Self-organized Operation and Stability Control of Electronized Power Networks

Xiaochao Hou · Yao Sun · Siqi Fu · Shimiao Chen ·
Mei Su

Self-organized Operation and Stability Control of Electronized Power Networks

Voltage Magnitude-Frequency Dynamics Perspective

Springer

Xiaochao Hou
Central South University
Changsha, China

Yao Sun
Central South University
Changsha, China

Siqi Fu
Changsha University
Changsha, China

Shimiao Chen
Central South University
Changsha, China

Mei Su
Central South University
Changsha, China

ISBN 978-981-96-9066-4 ISBN 978-981-96-9067-1 (eBook)
https://doi.org/10.1007/978-981-96-9067-1

This Springer imprint is published by the registered company Springer Nature Singapore Pte Ltd.
The registered company address is: 152 Beach Road, #21-01/04 Gateway East, Singapore 189721, Singapore

Preface

With the development of power electronic technology, the proportion of power electronic devices in power system is increasing. Modern power electronic technology has greatly enhanced the controllability and flexibility of power system and has been widely used in power system generation, transmission, distribution, utilization, storage and other applications. The power electronized power network is the inevitable trend, resulting in profound changes in the dynamic characteristics, control profile, and operation framework.

Nowadays, the power coordination operation and dynamic stability control are moving from centralized to decentralized manner. The traditional centralized control mode, which is highly dependent on high energy storage configuration and high bandwidth communication, cannot meet the development needs of flexible and reliable, plug and play, and variable operation of the system. It is urgent to develop a new simple control rule to manage the complex electronized power networks.

Self-organized operation is the fundamental mechanism for building networked system structure self-similarity and achieving power energy self-balance. It emphasizes that without external instructions, each subsystem within the system can form a certain structure and function according to certain rules. It belongs to the autonomous decision-making adjustment in the system that tends to guide rather than force control. Under the self-organized operation framework of the power electronized power networks, its core is to reduce the dependence on high-bandwidth communication and high-proportion energy storage and fully tap the flexible controllability of multi-layer converters. The core includes self-synchronization of AC-dominated System, Self-Adaptation to Variable Conditions, Self-Optimization Economic Regulation, and Self-Disciplined Robust Stability.

Power electronic device is a strong nonlinear time-varying system, which has the characteristics of low inertia, fast dynamic response, wide-frequency oscillation, weak overload capacity and toughness. In particular, the dynamics of the power electronic equipment and the interaction between the devices as well as the interaction with the characteristics of the power grid may bring major challenges to the safe and stable operation. In the past decades, although extensive researches have been done on control and stability analysis of electronized power systems, there are still many new issues including severe power oscillation, synchronous instability, poor dynamic response performance caused by multi-converter interaction and low reliability caused by communication dependency, and so on. Without a doubt, advanced self-organized control and synchronization stability from the amplitude-frequency dynamic perspective meet the development trends of electronized power systems. However, advanced self-organized control and synchronization stability technology has not been systematically introduced to the researchers and other readers.

This book gives a comprehensive and in-depth introduction into the development of self-organized operation and stability control of Electronized Power Networks. It presents modeling method from the amplitude-frequency dynamic perspective, oscillation suppression control, transient synchronous stability and self-organized operation of the electronized power systems. For each topic, a theoretical introduction and overview are backed by very concrete programming examples that enable the reader to not only understand the topic but to develop microgrid simulation models. This book will serve as an invaluable tool for researchers, engineers, and designers in the field of the electronized power systems who are involved in the cooperative control of high renewable based power converter networks. With this book, we intend to enable the reader to: (1) the first book available on system modeling from the amplitude-frequency dynamic perspective, oscillation suppression control, transient synchronous stability control and self-organized control of the electronized power systems; (2) help researchers, engineers, and designers with properly control design to improve dynamic response performance and improve system stability; (3) the book also enables the students to conduct the simulation and help them to start the research in this area.

The main research results of this book have been originally taken from the authors who carried out the related research together for almost ten years, which is a comprehensive summary of the authors' latest research results. This book is likely to be of interest to university researchers, R&D engineers, and graduate students in electrical engineering who wish to learn the core principles, control methods, and applications of Electronized Power Networks.

Changsha, China

Xiaochao Hou
Yao Sun
Siqi Fu
Shimiao Chen
Mei Su

Acknowledgements This book is supported by the Nature Science Foundation of China (NSFC) under Grant 62125308, Grant 52337008, and Grant 52307232, the Hunan Provincial Natural Science Foundation of China under Grant 2024JJ4055, and the Hunan Provincial Key Laboratory of Power Electronics Equipment and Gird.

The authors would like to thank postgraduate students Mr. Heng Ma, Mr. Zexiong Wei, Mr. Xinyu He, Ms. Jiatong He, Mr. Yuejiang Li, Mr. Hangrui Zhang, Mr. Yuxuan Li, Mr. Jiawei Tan, Ms. Xiaonuo Dong and Mr. Hongpeng Li for their contributions and proofreading. Special thanks for the academic support and contributions go to Prof. Hua Han from Central South University and Prof. Kai Sun from Tsinghua University and Prof. Xin Zhang from Zhejiang University. Finally, the authors would like to thank the long-term support and encouragement from their colleague, friends, and families.

Competing Interests The authors have no competing interests to declare that are relevant to the content of this manuscript.

Contents

Part II Oscillation Suppression Control

About the Authors

Xiaochao Hou received the B.S., M.S., and Ph.D. degrees in control science and engineering from the School of Automation, Central South University, Changsha, China, in 2014, 2017, and 2020, respectively. From September 2018 to September 2019, he was Joint Ph.D. Student with the School of Electrical and Electronic Engineering of Nanyang Technological University, Singapore. From July 2020 to December 2022, he was Postdoctoral Research Fellow with the Department of Electrical Engineering of Tsinghua University, China. From April 2023, he is Associate Professor with the School of Electronic Information, Central South University, Changsha, China. He was Recipient of the Outstanding Reviewers Award of IEEE Transactions on Sustainable Energy in 2021 and the Paper Award of IEEE Transactions on Power Electronics in 2022. His research interests include nonlinear control and stability of power electronic networks, grid-forming distributed PV/ESS/EV microgrid, high renewable distribution network.

Yao Sun is currently Professor at the School of Automation, Central South University, China. He received B.S., M.S., and Ph.D. degrees from the School of Information Science and Engineering, Central South University, in 2004, 2007, and 2010, respectively. Professor Sun has worked on microgrids, coordinated control, and power electronic converters for more than ten years, and he has published his research results in more than 200 journals and conferences. He was Recipient of the first prize of the 2020 Hunan Natural Science Award of China. His research interests include matrix converter, microgrid, and wind energy conversion systems.

Siqi Fu received the B.S. degree in electrical engineering and the automatization from the HeFei University of Technology, Anhui, China, in 2018. She received the Ph.D. degree in control science and engineering from the School of Automation, Central South University, Changsha, China, in 2024. She is currently Lecturer with the College of Electronic Information and Electrical Engineering, Changsha University, Changsha. Her research interests include control and stability of distributed microgrid and power-electronic-enabled power network.

Shimiao Chen received the B.S. degree in electrical engineering from Central South University, Changsha, Hunan, China, in 2019, where he is currently pursuing a Ph.D. degree in power electronics at the College of Automation. His research interests include renewable energy, power electronic transient stability analysis, and virtual synchronous generator (VSG).

Mei Su received the B.S. degree in automation, M.S., and Ph.D., degrees in control engineering from the School of Information Science and Engineering, Central South University, Changsha, China, in 1989, 1992, and 2005, respectively. She has been Full Professor with the School of Automation, Central South University. Her research interests include matrix converter, adjustable speed drives, and wind energy conversion. Professor Su is currently Associate Editor of the *IEEE Transactions on Power Electronics* and *IEEE Transactions on Sustainable Energy*.

Symbols

S_i	The Converter output apparent power of DG-i
P_i	The active power of DG-i
$P_{\max}$	The maximum values of the allowable active power
$P_{\min}$	The minimum values of the allowable active power
Q_i	The reactive power of DG-i
$Q_{\max}$	The maximum values of the allowable reactive power
$Q_{\min}$	The minimum values of the allowable reactive power
V_{pcc}	The amplitude of common ac bus voltage
θ_{pcc}	The common ac bus voltage angle
X_{line}	The line impedance
R_{line}	The line resistance
Z_{line}	The impedance between converter and the common bus
Z_{load}	The common bus load
V_g	The amplitude of grid voltage
δ_g	The grid voltage angle
δ_i	The angle of output voltage of DG-i
V_i	The amplitude of output voltage of DG-i
ω_i	The angular frequency of DG-i
$\omega_{\max}$	The maximum values of the allowable angular frequency
$\omega_{\min}$	The minimum values of the allowable angular frequency
m_i	The active droop coefficient
n_i	The reactive droop coefficient
ω^*	The nominal angular frequency value
V^*	The nominal voltage value
T_s	The sampling period
$C_i(P_i)$	The cost function
P_L	The total load
$P_{L,\max}$	The upper bound of load power
f_i	The DG-i actual frequency
f^*	The DG-i reference frequency
ζ	The damping factor

sgn ()	The sign function
$\hat{\cdot}$	A small perturbation around the steady state points
L_f	The filter inductance
R_f	The filter resistance
C_f	The filter capacitance
i	The grid current
i^*	The reference grid current
ω_c	The cut-off frequency

Chapter 1
Towards a Distributed, Electronized and Self-Organized Power Systems

1.1 Power Electronized Power Systems

1.1.1 Development Profile of Electronized Power Systems

Modern power electronic technology has greatly enhanced the controllability and flexibility of power system, and has been widely used in power system generation, transmission, distribution, use, storage and other applications [1–6], as shown in Fig. 1.1 Renewable energy generation needs to rely on power electronic converter devices for accurate and fast control. In the transmission link, HVDC transmission technology and flexible AC transmission technology based on high-power power electronic devices have played an important role in major projects. In the distribution link, the controllability of the power electronic converter device can effectively improve the active power peak regulation and reactive power support ability of the power system, and ensure the high quality and economic operation of the power grid. In the link of electricity consumption, the diversified load demand and high-quality power supply are inseparable from power electronic devices. In the energy storage part, the power electronic device is the hardware basis to play the role of peak cutting and valley filling. With the development of power electronic technology, the proportion of power electronic devices in power system is increasing, and the power electronized power systems is the inevitable trend of the development of the power system [7, 8].

1.1.2 Challenges for Power Electronized Power Systems

There are essential differences between new renewable energy generation and traditional synchronous generators, flexible AC-DC power transmission and traditional AC power transmission, resulting in profound changes in the dynamic characteristics of the system, and new power systems are facing new challenges [9–13].

X. Hou et al., *Self-organized Operation and Stability Control of Electronized Power Networks*, https://doi.org/10.1007/978-981-96-9067-1_1

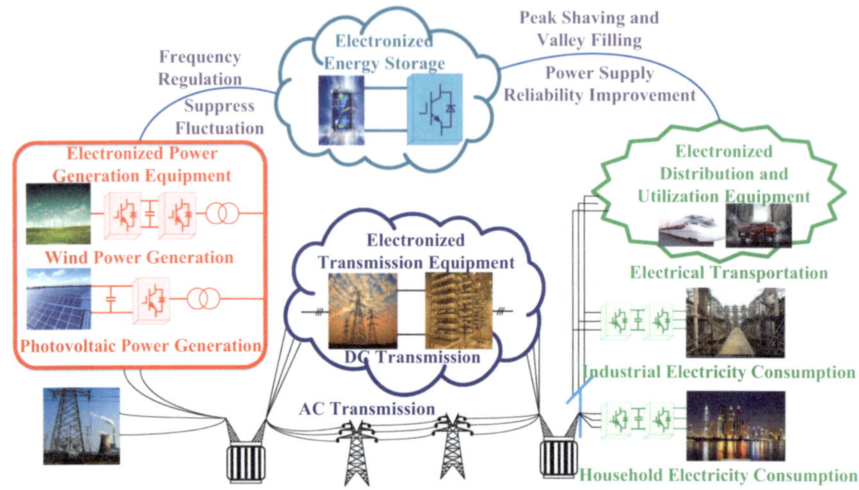

Fig. 1.1 New-type power systems: power electronized power systems

Equipment-to-Network Dynamic Behavior Modeling: Power electronic converters have the characteristics of multi-time scale, nonlinear, high order, coupling, etc., which makes the interaction characteristics of the system more complicated. How to establish equipment multi-time scale model by taking into account the nonlinear links is a new problem for modeling. The analysis of system dynamic characteristics needs to study the characteristics of each scale and the interaction of different scales at the same time.

Low Inertia and Wide-frequency Oscillation [14–16]: With the increasing penetration rate of Renewable energy and power electronic devices, the risk and influence range of broadband oscillation problem are also expanding. The frequency range of wide-band oscillation involves a number of Hz to a number of kHz, and the harm caused by oscillation does not decrease with the increase of frequency, resulting in increased security risks in the transmission network and even the large power grid. This kind of wide-band oscillation caused by the interaction between power electronic equipment and the grid poses a great threat to the Renewable energy grid-connected equipment, the station and the power system.

Transient Synchronous Stability Issue [17–19]: The characteristics of the electronic power system are that the topology of the system changes with the switching action of the power electronic device, the whole system is time-varying (non-autonomous), and there are multi-time scale control interactions. In addition, the power electronic converter itself has structural nonlinearity and complexity, and there are over modulation and limiting when using pulse width modulation, so it has saturated nonlinearity. It brings difficulties to the stability analysis of the system, especially the transient stability analysis.

Self-optimizing Control under the Weak-Centralized Framework [20, 21]: The traditional centralized control mode is highly dependent on high energy storage configuration and high bandwidth communication. It cannot meet the development needs of flexible and reliable system, plug and play, and variable operation. The real-time power balance and dynamic stability control of system under the trend of flat physical structure and weak-centralized framework are a major challenge.

Multi-scale Collaborative Planning under Strong Power Uncertainty [13, 22, 24]: Under the basic characteristics of high proportion penetration of Renewable energy, diversified load superposition and flexible resource application, how to adapt to the strong uncertainty on both sides of supply and demand, and effectively integrate multiple types of flexible resources to achieve accurate matching between supply and demand is the main challenge.

1.2 Issues and Methods of System Stability Analysis

1.2.1 Classification of Power Electronized System Stability

In 2004, the Institute of Electrical and Electronics Engineers (IEEE) and the International Council on Large Electric systems (CIGRE) jointly published the report "Definition and Classification of Power System Stability" [25], which divided power system stability into power angle stability, voltage stability and frequency stability, and then subdivided each stability problem according to the disturbance size and time span.

With the increasing penetration rate of renewable energy, high proportion of renewable energy and high proportion of power electronic equipment significantly affect the steady-state and transient characteristics of power system [26]. In recent years, the new form stability problems related to power electronics have become prominent, causing serious consequences such as generator torsional vibration, equipment damage, Renewable energy off-grid, and even causing malignant accidents such as large-scale power outages [27–29]. It seriously threatens grid security and even social public safety, and bring new power supply stability problems [30].

Thus, IEEE and CIGRE published the report "Dynamic Behavior Characteristics and Stability Definition of Power Systems with High-permeability power electronic Interface Devices" in 2020 [31], which further added resonance stability and converter drive stability. On this basis, power electronized power system stability is divided systematically into two categories [32]: power frequency stability dominated by electromechanical dynamics and non-power frequency stability dominated by electromagnetic dynamics, as shown in Fig. 1.2. The former is mainly aimed at the ability to operate stably at rated frequencies and provide high-quality electrical power, while the latter is mainly aimed at the ability to suppress unnecessary

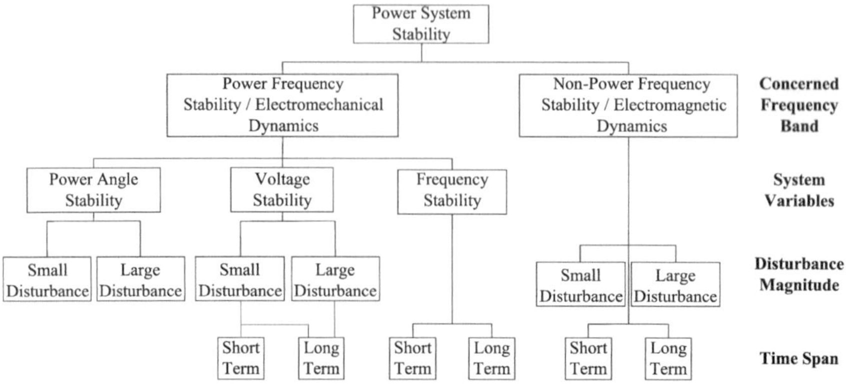

Fig. 1.2 Classification of power system stability

non-power frequency electromagnetic components. Among them, the newly added non-power frequency stability can also be subdivided by small-signal /large-signal stability and short/long term [32].

1.2.2 Small-Signal Stability Analysis

The methods of oscillation and stability analysis based on small signal linearization model of converter system can be divided into two categories: frequency domain analysis and time domain analysis. The two focus on the characteristics of the system model from different angles. Time domain analysis is mainly eigenvalue analysis, which focuses on measuring system stability through damping of eigenvalue. Frequency domain analysis is mainly based on the system transfer function analysis, the existing frequency domain analysis can be divided into the following three categories: damping torque analysis, impedance analysis and P–ω power frequency analysis.

1.2.2.1 Eigenvalue Analysis [33–40]

Eigenvalue analysis method is based on the state-space model of the system, which covers all internal dynamics of each part of the system, and its eigenvalue directly reflects the oscillation mode of the system. It can deeply analyze the relationship between each system state variable and some oscillation modes through the participation factor, and quantify the corresponding participation degree of system variable in the mode. By looking for the physical state variables most relevant to the oscillation, the weak link of the system is located. In addition, the influence of each control parameter on the oscillation mode can be easily analyzed based on the state-space model.

Although the eigenvalue analysis covers all the state variable dynamics of the system, it can analyze the influence of each system state variable and control variable on some oscillation modes in detail. But it relies on the hardware design and detailed control algorithms of each converter, and power electronic converter algorithms are usually not informed by the manufacturer. With the expansion of the system scale, the excessive order of the model may cause "dimensional disaster" and make it difficult to solve the eigenvalues. Moreover, the time-domain analysis method based on state-space model has poor scalability, and the model needs to be rebuilt when the system structure changes.

1.2.2.2 Damping Torque Analysis Method [41–47]

The damping torque analysis method is a common method to study the low frequency oscillation of power system. Damping torque analysis method. Based on the single-machine infinite bus system Philipps-Heffron model, the method is used to study the contribution of specific sources to the electric torque of the electromechanical oscillation circuit of synchronous generator in a power system, which is divided into synchronous torque and damping torque. The core idea is to analyze the positive and negative damping torque provided by the control link near the low frequency oscillation frequency to judge its influence on the stability of the system. Providing positive damping torque is conducive to enhancing system stability, whereas negative damping torque is not conducive to system stability.

The damping torque analysis method is mainly used to analyze the local oscillation mode of the single-machine infinite power system, which has clear physical meaning and is easy to understand. However, the positive and negative damping torque can only be used to judge the stability of the system in this case. This stability judgment method has not been strictly proved in the multi-machine system, and its scalability is poor.

1.2.2.3 Impedance Analysis Method [48–66]

Impedance analysis is based on the impedance model of the system and is essentially a transfer function between voltage and current. The impedance model can characterize the response characteristics of power electronic converters under small signal disturbance in different frequency bands, and the stability can be judged by pole diagram or Nyquist criterion. Because each subsystem model is characterized independently by port input/output characteristics, it is easier to analyze large-scale systems, modular and easy to expand. In addition, when the system parameters are unknown, the frequency response characteristics of the system can be measured experimentally to obtain the impedance model, which has distinct physical significance.

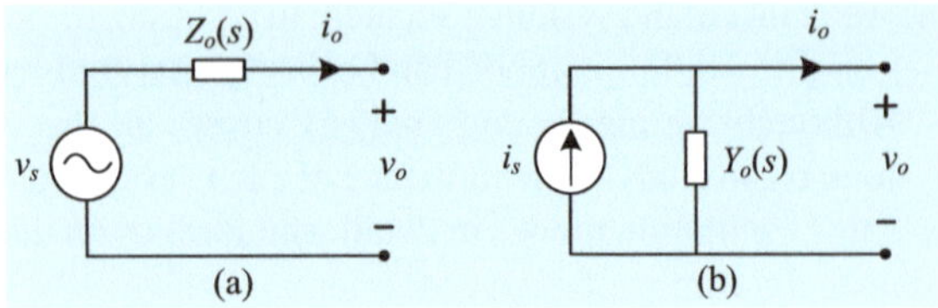

Fig. 1.3 Single-port system equivalent circuit: **a** Thevenin equivalent circuit and **b** Norton equivalent circuit

For the single-port source subsystem, when the converter is controlled as a voltage source, the output current is taken as the input variable, and the port voltage is taken as the output variable. According to the Thevenin theorem, the subsystem is equivalent to the ideal voltage source with serial output impedance $Z_o(s)$, as shown in Fig. 1.3(a) It is usually designed to maintain stability under the current source load in island mode i.e. $Z_o(s)$ is stable When the converter is a controlled as a current source, the port voltage is taken as the input variable and the output current is taken as the output variable. According to Norton theorem, the subsystem is equivalent to the ideal current source with parallel output admittance $Y_o(s)$ form, as shown in Fig. 1.3(b) It is usually designed to be stable under the ideal voltage source i.e. $Y_o(s)$ is stable.

Impedance analysis has been applied to the stability evaluation of power electronized power systems dominated by power electronic converters because of its modularity, easy expansion and measurable advantages. However, with voltage and current as the observed variable, it is not convenient to directly analyze power, frequency, voltage amplitude and their dynamic response.

1.2.2.4 *P–ω* Analysis Method [67–71]

Active power (P), reactive power (Q), voltage (V), frequency (f) and other state variables and their dynamic responses are the key characteristic variables that microgrids and even power systems pay attention to. In order to focus on the dynamic responses of active power (P) and angular frequency (ω), some scholars have introduced a P–ω modeling method, which can directly describe the dynamic responses of the output power and frequency of the converter under disturbance. It is convenient to analyze the power and frequency oscillation characteristics of the system. Based on the P–ω modeling method, the single-port source subsystem can be equivalent to the power loop two-port network shown in Fig. 1.4.

P–ω modeling method inherits the advantages that impedance model can be modularized, measured, and easily extended to multi-machine systems. However, the model ignores the influence of voltage fluctuation and reactive power control loop, and cannot analyze the coupling of reactive power and active power. The model has high accuracy only when the line impedance is highly inductive.

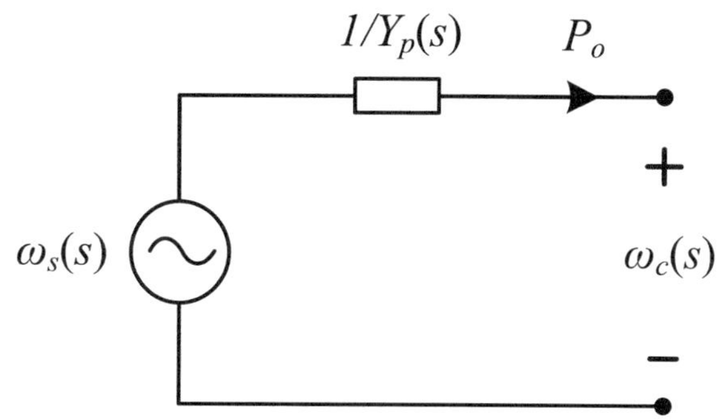

Fig. 1.4 Single-port system equivalent circuit based on P–ω modeling method

1.2.3 Large-Signal Stability Analysis

Compared with the small-signal stability problems such as wide-band oscillation described in Sect. 1.2.2, converters exhibit more complex nonlinear characteristics during transient faults. Due to the multi-time scale coupling and control mode switching of the converter, it is difficult to establish a complete model to analyze the transient characteristics during the fault period. So far, no complete system has been formed at home and abroad.

The essence of transient instability is that the grid-connected synchronous control strategy of converter loses its ability to keep pace with the power grid after experiencing large disturbance. Therefore, scholars began to try to study the "transient synchronous stability" of this grid-connected synchronization link, and the analysis methods can be mainly divided into time domain simulation analysis and quasi-steady state analysis. The quasi-steady-state analysis assumes that all the links of grid-connected synchronous control are in steady state, including equal area criterion (EAC) analysis, Lyapunov energy function method, phase plane analysis, etc.

1.2.3.1 Time Domain Simulation Analysis [72–74]

The time-domain simulation method solves the differential algebraic equations of the system to obtain the trajectory of the system state quantity and the number of generations over time. In the traditional AC power system, the time domain simulation method determines the transient stability of the system by the maximum power angle difference between the generators. As the most reliable evaluation method, time domain simulation can realize arbitrary complex system models and control strategies, and is often used as a standard for other transient stability analysis methods.

Time domain analysis method also has some shortcomings. The numerical integration of dynamic equations is slow. With the increase of the order of system state variables, the calculation amount of time domain simulation will be greatly increased, and the calculation speed cannot meet the needs of online monitoring and control.

Besides, it does not provide information about the stability degree of the system, and it is impossible to explore the stability mechanism.

1.2.3.2 Equal Area Criterion Analysis [75, 76]

Equal area criterion and extended equal area criterion (EEAC) are effective methods to analyze the transient stability of traditional power systems. Due to the inherent similarity of synchronization mechanism, it has been extended to the study of transient synchronous stability of converters, and the mechanism of transient instability of netted converters can be intuitively revealed. However, this method ignores the influence of the damping of the netted converter and is difficult to be used to quantitatively estimate the size of the stability region or the limit excision time CCT.

1.2.3.3 Transient Energy Function Method [77, 78]

Transient energy function method determines the transient stability of the system by constructing the Lyapunov energy function and comparing it with the maximum transient energy (called critical energy) that can be absorbed by the system. It can not only judge the stability of the system qualitatively, but also obtain the stability margin of the system and analyze the transient stability of the system quantitatively. The calculation of the critical energy in the traditional AC power system, that is, the boundary calculation of the attractive region, is the most difficult step of the direct method. The main methods include the unstable equilibrium point (UEP) method, the dominant UEP method and the persistent fault method.

Compared with traditional AC power systems, electronic power systems have the characteristics of time-varying topology, fast transient process, strong nonlinearity and high order of system state variables, which bring many challenges to transient stability analysis by direct method. Even if the fluctuation characteristics of Renewable energy are not considered, the system model will become very complex when all kinds of power electronic converters operate in combination. How to simplify the analysis process according to some characteristics of the electric power system is the key to use the direct method

1.2.3.4 Phase Plane Analysis [79, 80]

Phase plane analysis has obvious advantages in the study of dynamic trajectories of second-order systems. According to the phase diagram, the attractive range of the stable equilibrium point can be studied to predict the stability of the system after experiencing large perturbations. This method is called the phase plane/phase diagram analysis method, which has certain engineering value, and can be used to study the parameter-oriented design for a specific system to obtain better parameters.

However, this method is still a numerical solution in essence, which can only be quantitatively analyzed for a specific set of parameters, and can not get a universal quantitative analysis conclusion.

1.3 Self-Organized Operation Control

Renewable energy microgrids and their clusters are important basic units for the construction of power electronized power systems and future energy internet, as shown in Fig. 1.5 It has the essential characteristics of high proportion of renewable energy and power electronics, and present new characteristics such as large random fluctuations, weak networking support, and hard collaborative control. On the one hand, the dynamic characteristics of the diversified distributed source and load are different. The system operation is highly dependent on the high-proportion energy storage capacity unit to stabilize voltage and frequency, which reduces the economy. On the other hand, the system networking mode and operating conditions are changeable.

The power coordination and stability control at the micro source level, bus level and cluster level rely heavily on the centralized control under the master-slave framework. There are major hidden dangers of insufficient flexible expansion and single point failure of high-bandwidth communication, and it limits the scale development and reliability improvement of the system. At present, it is increasingly difficult to control the real-time power balance and dynamic stability of microgrid clusters under the trend of flat physical structure and weak-centralized framework [81]. The traditional centralized control mode, which is highly dependent on high energy storage configuration and high bandwidth communication, cannot meet the development needs of flexible and reliable, plug and play, and variable operation of the system. It

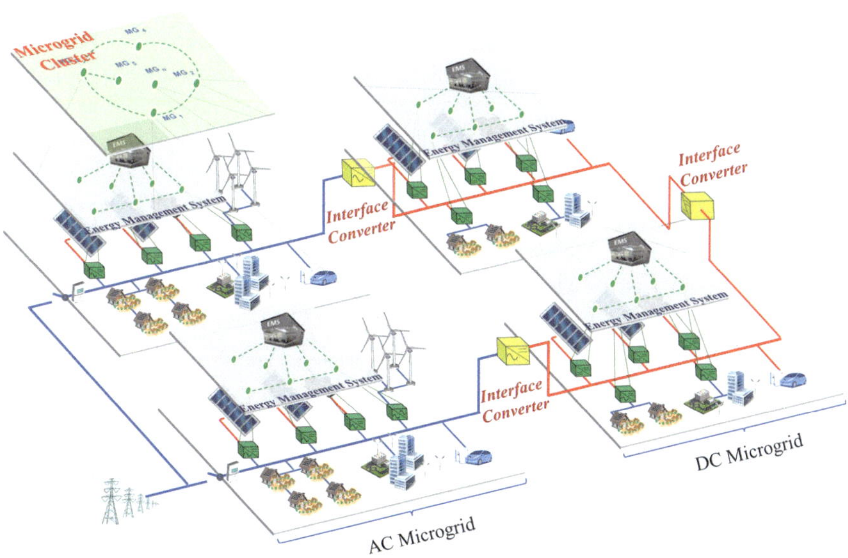

Fig. 1.5 The multi-level morphological structure of interconnected microgrid group

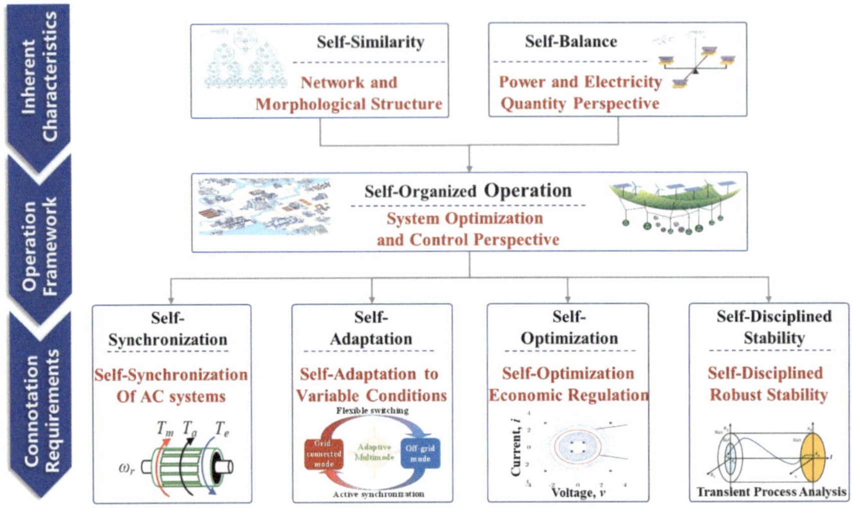

Fig. 1.6 Self-organized operation framework based on interconnected microgrid group

is urgent to develop a new microgrid control mode with self-organized operation as the core principle.

Self-organized operation is the fundamental mechanism for building microgrid group structure self-similarity and achieving power and quantity self-balance. It emphasizes that without external instructions, each subsystem within the system can form a certain structure and function according to certain rules, which belongs to the autonomous decision-making adjustment in the system that tends to guide rather than force control. Under the self-organized operation framework of the power electronized power system, its core is to reduce the dependence on high-bandwidth communication and high-proportion energy storage, and fully tap the flexible controllability of multi-level interconnections such as micro-source interface converters, bus interlinked converters, and microgrid interlinked converters. The specific requirements as shown in Fig. 1.6 include AC leading self-synchronization, variable operating condition self-adaptation, economic regulation self-optimization, stable operation self-stability and extreme elastic self-recovery.

1.4 Organization of the Book

According to the preliminary study [81–95], this book would bring latest research result on advanced self-organized control and synchronization stability of electronized power systems from voltage amplitude-frequency dynamics. This book can be divided into four parts. The first part is focused on novel amplitude-frequency modeling method, which spans Chaps. 2 to 3. The second part presents oscillation

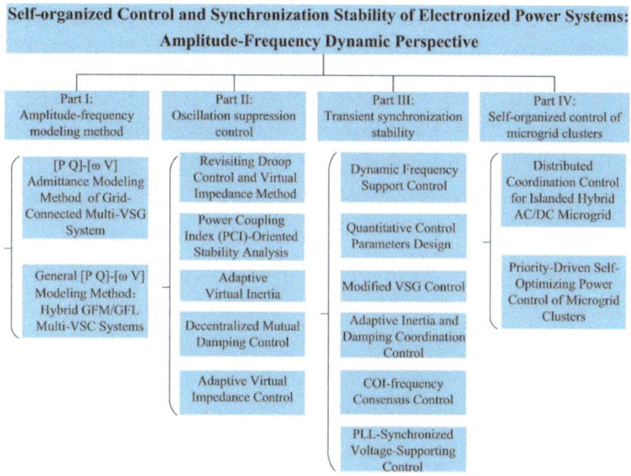

Fig. 1.7 The overall structure of the book

suppression control, which are included from Chaps. 4–8. The third part introduces transient synchronization stability, which is from Chaps. 9–14. And the forth part presents self-organized control of microgrid clusters, which are included from Chaps. 15–16. The overall structure of the book is shown in Fig. 1.7.

Part I introduces novel amplitude-frequency modeling method for power electronized power system. Chapter 2 presents a $[P\ Q]$–$[\omega\ V]$ “admittance” model of multi-VSG grid-connected systems and analyzes the power, frequency and voltage oscillation of the system. Based on the model, the stability criterion of the system is obtained and the influence of key parameters on the power coupling and oscillation characteristics is analyzed. Chapter 3 introduces a general $[P\ Q]$–$[\omega\ V]$ modeling method of hybrid GFM/GFL multi-VSC systems. It could serve as a standard model that provides a convenient and universal tool for both the low and medium frequency (a few tenths to tens Hz) power oscillation analysis and stability analysis.

In part II, a series of oscillation suppression control are presented. Chapter 4 compares basic droop control with virtual impedance methods from a new analogy perspective. By combing both of them, a unified multi-degree-of-freedom droop control is introduced with systematic consideration of steady-state power-sharing performance, dynamic performance, and universal applicability of complex line impedance. Chapter 5 presents a power coupling index (PCI)-oriented stability analysis and $[P\ Q]$–$[\omega\ V]$ interactive control design method for inverter-based microgrid, which makes inverter universal applicability under complex line impedances and effectively suppresses system power medium frequency oscillation. To suppress the low-frequency power oscillation in the multi-VSG grid, Chaps. 6–8 presents oscillation suppression control methods from different angles. Chapter 6 introduces a distributed adaptive virtual inertia control method. The basic principle of the method is that when the local frequency of a VSG deviates from the average frequency of its neighbors, the adaptive inertia control will exhibit a large inertia and when the local frequency starts to return, a small inertia is shaped to accelerate the convergence to the average frequency. Chapter 7 presents a decentralized mutual damping control.

The difference between the local angular frequency and other angular frequencies is obtained indirectly by the derivative of the local active power, and forming a mutual damping term that promotes frequency convergence in the dynamic process. Chapter 8 offers an adaptive virtual impedance method to suppress the power oscillation, which endows the virtual impedance a new function. Stability analysis of the system based on a small-signal model and guidelines for designing key parameters are provided.

Our researches for system transient synchronization stability are introduced in Part III. Chapter 9 presents a systematic analysis of the influence of virtual inertia and damping coefficient on the dynamic frequency characteristic among parallel VSG systems and a coordinated control for virtual inertia and damping to improve frequency stability and achieve a better dynamic performance of parallel VSG systems. Chapter 10 introduces a systematic feasible parameters region that satisfies grid-code-based transient stability and rate-of-change-of-frequency (RoCoF) requirement under large disturbances. provides a practical reference for control parameters design and engineering application of VSG. Chapter 11 presents a modified VSG control scheme which solves the conflict caused by virtual resistance between small-signal stability and transient stability. A proportional integral controller is added at the forward loop to control the real active power to track the power reference. Chapter 12 introduces an adaptive inertia and damping coordination (AIDC) control strategy for grid-forming VSGs to improve transient stability. It adaptively adjusts the virtual inertia and damping coefficients based on real-time conditions of operation frequency deviation and its rate of change of frequency (RoCoF). Chapter 13 presents a distributed control method via COI frequency for multi-VSG grids. The stability of the control method is proved and the feasible range of the control parameter is presented. Chapter 14 introduces a new first-order phase-lock-loop (PLL)-synchronized voltage-supporting control method for the islanded operation of parallel inverters in UPS systems. The first-order PLL is employed for frequency synchronization in islanded operation. Meanwhile, voltage support and power sharing can be guaranteed with modified current control.

Part IV introduces our researches for self-organized control of microgrid clusters. Chapter 15 presents a distributed normalized power coordination (NPC) embedded with virtual synchronous generator (VSG) for hybrid microgrid. The NPC controller can achieve cross inertia support, accurate global power sharing and frequency/voltage recovery among dispatchable DGs simultaneously. Chapter 16 introduces a decentralized self-optimizing power control scheme for interlinking converters (ILCs) to realize optimal power exchanges among microgrid clusters. It has the features of decreased cost, increased scalability, reduced geographic restrictions and high resilience and robustness in terms of communication faults.

References

1. D. Boroyevich, R. Brasov, I. Cvetkovic, D. Dong, R. Burgos, F. Wang, F. Lee, Future electronic power distribution systems: a contemplative view. in *Proceedings of 12th IEEE International Conference on OPTIM* (2010), pp. 1369–1380
2. W. Said, A fight to quality. IEEE Power Energy Mag. **9**(4), 38–45 (2011)
3. C.C. Davidson, Power transmission with power electronics, in *Proceedings of IEEE 14th European Conference on Power Electronics and Applications* (2011), pp. 1–10
4. R.S. Whitehouse, Technical challenges of realising multi-terminal networking with VSC. in *Proceedings of IEEE 14th European Conference Power Electronics and Applications* (2011), pp. 1–12
5. G. Tang, X. Luo, X. Wei, Multi-terminal HVDC and DC-grid technology. Proc. CSEE **33**(10), 8–17 (2013)
6. D. Boroyevich, I. Cvetkovic, R. Burgos et al., Intergrid: a future electronic energy network. IEEE J. Emerg. Select. Top. Power Electron. **1**(3), 127–138 (2013)
7. X. Yuan, S. Cheng, J. Hu, Multi-time scale voltage and power angle dynamics in power electronics dominated large power systems. Proc. CSEE **36**(19), 5145–5154 (2016)
8. J. Hu, X. Yuan, S. Cheng, Multi-time scale transients in power-electronized power systems considering multi-time scale switching control schemes of power electronics apparatus. Proc. CSEE **39**(18), 5457–5467 (2019)
9. M. Zhao, X. Yuan, J. Hu et al., Voltage dynamics of current control time-scale in a VSC-connected weak grid. IEEE Trans. Power Syst. **31**(4), 2925–2937 (2016)
10. M. Li, Z. Yu, T. Xu, et al., Study of complex oscillation caused by renewable energy integration and its solution. Power Syst. Technol. **41**(4), 1035–1042 (in Chinese, 2017)
11. G. Chen, M. Li, T. Xu, et al., Study on technical bottleneck of new energy development. Proc. CSEE, **37**(1), 20–26 (in Chinese, 2017)
12. M. Farhadi, O. Mohammed, Energy storage technologies for high-power applications. IEEE Trans. Ind. Appl. **52**(3), 1953–1961 (2016)
13. A. Ortega, F. Milano, Modeling, simulation, and comparison of control techniques for energy storage systems. IEEE Trans. Power Syst. **32**(3), 2445–2454 (2017)
14. J. Adams, C. Carter, S.H. Huang, ERCOT experience with sub-synchronous control interaction and proposed remediation. in *Proceedings of Transmission & Distribution Conference & Exposition, Orlando, USA* (unclear page range) (2012)
15. X. Xie, X. Zhang, H. Liu, H. Liu, Y. Li, C. Zhang, Characteristic analysis of subsynchronous resonance in practical wind farms connected to series-compensated transmissions. IEEE Trans. Energy Convers. **32**(3), 1117–1126 (2017)
16. C. Buchhagen, C. Rauscher, A. Menze, J. Jung, "BorWin1-first experiences with harmonic interactions in converter dominated grids. in *Proceedings of International ETG Congress, Die Energiewende-Blueprints for the new energy age* (Bonn, Germany, 2015), p. 2015
17. J. Zhao, M. Huang, H. Yan, C.K. Tse, X. Zha, Nonlinear and transient stability analysis of phase-locked loops in grid-connected converters. IEEE Trans. Power Electron. **36**(1), 1018–1029 (2021)
18. X. Xiong, C. Wu, B. Hu, D. Pan, F. Blaabjerg, Transient damping method for improving the synchronization stability of virtual synchronous generators. IEEE Trans. Power Electron. **36**(7), 7820–7831 (2021)
19. C. Wu, X. Xiong, M.G. Taul, F. Blaabjerg, On the equilibrium points in three-phase PLL based on the d-axis voltage normalization. IEEE Trans. Power Electron. **36**(11), 12146–12150 (2021)
20. C. Jin, P.C. Loh, P. Wang, Y. Mi, F. Blaabjerg, Autonomous operation of hybrid AC-DC microgrids. in *Proceedings of 2010 IEEE International Conference on Sustainable Energy Technologies (ICSET 2010)* (2010)
21. P.C. Loh, D. Li, Y.K. Chai, F. Blaabjerg, Autonomous operation of hybrid microgrid with AC and DC subgrids. IEEE Trans. Power Electron. **28**(5), 2214–2223 (2013)
22. X. Suo, S. Zhao, Y. Ma, L. Dong, New energy wide area complementary planning method for multi-energy power system. IEEE Access **9**, 157295–157305 (2021)

23. A. van Stiphout, K. De Vos, G. Deconinck, The impact of operating reserves on investment planning of renewable power systems. IEEE Trans. Power Syst. **32**(1), 378–388 (2017)
24. D. Ciechanowicz, D. Pelzer, B. Bartenschlager, A. Knoll, A modular power system planning and power flow simulation framework for generating and evaluating power network models. IEEE Trans. Power Syst. **32**(3), 2214–2224 (2017)
25. P. Kundur et al., Definition and classification of power system stability IEEE/CIGRE joint task force on stability terms and definitions. IEEE Trans. Power Syst. **19**(3), 1387–1401 (2004)
26. C. Zhang, X. Cai, A. Rygg, M. Molinas, Sequence domain SISO equivalent models of a grid-tied voltage source converter system for small-signal stability analysis. IEEE Trans. Energy Convers. **33**(2), 741–749 (2018)
27. R. Liu, J. Yao, X. Wang, P. Sun, J. Pei, J. Hu, Dynamic stability analysis and improved LVRT schemes of DFIG-based wind turbines during a symmetrical fault in a weak grid. IEEE Trans. Power Electron. **35**(1), 303–318 (2020)
28. H. Nian, B. Pang, Stability and power quality enhancement strategy for DFIG system connected to harmonic grid with parallel compensation. IEEE Trans. Energy Convers. **34**(2), 1010–1022 (2019)
29. Y. Song, E. Ebrahimzadeh, F. Blaabjerg, Analysis of high-frequency resonance in DFIG-based offshore wind farm via long transmission cable. IEEE Trans. Energy Convers. **33**(3), 1036–1046 (2018)
30. C. Zou et al., Analysis of resonance between a VSC-HVDC converter and the AC Grid. IEEE Trans. Power Electron. **33**(12), 10157–10168 (2018)
31. N. Hatziargyriou, J.V. Milanovic, C. Rahmann, et al., Stability definitions and characterization of dynamic behavior in systems with high penetration of power electronic interfaced technologies, in *PES-TR77, Piscataway* (IEEE, 2020)
32. X. Xie, J. He, H. Mao, et al., New issues and classification of power system stability with high shares of renewables and power electronics. Proc. CSEE **41**(2), 461–474 (in Chinese, 2021)
33. J. Liu, Y. Miura, H. Bevrani, T. Ise, Enhanced virtual synchronous generator control for parallel inverters in microgrids. IEEE Trans. Smart Grid **8**(5), 2268–2277 (2017)
34. Z. Shuai, W. Huang, Z.J. Shen, A. Luo, Z. Tian, Active power oscillation and suppression techniques between two parallel synchronverters during load fluctuations. IEEE Trans. Power Electron. **35**(4), 4127–4142 (2020)
35. F. Katiraei, M.R. Iravani, P.W. Lehn, Small-signal dynamic model of a micro-grid including conventional and electronically interfaced distributed resources. IET Gener. Transm. Distrib. **1**(3), 369–378 (2007)
36. N. Pogaku, M. Prodanovic, T.C. Green, Modeling, analysis and testing of autonomous operation of an inverter-based microgrid. IEEE Trans. Power Electron. **22**(2), 613–625 (2007)
37. R. Majumder, B. Chaudhuri, A. Ghosh, Improvement of stability and load sharing in an autonomous microgrid using supplementary droop control loop. IEEE Trans. Power Syst. **25**(2), 796–808 (2010)
38. R. Majumder, Some aspects of stability in microgrids. IEEE Trans. Power Syst. **28**(3), 3243–3252 (2013)
39. E. Barklund, N. Pogaku, M. Prodanovic, C. Hernandez-Aramburo, T.C. Green, Energy management in autonomous microgrid using stability-constrained droop control of inverters. IEEE Trans. Power Electron. **23**(5), 2346–2352 (2008)
40. X. Guo, *Dynamic Phasors-Based Modeling and Stability Analysis of Droop-Controlled Inverters for Microgrid Applications* (2014)
41. Z. Qu, J.C.H. Peng, H. Yang et al., Modeling and analysis of inner controls effects on damping and synchronizing torque components in VSG-controlled converter. IEEE Trans. Energy Convers. **36**(1), 488–499 (2020)
42. J. Liu, W. Yao, J. Wen, J. Fang, L. Jiang, H. He, S. Cheng, Impact of power grid strength and PLL parameters on stability of grid-connected DFIG wind farm. IEEE Trans. Sustain. Energy **11**(1), 545–557 (2019)
43. L. Huang, Damping low-frequency oscillations through VSC-HVDC stations operated as virtual synchronous machines. IEEE Trans. Power Electron. **34**(6), 5803–5818 (2018)

44. L. Harnefors, Proof and application of the positive-net-damping stability criterion. IEEE Trans. Power Syst. **26**(1), 481–482 (2010)
45. W. Du, J. Bi, J. Cao, H.F. Wang, A method to examine the impact of grid connection of the DFIGs on power system electromechanical oscillation modes. IEEE Trans. Power Syst. **31**(5), 3775–3784 (2016)
46. B. Ren, W. Du, H. Wang, Damping torque analysis of electromechanical oscillation modes for power system affected by grid-connected UPFC. Autom. Elect. Power Syst. **42**(17), 18–30 (2018)
47. B. Ren, W. Du, H. Wang, Damping channel selection based on damping torque analysis and coordinated design of auxiliary stabilizer for UPFC. Electr. Power Autom. Equip. **37**(5), 18–24 (2017)
48. J. Sun, Impedance-based stability criterion for grid-connected inverters. IEEE Trans. Power Electron. **26**(11), 3075–3078 (2011)
49. H. Liu, S. Shah, J. Sun, An impedance-based approach to HVDC system stability analysis and control development, in *Proceedings of 2014 International Power Electronics Conference, Japan* (2014)
50. L. Hong, W. Shu, J. Wang, R. Mian, Harmonic resonance investigation of a multi-inverter grid-connected system using resonance modal analysis. IEEE Trans. Power Deliver. **34**(1), 63–72 (2018)
51. K. Sun, W. Yao, J. Fang, X. Ai, J. Wen, S. Cheng, Impedance modeling and stability analysis of grid-connected DFIG-based wind farm with a VSC-HVDC. IEEE J. Emerg. Select. Top. Power Electron. **8**(2), 1375–1390 (2020)
52. R.D. Middlebrook, Input filter considerations in design and application of switching regulators. in *Proceedings, Record of 1976 IEEE Industry Applications Society Annual Meeting* (1976), pp. 366–382
53. J. Sun, Impedance-based stability criterion for grid-connected inverters. IEEE Trans. Power Electron. **26**(11), 3075–3078 (2011)
54. X. Liu, A.J. Forsyth, A.M. Cross, Negative input-resistance compensator for a constant power load. IEEE Trans. Ind. Electron. **54**(6), 3188–3196 (2007)
55. M. Cespedes, L. Xing, J. Sun, Constant-power load system stabilization by passive damping. IEEE Trans. Power Electron. **26**(7), 1832–1836 (2011)
56. H. Liu, S. Shah, J. Sun, An impedance-based approach to HVDC system stability analysis and control development, in *Proceedings of International Power Electronics Conference, Hiroshima, Japan* (2014)
57. M. Amin, A. Rygg, M. Molinas, Self-synchronization of wind farm in an MMC-based HVDC system: a stability investigation. IEEE Trans. Energy Convers. **32**(2), 458–470 (2017)
58. H. Liu, J. Sun, Voltage stability and control of offshore wind farms with AC collection and HVDC transmission. IEEE J. Emerg. Sel. Topics Power Electron. **2**(4), 1181–1189 (2014)
59. M. Cespedes, J. Sun, Impedance modeling and analysis of grid-connected voltage-source converters. IEEE Trans. Power Electron. **29**(3), 1254–1261 (2014)
60. B. Wen, D. Boroyevich, R. Burgos, P. Mattavelli, Z. Shen, Impedance-based analysis of gridsynchronization stability for threephase paralleled converters. IEEE Trans. Power Electron. **31**(1), 26–38 (2016)
61. S. Lissandron, L. Dalla Santa, P. Mattavelli, B. Wen, Experimental validation for impedance-based small-signal stability analysis of singlephase interconnected power systems with grid-feeding inverters. IEEE J. Emerg. Sel. Topics Power Electron. **4**(1), 103–115 (2016)
62. B. Wen, D. Boroyevich, R. Burgos, P. Mattavelli, Z. Shen, Inverse nyquist stability criterion for grid-tied inverters. IEEE Trans. Power Electron. **32**(2), 1548–1556 (2017)
63. M. Amin, M. Molinas, Understanding the origin of oscillatory phenomena observed between wind farms and HVDC systems. IEEE J. Emerg. Sel. Topics Power Electron. **5**(1), 378–392 (2017)
64. B. Wen, D. Boroyevich, R. Burgos, P. Mattavelli, Z. Shen, Small-signal stability analysis of three-phase AC systems in the presence of constant power loads based on measured d-q frame impedances. IEEE Trans. Power Electron. **30**(10), 5952–5963 (2015)

65. A.A.A. Radwan, Y.A.R.I. Mohamed, Linear active stabilization of converter-dominated DC microgrids. IEEE Trans. Smart Grid **3**(1), 203–216 (2012)
66. B. Wen, D. Boroyevich, R. Burgos, P. Mattavelli, Z. Shen, Analysis of D-Q small-signal impedance of grid-tied inverters. IEEE Trans. Power Electron. **31**(1), 675–687 (2016)
67. J. Jia, X. Yan, B. Qin, B. Zhang, Modeling and analysis of the torque-frequency dynamics for multi-VSC parallel system based on the equivalent admittance. IEEE Trans. Power Deliver. **37**(5), 3597–3607 (2021)
68. B. Qin, Y. Xu, C. Yuan, J. Jia, A unified method of frequency oscillation characteristic analysis for multi-VSG grid-connected system. IEEE Trans. Power Deliver. **37**(1), 279–389 (2021)
69. P. Zhang, H. Cai, J. Shi, Stability analysis of parallel inverter systems using a P/ω(Q/E) Admittance. Proc. CSEE **36**(9), 2486–2493 (2016)
70. B. Qin, Y. Xu, J. Jia, Modeling and analysis of SGs/VSGs torque-frequency dynamics based on mechanical admittance method. Proc. CSEE **40**(21), 6903–6913 (2020)
71. B. Qin, Y. Xu, C. Yong, J. Jia, The P/ω admittance modeling and power-frequency oscillation analysis of multi-VSGs grid-connected systems. Proc. CSEE **40**(9), 2932–2942 (2020)
72. A.A. van der Meer, J. Rueda-Torres, F.F. da Silva, et al., Computationally efficient transient stability modeling of multi-terminal VSC-HVDC, in *Proceedings of 2016 IEEE Power and Energy Society General Meeting, Boston, MA* (IEEE, 2016), pp. 1–5
73. R. Reza, D. Sudarmadi, F.A. Viawan, et al., Dynamic stability of power systems with power electronic interfaced DG, in *Proceedings of 2016 IEEE PES Power Systems Conference and Exposition, Atlanta, GA* (IEEE, 2006), pp. 1423–1428
74. A.M. Azmy, I. Erlich, Impact of distributed generation on the stability of electrical power system, in Proceedings of 2005 IEEE Power Engineering Society General Meeting, San Francisco, CA, USA (IEEE, 2005), pp. 1056–1063
75. G.H. Geng, L. Liu, R.Q. Li, Synchronization and reactive current support of PMSG-based wind farm during severe grid fault. IEEE Trans. Sustain. Energy **9**(4), 1596–1604 (2018)
76. H. Wu, X.F. Wang, Transient stability impact of the phase-locked loop on grid-connected voltage source converters. in *2018 International Power Electronics Conference (International Conference on Power Electronics* (2018), pp. 2673–2680
77. Z. Shuai, S. Member, C. Shen, X. Liu, Z. Li, S. Member, Transient angle stability of virtual synchronous generators using lyapunov's direct method. IEEE Trans. Smart Grid **10**(4), 4648–4661 (2019)
78. M. Choopani, S.H. Hosseinian, B. Vahidi, New transient stability and LVRT improvement of multi-VSG grids using the frequency of the center of inertia. IEEE Trans. Power Syst. **35**(1), 527–538 (2020)
79. D. Pan, X. Wang, F. Liu, R. Shi, Transient stability of voltage-source converters with grid-forming control: a design-oriented study. IEEE J. Emerg. Select. Top. Power Electron. **8**(2), 1019–1033 (2020)
80. K. Wang, C. Qi, X. Huang, G. Li, Large disturbance stability evaluation of interconnected multi-inverter power grids with VSG model. J. Eng. **2017**(13), 2483–2488 (2017)
81. X. Hou, K. Sun, N. Zhang, F. Teng, X. Zhang, T.C. Green, Priority-driven self-optimizing power control scheme for interlinking converters of hybrid AC/DC microgrid clusters in decentralized manner. IEEE Trans. Power Electron. **37**(5), 5970–5983 (2021)
82. S. Fu, Y. Sun, J. Lin, S. Chen, M. Su, P/Q-ω/V admittance modeling and oscillation analysis for multi-VSG grid-connected system. IEEE Trans. Power Syst. **38**(6), 5849–5859 (2022)
83. Y. Sun, X. Hou, S. Chen, G. Shi, M. Su, A general [P Q]-[ω V] model of hybrid GFM/GFL multi-VSC systems: power oscillation analysis and suppression method. IEEE J. Emerg. Select. Top. Ind. Electron. (to be updated with page range if available) (2024)
84. L. Wang, T. Li, Z. Cheng, X. Hu, Z. Li, Z. Liu, J. Huang, X. Hou, A unified droop control of AC microgrids under different line impedances: Revisiting droop control and virtual impedance method. Front. Energy Res. **11**, 1190833 (2023)
85. S. Fu, Y. Sun, X. Hou, F. Zhou, M. Su, X. Zhang, K. Sun, Power coupling index (PCI)-oriented stability analysis and interaction control design for inverter-based microgrid under complex line impedance. IEEE Trans. Power Electron. (2024)

86. S. Fu, Y. Sun, Z. Liu, X. Hou, H. Han, M. Su, Power oscillation suppression in multi-VSG grid with adaptive virtual inertia. Int. J. Electr. Power Energy Syst. **135**, 107472 (2022)
87. S. Fu, Y. Sun, L. Li, Z. Liu, H. Han, M. Su, Power oscillation suppression of multi-VSG grid via decentralized mutual damping control. IEEE Trans. Industrial Electron. **69**(10), 10202–10214 (2022)
88. S. Fu, Y. Sun, L. Li, Z. Liu, H. Han, M. Su, Power oscillation suppression in multi-VSG grid by adaptive virtual impedance control. IEEE Syst. J. **16**(3), 4744–4755 (2022)
89. S. Chen, Y. Sun, H. Han, G. Shi, Y. Guan, J.M. Guerrero, Dynamic frequency performance analysis and improvement for parallel VSG systems considering virtual inertia and damping coefficient. IEEE J. Emerg. Select. Top. Power Electron. **11**(1), 478–489 (2022)
90. S. Chen, Y. Sun, X. Hou, H. Han, S. Fu, M. Su, Quantitative parameters design of VSG oriented to transient synchronization stability. IEEE Trans. Power Syst. (2023)
91. S. Chen, Y. Sun, H. Han, S. Fu, S. Luo, G. Shi, A modified VSG control scheme with virtual resistance to enhance both small-signal stability and transient synchronization stability. IEEE Trans. Power Electron. **38**(5), 6005–6014 (2023)
92. L. Wang, H. Zhou, X. Hu, X. Hou, C. Su, K. Sun, Adaptive inertia and damping coordination (AIDC) control for grid-forming VSG to improve transient stability. Electronics **12**(9), 2060 (2023)
93. S. Chen, Y. Sun, H. Han, Z. Luo, G. Shi, L. Yuan, J.M. Guerrero, Active power oscillation suppression and dynamic performance improvement for multi-VSG grids based on consensus control via COI frequency. Int. J. Electr. Power Energy Syst. **147**, 108796 (2023)
94. S. Chen, Y. Sun, X. Hou, X. Zheng, M. Su, K. Sun, A first-order PLL-synchronized voltage-supporting control method for parallel inverters with near-zero line impedance, IEEE Trans. Power Electron. (2024)
95. K. Zhang, M. Su, Z. Liu, H. Han, X. Zhang, P. Wang, A distributed coordination control for islanded hybrid AC/DC microgrid. IEEE Syst. J. **17**(2), 1819–1830 (2023)

Part I
Novel Amplitude-Frequency Modeling

Chapter 2
[P Q]–[ω V] Admittance Modeling Method of Multi-VSG System

2.1 Small-Signal Modeling Method

Affected by different VSG controller designs, line impedance mismatch and other factors, the multi-VSG system faces the risk of the power, frequency and voltage oscillation [1, 2], which will threaten the security and stability of system [3, 4]. As a result, a multi-VSG grid-connected system model must be built in order to investigate the oscillation features and the relevant affecting factors.

Currently, there are two types of small-signal modeling method utilized to investigate the aforementioned oscillation problem [5]: the time-domain modeling method, and the frequency domain modeling method. The time-domain modeling method obtains all the state information of the system by establishing the time domain equation relationship between the variables of the system. They have a high degree of accuracy in estimating system stability. Whereas, they will be quite computationally intensive and complicated as the number of parallel inverters increases, which results in poor application flexibility [6]. Frequency domain modeling relies on the system transfer function. The well-known impedance model based on the transfer function between voltage and current is widely used in grid-connection stability analysis [7, 8] and harmonic resonance analysis [9, 10]. However, from a power system point of view, the four variables [P Q]–[ω V] are the key characteristic indicators that are paid attention to, which are usually used to describe the nodal dynamic characteristics, instead of the traditional impedance model. Thus, this chapter introduces a [P Q]–[ω V] admittance modeling method of grid-connected multi-VSG system.

2.2 [P Q]–[ω V] Admittance Model of Multi-VSG System

2.2.1 Modeling of Three-Phase VSG

The principle diagram of the VSG is shown in Fig. 2.1 [11]. P_i^* and Q_i^* are the active and reactive power references. ω^* and E^* are the system angular frequency

X. Hou et al., *Self-organized Operation and Stability Control of Electronized Power Networks*, https://doi.org/10.1007/978-981-96-9067-1_2

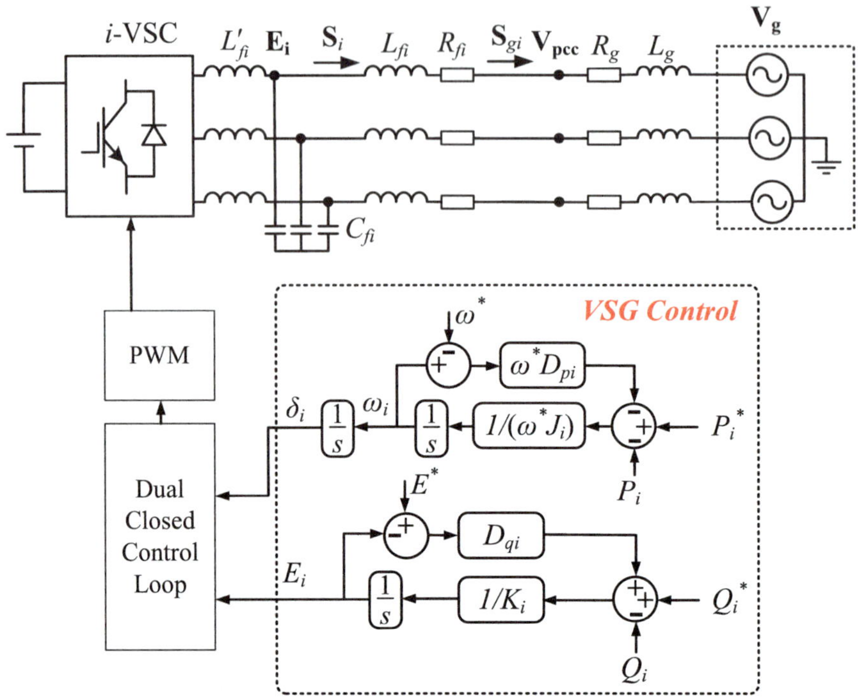

Fig. 2.1 Conventional topology and control scheme of a grid-connected VSG

and voltage reference. J_i and D_{pi} represent the inertia and damping coefficient. D_{qi} represents the voltage-drooping coefficient. C_{fi}, L_{fi} (L'_{fi}) and R_{fi} represent the capacitor filter, the inductor filter and parasitic resistance. L_g and R_g represent the grid inductance and grid resistance, respectively. $\mathbf{E}_i = E_i e^{j\delta_i}$ represents the space vector of the output voltage of the i-VSG. $\mathbf{V}_{pcc} = V_p e^{j\delta_p}$ and $\mathbf{V}_g = V_g e^{j\delta_g}$ are the space vector of the PCC voltage and the grid voltage, where δ_p and δ_g are the PCC voltage phase and grid voltage phase. $\mathbf{S}_i$ represents the apparent power.

In the rotating reference frame, aligning $\mathbf{V}_{pcc}$ with axis d, then $\mathbf{V}_{pdq} = V_{pcc}$, $\mathbf{E}_{idq} = E_i e^{j\delta_i - \delta_p}$ and $\mathbf{V}_{gdq} = V_g e^{j\delta_p - \delta_g}$. $\mathbf{V}_{pdq}$, $\mathbf{E}_{idq}$ and $\mathbf{V}_{gdq}$ are defined as the voltage vectors of $\mathbf{V}_{pcc}$, $\mathbf{E}_i$ and $\mathbf{V}_g$ in dq synchronous frame.

According to Fig. 2.1, in the dq synchronous frame, the main circuit dynamics can be approximately described as

$$\begin{cases} L_{fi}\frac{di_{id}}{dt} = E_{id} - V_{pd} - i_{id}R_{fi} + \omega_n L_{fi} i_{iq} \\ L_{fi}\frac{di_{iq}}{dt} = E_{iq} - V_{pq} - i_{iq}R_{fi} - \omega_n L_{fi} i_{id} \end{cases} \tag{2.1}$$

where i_{id} and i_{iq} represent d-axis and q-axis component output current of the i-VSG. Applying a Laplace transform to (2.1), yields

$$\begin{cases} i_{id} = \frac{(R_{fi}+L_{fi}s)(E_i\cos\delta_{ip}-V_p)+X_{fi}E_i\sin\delta_{ip}}{(R_{fi}+L_{fi}s)^2+X_{fi}^2} \\ i_{iq} = \frac{-X_{fi}(E_i\cos\delta_{ip}-V_p)+(R_f+L_{fi}s)E_i\sin\delta_{ip}}{(R_{fi}+L_{fi}s)^2+X_{fi}^2} \end{cases} \tag{2.2}$$

where $X_{fi} = \omega_n L_{fi}$ and $\delta_{ip} = \delta_i - \delta_p$.

The active and reactive power can be obtained in dq frame:

$$\begin{cases} P_i = 1.5\left(E_{id}i_{id} + E_{iq}i_{iq}\right) \\ Q_i = 1.5\left(-E_{id}i_{iq} + E_{iq}i_{id}\right) \end{cases} \tag{2.3}$$

Substituting (2.2) into (2.3),

$$\begin{cases} P_i = \frac{1.5\left[(R_{fi}+L_{fi}s)(E_i^2-E_iV_p\cos\delta_{ip})+X_{fi}E_iV_p\sin\delta_{ip}\right]}{(R_{fi}+L_{fi}s)^2+X_{fi}^2} \\ Q_i = \frac{1.5\left[X_{fi}E_i(E_i-V_p\cos\delta_{ip})-(R_{fi}+L_{fi}s)E_iV_p\sin\delta_{ip}\right]}{(R_{fi}+L_{fi}s)^2+X_{fi}^2} \end{cases} \tag{2.4}$$

Linearizing (2.4) around the equilibrium points,

$$\tilde{\mathbf{S}}_i = \mathbf{C}_i(s)\,\tilde{\mathbf{v}}_i - \mathbf{D}_i(s)\,\tilde{\mathbf{v}}_p \tag{2.5}$$

where $\tilde{\mathbf{S}}_i = [\tilde{\mathbf{P}}_i\ \tilde{\mathbf{Q}}_i]^T$, $\tilde{\mathbf{v}}_i = [\tilde{\boldsymbol{\omega}}_i\ \tilde{\mathbf{E}}_i]^T$ and $\tilde{\mathbf{v}}_p = [\tilde{\boldsymbol{\omega}}_p\ \tilde{\mathbf{V}}_p]^T$, Similarly, the expression for grid-connected power $\tilde{\mathbf{S}}_{gi}$ can be obtained:

$$\tilde{\mathbf{S}}_{gi} = \mathbf{C}_{ai}(s)\,\tilde{\mathbf{v}}_i - \mathbf{D}_{ai}(s)\,\tilde{\mathbf{v}}_p \tag{2.6}$$

The transfer function matrices $\mathbf{C}_i$, $\mathbf{D}_i$, $\mathbf{C}_{ai}$ and $\mathbf{D}_{ai}$ are given in the Appendix (2.5).

Based on the assumption that the control bandwidth of the power loop is much smaller than that of the dual closed-loop, the VSG dynamics is expressed as

$$\tilde{\mathbf{v}}_i = \mathbf{B}_i(s)\,\tilde{\mathbf{S}}_i^* - \mathbf{B}_i(s)\,\tilde{\mathbf{S}}_i \tag{2.7}$$

where $\tilde{\mathbf{S}}_i^* = [\tilde{\mathbf{P}}_i^*\ \tilde{\mathbf{Q}}_i^*]^T$ and $\mathbf{B}_i(s) = diag\{\frac{1}{J_is+D_{pi}}\ \frac{1}{K_is+D_{qi}}\}$, Simplification is made in $\mathbf{B}_i(s)$ since the ω^* multiplied by J_i and D_{pi} is a constant.

Combining (2.5)–(2.7), the relationship between $\tilde{\mathbf{S}}_i$ and $\tilde{\mathbf{S}}_{gi}$ can be expressed as

$$\begin{aligned} \tilde{\mathbf{S}}_{gi} = {} & [\mathbf{I} - (\mathbf{C}_{ai}(s) - \mathbf{C}_i(s))\,\mathbf{B}_i(s)]\,\tilde{\mathbf{S}}_i \\ & + (\mathbf{C}_{ai}(s) - \mathbf{C}_i(s))\,\mathbf{B}_i(s)\,\tilde{\mathbf{S}}_i^* - (\mathbf{D}_{ai}(s) - \mathbf{D}_i(s))\,\tilde{\mathbf{v}}_p \end{aligned} \tag{2.8}$$

where $\mathbf{I}$ represents a 2×2 unit matrix.

From the inverter side, combining (2.5) and (2.7), yields

$$\tilde{\mathbf{S}}_i = \mathbf{G}_i(s)\,\tilde{\mathbf{S}}_i^* - \mathbf{Y}_{pi}(s)\,\tilde{\mathbf{v}}_p \tag{2.9}$$

where

$$\begin{cases} \mathbf{G}_i(s) = (\mathbf{I} + \mathbf{C}_i(s)\,\mathbf{B}_i(s))^{-1}\mathbf{C}_i(s)\,\mathbf{B}_i(s) \\ \mathbf{Y}_{pi}(s) = (\mathbf{I} + \mathbf{C}_i(s)\,\mathbf{B}_i(s))^{-1}\mathbf{D}_i(s) \end{cases} \tag{2.10}$$

Combining (2.8) and (2.9), the grid-connected power can be represented as:

$$\tilde{\mathbf{S}}_{gi} = \mathbf{G}_{bi}(s)\,\tilde{\mathbf{S}}_i^* - \mathbf{Y}_{bi}(s)\,\tilde{\mathbf{v}}_p \tag{2.11}$$

where

$$\begin{cases} \mathbf{G}_{bi}(s) = [\mathbf{I} - (\mathbf{C}_{ai}(s) - \mathbf{C}_i(s))\,\mathbf{B}_i(s)]\,\mathbf{G}_i(s) + (\mathbf{C}_{ai}(s) - \mathbf{C}_i(s))\,\mathbf{B}_i(s) \\ \mathbf{Y}_{bi}(s) = [\mathbf{I} - (\mathbf{C}_{ai}(s) - \mathbf{C}_i(s))\,\mathbf{B}_i(s)]\,\mathbf{Y}_{pi}(s) + (\mathbf{D}_{ai}(s) - \mathbf{D}_i(s)) \end{cases} \tag{2.12}$$

For a grid-connected inverter worked as a current source, the relation between the voltage and the feeding current is expressed as [7]

$$i = G(s)\,i_{ref} - Y_s(s)\,u_P \tag{2.13}$$

Comparing (2.11) with (2.13), they have the same mathematical expression form that indicates they share some physical principles [13]. Therefore, the Norton equivalent model of the VSG-controlled inverter can be obtained as shown in Fig. 2.2. The reason for choosing power and frequency/voltage as variables is that it is more convenient to analyze power, frequency and voltage oscillation, and influencing factors under disturbances.

By analogy with the derivation process of (2.1)–(2.5), at the grid side, the relation between feeding power $\tilde{\mathbf{S}}_{gi}$ and the grid $\tilde{\mathbf{v}}_g$ ($\tilde{\mathbf{v}}_g = [\tilde{\boldsymbol{\omega}}_g\ \tilde{\mathbf{V}}_g]^T$) is derived as

$$\tilde{\mathbf{S}}_{gi} = \mathbf{C}_g(s)\,\tilde{\mathbf{v}}_p - \mathbf{D}_g(s)\,\tilde{\mathbf{v}}_g \tag{2.14}$$

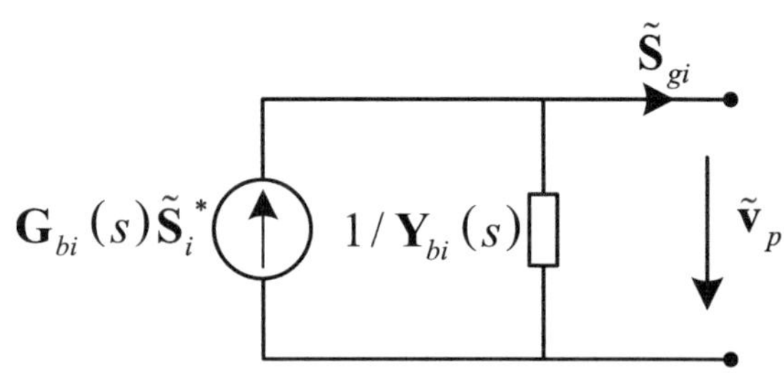

Fig. 2.2 Equivalent Norton model

Further,

$$\tilde{\mathbf{v}}_p = \mathbf{Y}^{-1}{}_g(s)\,\tilde{\mathbf{S}}_{gi} + \mathbf{G}_g(s)\,\tilde{\mathbf{v}}_g \tag{2.15}$$

where $\mathbf{Y}_g(s) = \mathbf{C}_g(s)$ and $\mathbf{G}_g(s) = \mathbf{Y}_g^{-1}(s)\mathbf{D}_g(s)$. The transfer function matrices $\mathbf{C}_g$ and $\mathbf{D}_g$ are shown in the Appendix (2.5).

2.2.2 Modeling of Parallel Three-Phase VSGs

Combining (2.15) and Fig. 2.2, we obtain the $P/Q-\omega/V$ admittance model of n-parallel VSGs system as Fig. 2.3 shows. The simplified Norton equivalent model is shown in Fig. 2.4.

From Fig. 2.4 and (2.14), according to Kirchhoff's law, the following equation is satisfied

$$\sum_{c=1}^{n}\mathbf{G}_{bc}(s)\,\tilde{\mathbf{S}}_c^* - \sum_{c=1}^{n}\mathbf{Y}_{bc}(s)\tilde{\mathbf{v}}_p = \mathbf{C}_g(s)\,\tilde{v}_p - \mathbf{D}_g(s)\,\tilde{v}_g \tag{2.16}$$

Further, the frequency/voltage $\tilde{\mathbf{v}}_p$ of PCC can be expressed as:

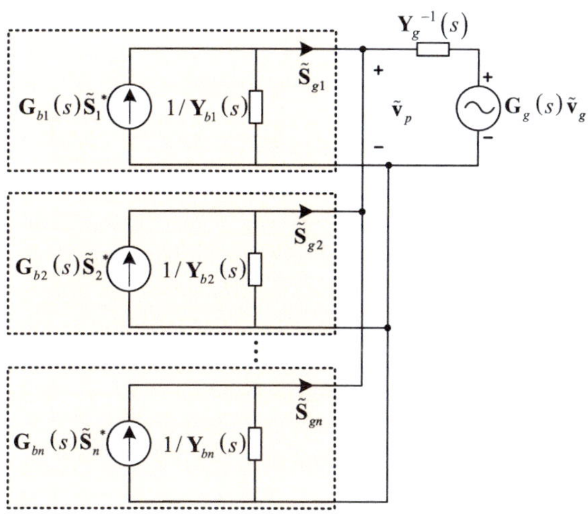

Fig. 2.3 P/Q–ω/V "admittance" model of n-parallel VSG system

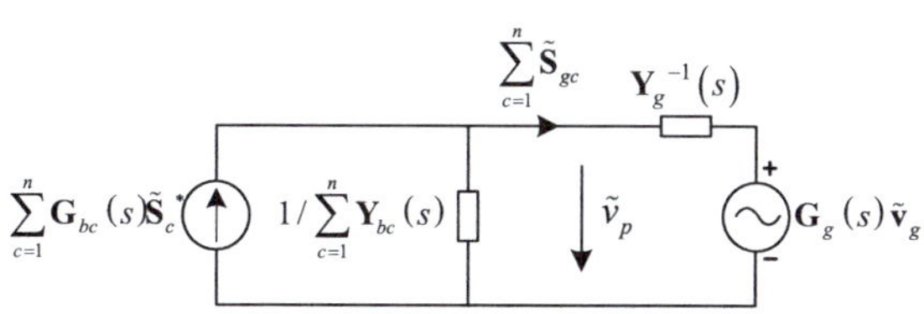

Fig. 2.4 Equivalent Norton model

$$\tilde{\mathbf{v}}_p = \left(\mathbf{Y}_g(s) + \sum_{c=1}^{n} \mathbf{Y}_{bc}(s)\right)^{-1} \left[\sum_{c=1}^{n} \mathbf{G}_{bc}(s)\tilde{\mathbf{S}}_c^* + \mathbf{D}_g(s)\tilde{\mathbf{v}}_g\right] \tag{2.17}$$

As depicted in (2.17), when the reference power of any source fluctuates or grid frequency/voltage fluctuates, frequency and voltage fluctuations at PCC will occur.

2.3 [P Q]–[ω V] Dynamic Response and Stability Analysis

2.3.1 Power, Frequency and Voltage Dynamic Response

By substituting (2.17) into (2.11), the output power of i-VSG can be expressed as

$$\tilde{\mathbf{S}}_i = \mathbf{H}_i(s)\tilde{\mathbf{S}}_i^* - \sum_{t\neq i, t=1}^{n} \mathbf{F}_{i,t}(s)\tilde{\mathbf{S}}_t^* - \mathbf{R}_i(s)\tilde{\mathbf{v}}_g \tag{2.18}$$

where

$$\begin{cases} \mathbf{H}_i(s) = \mathbf{G}_i(s) - \mathbf{Y}_{pi}(s)\left(\mathbf{Y}_g(s) + \sum_{c=1}^{n} \mathbf{Y}_{bc}(s)\right)^{-1} \mathbf{G}_{bi}(s) \\ \mathbf{F}_{i,t}(s) = \mathbf{Y}_{pi}(s)\left(\mathbf{Y}_g(s) + \sum_{c=1}^{n} \mathbf{Y}_{bc}(s)\right)^{-1} \mathbf{G}_{bt}(s) \\ \mathbf{R}_i(s) = \mathbf{Y}_{pi}(s)\left(\mathbf{Y}_g(s) + \sum_{c=1}^{n} \mathbf{Y}_{bc}(s)\right)^{-1} \mathbf{D}_g(s) \end{cases} \tag{2.19}$$

The above (2.18) depicts the power dynamic response of the multi-VSG grid-connected system. It is observed that the output power response of i-VSG is mainly influenced by three parts. The first part, the influence of i-VSG power reference fluctuation on the own output power response is reflected in coefficient $\mathbf{H}_i(s)$. The second part, the influence of t-VSG power reference fluctuation on the i-VSG output power response is reflected in coefficient $\mathbf{F}_{i,t}(s)$. In other words, $\mathbf{F}_{i,t}(s)$ reflects the interaction with other VSG that further complicates the power oscillation characteristics of the system. The third part, the influence of grid frequency/voltage fluctuation on the i-VSG output power response is reflected in coefficient $\mathbf{R}_i(s)$. As a result, by analyzing the three coefficients $\mathbf{H}_i(s)$, $\mathbf{F}_{i,t}(s)$, $\mathbf{R}_i(s)$, and their influencing factors and variation rules, the power oscillation characteristics of the system can be studied. Based on (2.18), the matrices for apparent power of every VSG grid-connected can be expressed as

$$\begin{aligned}&\left[\tilde{\mathbf{S}}_1\ \tilde{\mathbf{S}}_2\ \cdots \tilde{\mathbf{S}}_{n-1}\ \tilde{\mathbf{S}}_n\right]^T\\&=\begin{bmatrix}\mathbf{H}_1 & -\mathbf{F}_{12} & \cdots & -\mathbf{F}_{1(n-1)} & -\mathbf{F}_{1n}\\ -\mathbf{F}_{21} & \mathbf{H}_2 & \cdots & -\mathbf{F}_{2(n-1)} & -\mathbf{F}_{2n}\\ \vdots & \vdots & \ddots & \vdots & \vdots\\ -\mathbf{F}_{(n-1)1} & -\mathbf{F}_{(n-1)2} & \cdots & \mathbf{H}_{n-1} & -\mathbf{F}_{(n-1)n}\\ -\mathbf{F}_{n1} & -\mathbf{F}_{n2} & \cdots & -\mathbf{F}_{n(n-1)} & \mathbf{H}_n\end{bmatrix}\begin{bmatrix}\tilde{\mathbf{S}}_1^*\\ \tilde{\mathbf{S}}_2^*\\ \vdots\\ \tilde{\mathbf{S}}_{n-1}^*\\ \tilde{\mathbf{S}}_n^*\end{bmatrix}-\begin{bmatrix}\mathbf{R}_1\\ \mathbf{R}_2\\ \vdots\\ \mathbf{R}_{n-1}\\ \mathbf{R}_n\end{bmatrix}\tilde{\mathbf{v}}_g\end{aligned} \tag{2.20}$$

Further, the matrix expression of frequency/voltage can be expressed as

$$\begin{aligned}&\left[\tilde{\mathbf{v}}_1\ \tilde{\mathbf{v}}_2\ \cdots\ \tilde{\mathbf{v}}_{n-1}\ \tilde{\mathbf{v}}_n\right]\\&=\mathbf{B}(s)\left[\tilde{\mathbf{S}}_1^*\ \tilde{\mathbf{S}}_2^*\ \cdots \tilde{\mathbf{S}}_{n-1}^*\ \tilde{\mathbf{S}}_n^*\right]^T-\mathbf{B}(s)\left[\tilde{\mathbf{S}}_1\ \tilde{\mathbf{S}}_2\ \cdots \tilde{\mathbf{S}}_{n-1}\ \tilde{\mathbf{S}}_n\right]^T\end{aligned} \tag{2.21}$$

where $\mathbf{B}(s) = \text{diag}\{\mathbf{B}_1(s)\ \mathbf{B}_2(s)\ ...\ \mathbf{B}_{n-1}(s)\ \mathbf{B}_n(s)\}$. Therefore, the frequency/voltage oscillation characteristics can be analyzed based on (2.21).

Combining (2.20) and (2.21), the power-frequency/voltage dynamic response of the multi-VSG grid-connected system is depicted. The structure diagram of the interaction between multi-VSG is shown in Fig. 2.5.

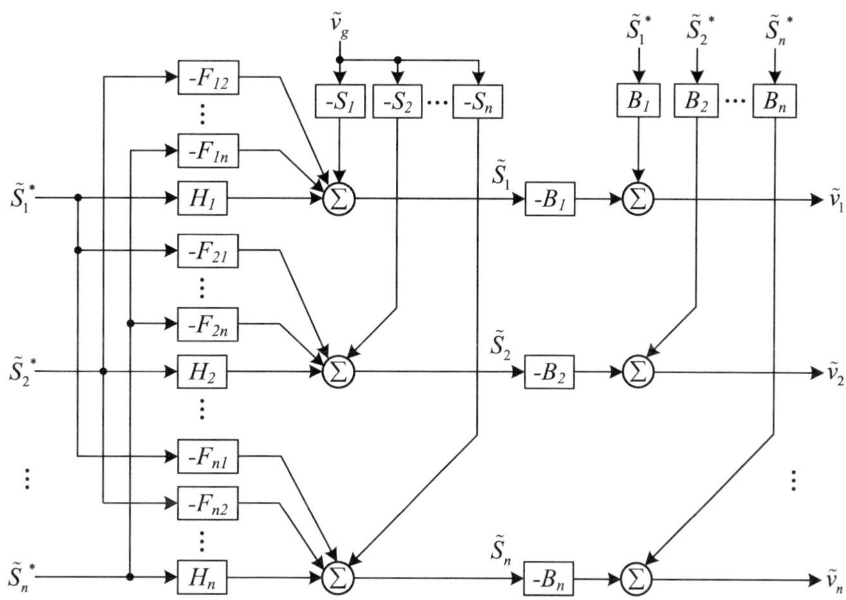

Fig. 2.5 Structure diagram of the interaction of multi-VSG

2.3.2 System Stability Analysis

Based on (2.20), the multi-VSG grid-connected system is stable if and if all poles (transfer functions) in the matrices $\mathbf{H}(s)$, $\mathbf{F}(s)$ and $\mathbf{R}(s)$ have negative real parts.

Assuming that the single VSG is stable, i.e. no right-plane poles from $\mathbf{G}_i(s)$ and $\mathbf{Y}_{pi}(s)$, since transfer function matrix $\mathbf{C}_i$, $\mathbf{D}_i$, $\mathbf{C}_{ai}$ and $\mathbf{D}_{ai}$ are stable, then $\mathbf{G}_{bi}(s)$ and $\mathbf{Y}_{bi}(s)$ are sTable From (2.17), the stability of the system will depend on the matrix $\mathbf{M}_n(s) = \left(\mathbf{Y}_g(s) + \sum_{c=1}^{n} \mathbf{Y}_{bc}(s)\right)^{-1}$

The matrix $\mathbf{M}_n(s)$ can be rebuilt as a determinant and an adjacent matrix:

$$\mathbf{M}_n(s) = \operatorname{adj}\left(\mathbf{Y}_g(s) + \sum_{c=1}^{n} \mathbf{Y}_{bc}(s)\right) \det\left(\mathbf{Y}_g(s) + \sum_{c=1}^{n} \mathbf{Y}_{bc}(s)\right)^{-1} \tag{2.22}$$

The admittance matrix $\mathbf{Y}_{bi}(s)$ of VSG has no right-plane poles, neither to the $\mathbf{Y}_g(s)$. Hence, there are no right-plane poles exist in $\operatorname{adj}\left(\mathbf{Y}_g(s) + \sum_{c=1}^{n} \mathbf{Y}_{bc}(s)\right)$. The system stability is determined by $L_n(s) = \det\left(\mathbf{Y}_g(s) + \sum_{c=1}^{n} \mathbf{Y}_{bc}(s)\right)$ via (2.22). The system is stable if and only if there are no right-plane zeros in $L_n(s)$, i.e. the Nyquist plot does not enclose the origin.

2.4 Oscillation Characteristics Analysis

This part is conducted to analyze the influence of the key parameters on power coupling, oscillation characteristics, and the interaction between multiple VSGs. The introduced model takes active power and reactive power as output variables that is beneficial to analyzing the dynamic power response characteristics of VSG. As (2.17) shows, the power coupling of the i-VSG can be described by transfer functions $\Delta P_i/\Delta Q_i^*$ and $\Delta Q_i/\Delta P_i^*$, i.e. $\mathbf{H}_{i\_12}(s)$ and $\mathbf{H}_{i\_21}(s)$. There is a correspondence between the time domain response curve and the amplitude-frequency characteristic curve. For example, the oscillation frequency and amplitude of the system correspond to the resonance frequency and resonance peak of the amplitude-frequency curve [12]. Such analysis can also be done on other elements of $\mathbf{H}_i(s)$, $\mathbf{F}_{i,t}(s)$ and $\mathbf{R}_i(s)$.

2.4.1 *Impact of Parameters on Power Coupling and Oscillation*

This case analyzes a VSG grid-connected system as Fig. 2.1 shows. Figure 2.6a shows the effect of reactive power disturbance on active power response with different voltage-drooping coefficient D_q. The amplitude-frequency curve exhibits attenuation characteristics at low-frequency bands, and the corresponding steady-state gain is 0. It indicates that the steady-state value of the output active power is not affected by the variation of reactive power. Besides, there are low-frequency (e.g., a few tenths to several Hz) $f_1 = 3.84$ Hz resonance peak and medium-frequency (e.g., tens to hundreds of Hz) $f_2 = 78.34$ Hz resonance peak, which indicates the reactive power reference disturbances cause active power oscillation. With the decrease of voltage-drooping coefficient, the medium-frequency resonant peak around f_2 increases significantly, which indicates the medium-frequency oscillation is more severe. As can be seen from Fig. 2.9a, when D_p decreases to a certain extent, the medium-frequency eigenvalues $\lambda_{1,2}$ cross the imaginary axis. Thus, the system will be unsTable The effect of active power disturbance on reactive power response is shown in Fig. 2.6b. Active power disturbance also causes active power oscillation. With the decrease of voltage-drooping coefficient, the low-frequency resonance peak around f1 slightly decreases. It indicates that the low-frequency oscillation is weakened. Figure 2.7 shows the bode plots of $\mathbf{H}_{i\_12}(s)$ (a) and $\mathbf{H}_{i\_21}(s)$ (b) with integral coefficient K. With the increase of the integral coefficient, the medium-frequency resonant peak decreases significantly, which helps to suppress the medium-frequency oscillation.

Figures 2.8 and 2.9 show the bode plots with different inertia coefficient J and with different damping coefficient D_p, respectively. It is observed that J and D_p mainly affect the low-frequency oscillation. When J increases, the system begins to show low-frequency oscillation, the oscillation frequency decreases, and the resonance peak increases. With the decrease of D_p, the low-frequency resonance peak increases.

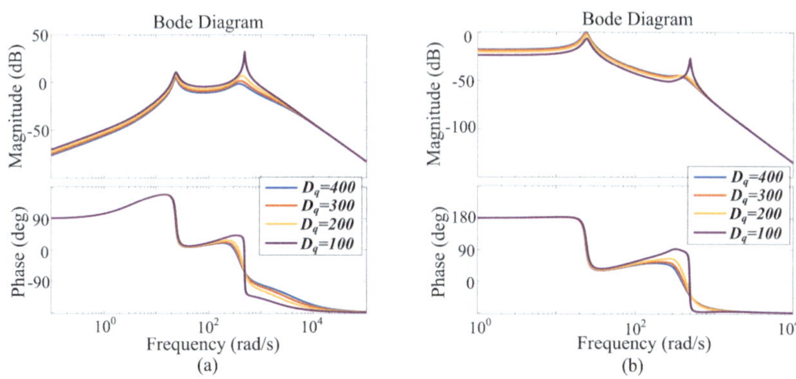

Fig. 2.6 Bode plots of: **a** $\mathbf{H}_{i\_12}(s)$ and **b** $\mathbf{H}_{i\_21}(s)$ with different D_q

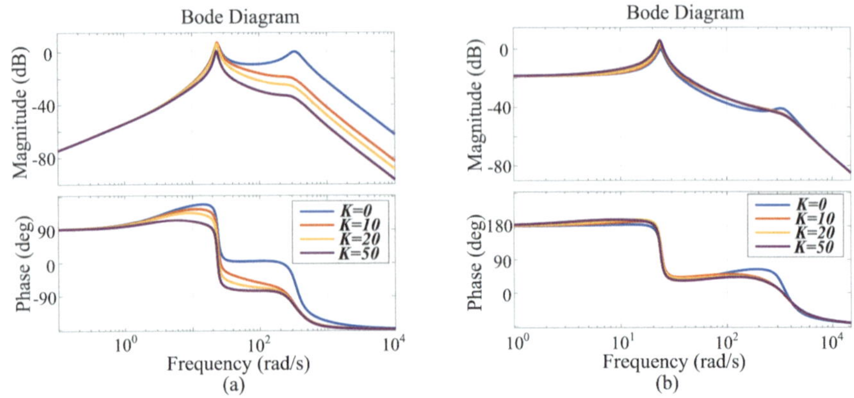

Fig. 2.7 Bode plots of: **a** $\mathbf{H}_{i\_12}(s)$ and **b** $\mathbf{H}_{i\_21}(s)$ with different K

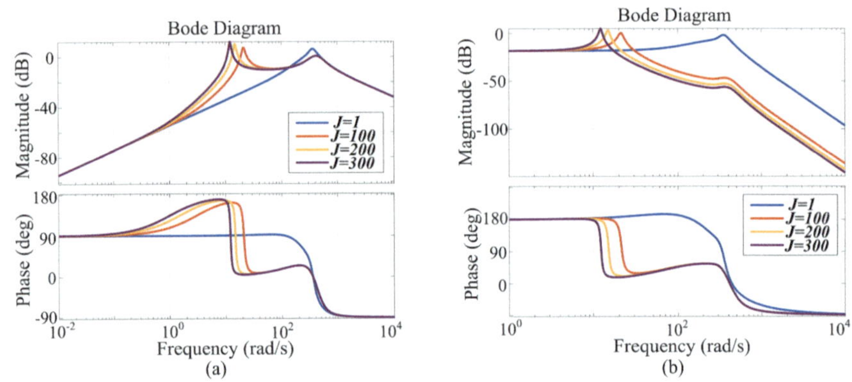

Fig. 2.8 Bode plots of: **a** $\mathbf{H}_{i\_12}(s)$ and **b** $\mathbf{H}_{i\_21}(s)$ with different J

As can be seen from Fig. 2.10a, when D_p decreases to a certain extent, the eigenvalues $\lambda_{3,4}$ of oscillation frequency around f_1 cross the virtual axis and the system becomes unsTable

Based on the above analysis, it is observed that active power loop control parameters have more influence on the low-frequency eigenvalues $\lambda_{1,2}$, and reactive power loop control parameters affect the medium-frequency eigenvalues $\lambda_{3,4}$ more.

2.4.2 *Impact of Grid Resistance*

Figure 2.11 shows the bode plots of $\mathbf{H}_{i\_12}(s)$ (a) and $\mathbf{H}_{i\_21}(s)$ (b) with different grid resistance R_g. It is observed that with the increase of R_g, the medium-frequency resonance peak decreases and the low-frequency resonance peak increases slightly. Besides, the low-frequency gain also increases, which indicates the coupling between

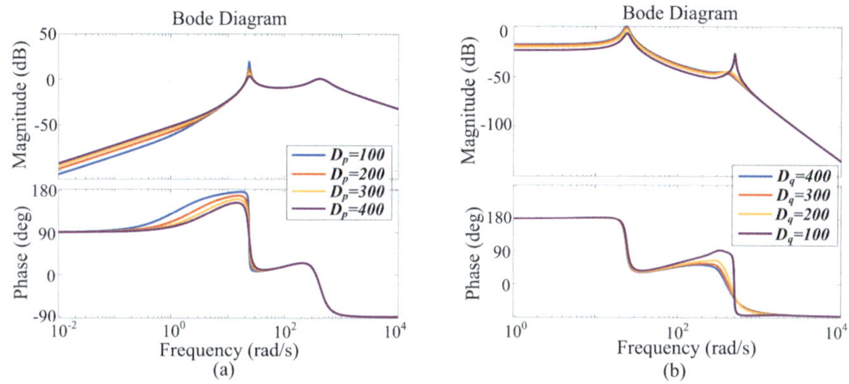

Fig. 2.9 Bode plots of: **a** $\mathbf{H}_{i\_12}(s)$ and **b** $\mathbf{H}_{i\_21}(s)$ with different damping coefficient D_p

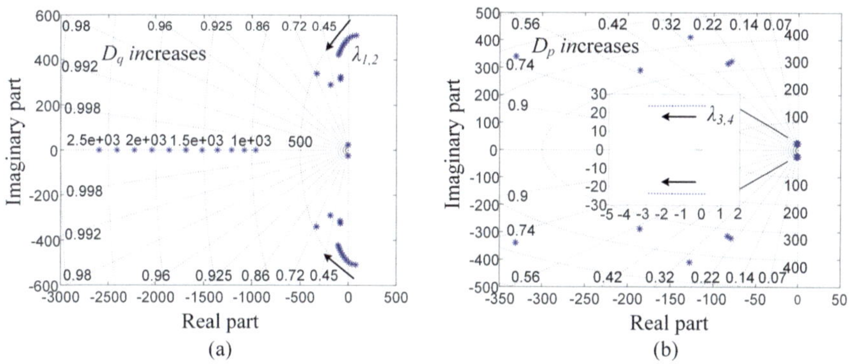

Fig. 2.10 Pole locus of $\mathbf{H}_i(s)$ with: **a** D_q[50, 200], and **b** D_p ∈[25, 300]

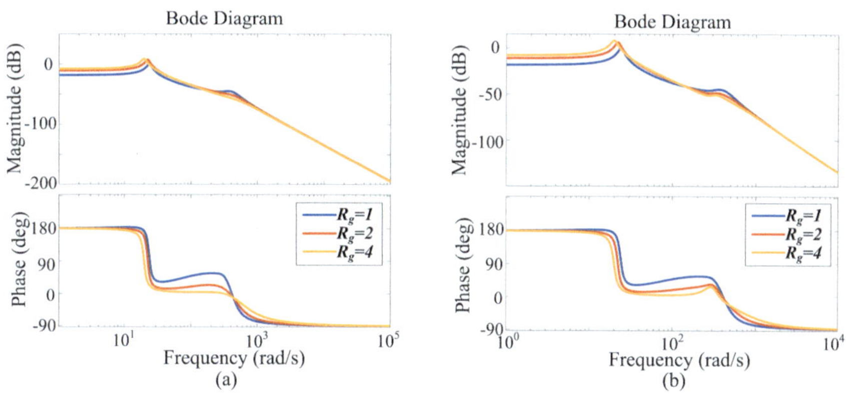

Fig. 2.11 Bode plots of: **a** $\mathbf{H}_{i\_12}(s)$ and **b** $\mathbf{H}_{i\_21}(s)$ with different R_g

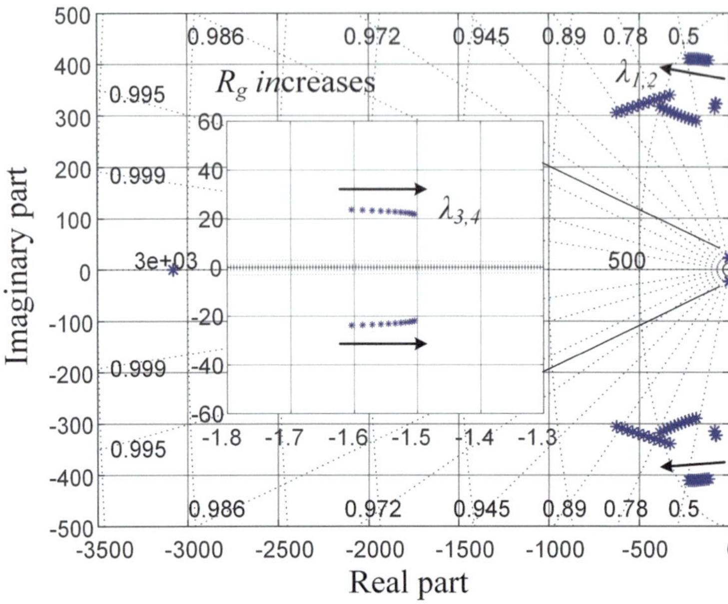

Fig. 2.12 Pole locus of $\mathbf{H}_i(s)$ with grid resistance $R_g \in [1, 2]$

reactive power and active power loop increases. As shown in Fig. 2.12, with the increase of R_g, eigenvalues $\lambda_{1,2}$ move away from the imaginary axis and $\lambda_{3,4}$ move closer to the imaginary axis.

2.4.3 Impact of Number on Power Coupling and Oscillation

Figure 2.13 shows the bode plots of $\mathbf{H}_{i\_12}(s)$ (a) and $\mathbf{H}_{i\_21}(s)$ (b) with different number n of VSGs in parallel. When n increases to 2, the system shows two low-frequency (f_1' and f_1'') resonance peaks, and the medium-frequency resonant peak is much larger. When n continues to increase, the low-frequency f_1' resonance peak is very small, and the resonance peaks of f_1'' and f_2 change not evidently. Figure 2.14a shows the pole locus of D_q changing from 50 to 160, when $n = 1, 2, 3$ and 4. It is observed that when $n > 1$, the damping ratio corresponding to eigenvalues$\lambda_{1,2}$ is weaker. When n continues to increase, the pole locus of eigenvalues $\lambda_{1,2}$ almost coincide. It indicates increasing n deteriorates the stability of eigenvalues $\lambda_{1,2}$. Figure 2.13 shows that the low-frequency resonance peak of the single grid-connected system is larger than the two low frequencies of the multi-grid-connected system. From Fig. 2.14b, with the same parameter, the low-frequency eigenvalues are closer to the virtual axis when $n = 1$ than when $n>1$.

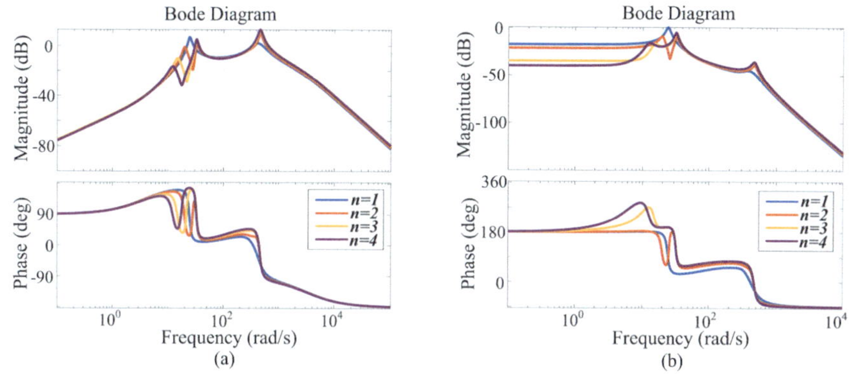

Fig. 2.13 Bode plots of: **a** $\mathbf{H}_{i\_12}(s)$ and **b** $\mathbf{H}_{i\_21}(s)$ with different number n of VSGs in parallel

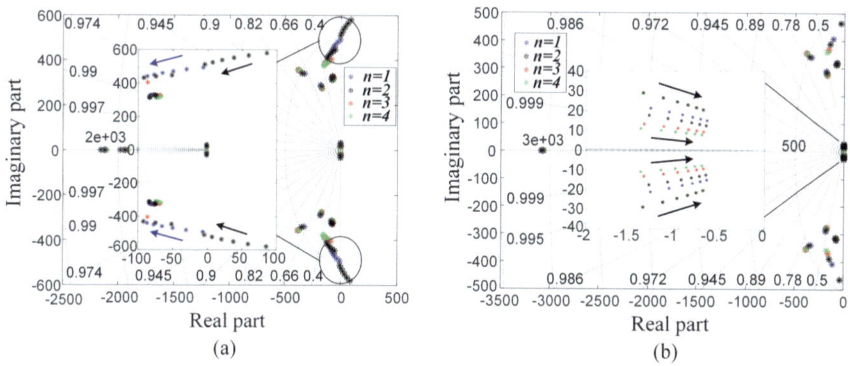

Fig. 2.14 Pole locus of $\mathbf{H}_i(s)$ with: **a** $D_q[50, 160]$ and **b** $D_p[25, 300]$

2.4.4 Interaction Analysis Between Two VSGs

To investigate the interaction between VSGs, the analysis in this case is based on a two-VSG grid-connected system. Figure 2.15a shows the bode plots of $\mathbf{H}_{1\_11}(s)$ and $\mathbf{F}_{12\_11}(s)$ (i.e. $\Delta P_1/\Delta P_1^*$ and $\Delta P_2/\Delta P_1^*$) and Fig. 2.15b shows the bode plots of $\mathbf{H}_{1\_22}(s)$ and $\mathbf{F}_{12\_22}(s)$ (i.e. $\Delta Q_1/\Delta Q_1^*$ and $\Delta Q_2/\Delta Q_1^*$). It is observed that there are two low-frequency resonance peaks and one medium-frequency (f_2) resonance peak. From Fig. 2.15a, under active reference disturbance, the output active power of two VSGs may both appear severe low-frequency oscillation, since $\mathbf{H}_{1\_11}(s)$ and $\mathbf{F}_{12\_11}(s)$ appear the low-frequency resonant peaks. The corresponding frequencies of the two low-frequency resonant peaks are $f_1' = 3.1$ Hz and $f_1'' = 5.2$ Hz respectively. Similarly, from Fig. 2.15b, under reactive reference disturbance, the output reactive power of two VSGs may both exhibit severe medium-frequency oscillation with the frequency of $f_2 = 73.6$ Hz. Besides, the low-band phase of $\mathbf{F}_{12\_22}(s)$ is 180°, which indicates that the reactive power of VSG2 changes in the direction opposite to that of the reactive power reference command ΔQ_1^*.

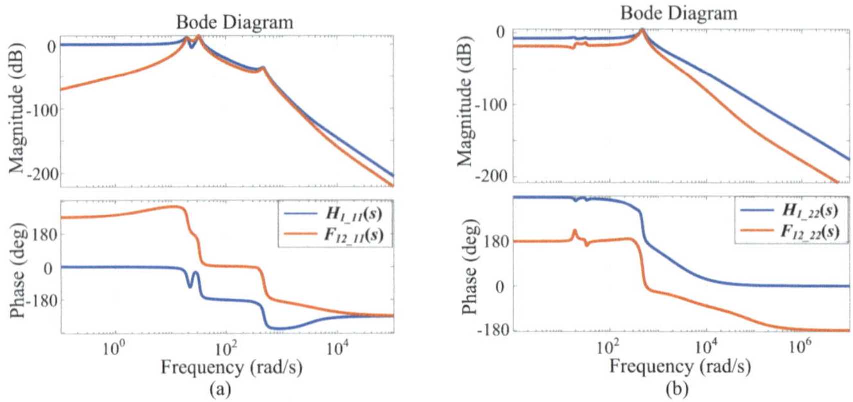

Fig. 2.15 Bode plots of: **a** $\mathbf{H}_{1\_11}(s)$ and $\mathbf{F}_{12\_11}(s)$, **b** $\mathbf{H}_{1\_22}(s)$ and $\mathbf{F}_{12\_22}(s)$

Table 2.1 CHIL parameter

Parameter	Value	Parameter	Value
Power reference P^*	11 kW	Filter inductance L_f	6 mH
Damping coefficient D_p	300	Filter resistance R_f	0.5 Ω
Inertia coefficient J	80	Line inductance L_g	4 mH
Droop coefficient D_q	300	Line resistance R_g	1 Ω
Integral coefficient K	0	Angular frequency ω^*	314 rad/s

The control-hardware-in-loop (CHIL) experimental results are displayed to validate the correctness of the introduced [P Q]–[ω V] "admittance" model and the oscillation characteristic analysis discussed in Sect. 2.4. The parameters adopted in the experiment are listed in Table 2.1. The main circuit is emulated in the OPAL-RT4510 simulator. The controller of the converter is implemented in the DSP control board. The experimental data is extracted from the oscilloscope.

2.4.5 *Power-Frequency/Voltage Response*

Figure 2.16 shows the active power and angular frequency response waveforms. At $t = 10$ s, the VSG is subjected to step perturbation of active power command ΔP^* = 1 kW. The blue line represents the estimated response waveform based on the model introduced in [12]. The model in [12] predicts the active power and frequency response well, when the system impedance is highly inductive and the voltage fluctuation is negligible. However, in practice, it is difficult to ensure that the grid impedance is fully inductive. Thus, the output active power and reactive power are controlled by

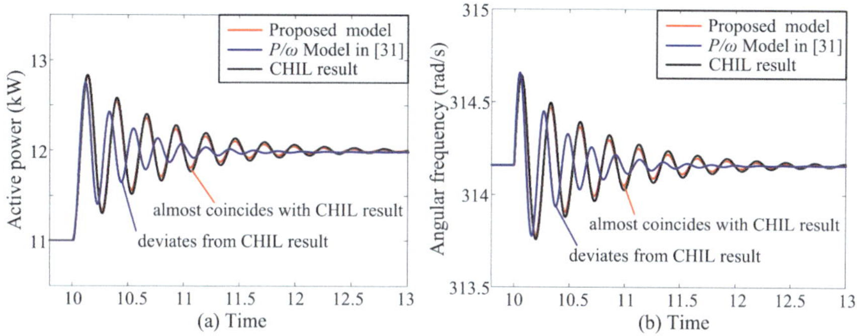

Fig. 2.16 Active power **a** frequency and **b** response waveform under disturbance

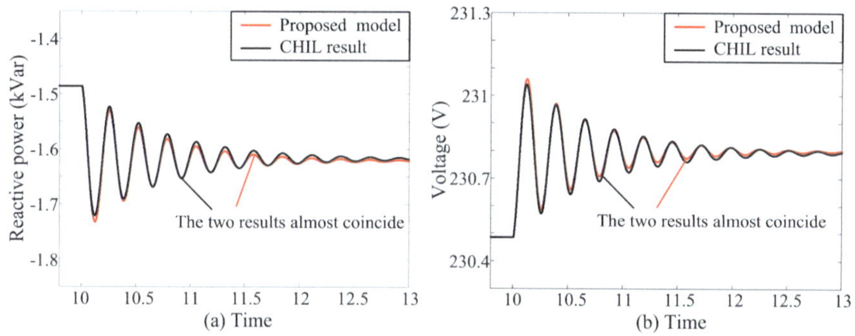

Fig. 2.17 Reactive power **a** voltage and **b** response waveform under disturbance

the power angle and voltage amplitude simultaneously, and their dynamic process is coupled with each other. In this situation, comparing with the model in [12], as shown in Fig. 2.16, the introduced $[P\ Q]$–$[\omega\ V]$ model is more accurate that almost unmistakably reflect the key information of system oscillation (oscillation damping, frequency and amplitude). As Fig. 2.17 shows, the introduced model also tracks the output reactive power and voltage of the inverter well. Consequently, the introduced $[P\ Q]$–$[\omega\ V]$ model is convenient to analyze power-frequency/voltage oscillation characteristics for the VSG system.

2.4.6 *Impact of Control Parameters on Oscillation*

This case is conducted to verify the influence of key control parameters on power coupling and system oscillation. As shown in Fig. 2.18, a step change $\Delta Q^* = 1$ kvar is applied to the VSG at 10 s. In the figures, MFO represents medium frequency oscillation and LFO represents low frequency oscillation. It is observed that there are

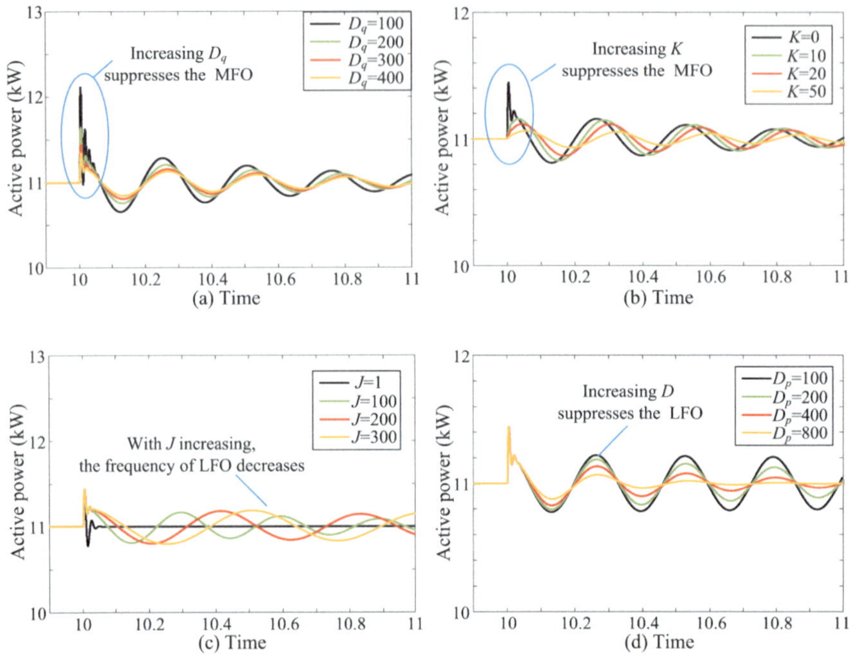

Fig. 2.18 Active power response waveform with different **a** D_q, **b** K, **c** J and **d** D_p under disturbance

two oscillation frequencies (f_1 and f_2) of active power and the medium-frequency (f_2) oscillation decay faster than the low-frequency (f_1) oscillation. As shown in Fig. 2.18a, as the reactive droop coefficient D_q decreases, the overshoot of the active power caused by medium-frequency oscillation increases, and the corresponding oscillation decay more slowly. As shown in Fig. 2.18b, when K increases, medium frequency oscillation is suppressed and the amplitude of low-frequency oscillations is reduced. As shown in Fig. 2.18c, d, the inertia coefficient J and the damping coefficient D_p have no effect on the overshoot of the active power caused by medium frequency oscillation. When J is very small, there is no low-frequency oscillation on active power. As J becomes larger, the oscillation frequency f_1 of active power becomes smaller. The damping coefficient D_p has little impact on the oscillation frequency, but more impact on the amplitude of low-frequency oscillation. When D_p increases, the amplitude of oscillation decreases.

2.4.7 Impact of Grid Resistance on Oscillation

The influence of grid resistance R_g on power coupling and system oscillation is verified in this case. A step change is applied to reactive power reference at 10 s.

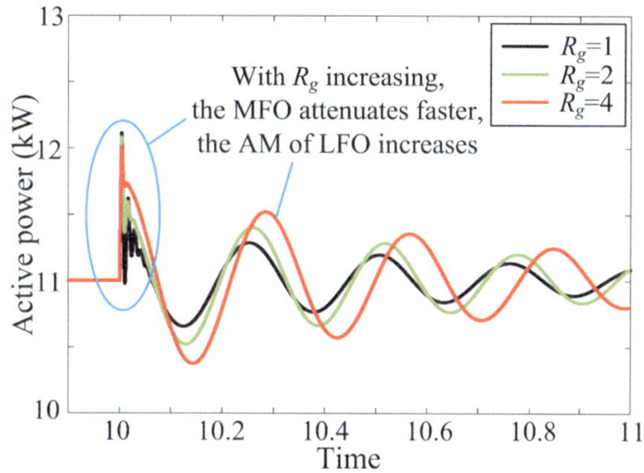

Fig. 2.19 Active power response waveform with different R_g under disturbance

As shown in Fig. 2.19 (AM represents amplitude), with the increase of R_g, the medium-frequency oscillation of the output active power attenuates faster. However, the low-frequency oscillation is more severe, and the amplitude of the low-frequency oscillations increased obviously.

2.4.8 *Impact of Parallel Number on Oscillation*

This case is conducted to verify the influence of the number of VSGs in parallel on power coupling and system oscillation. In this case, the difference of the parallel VSG is ignored, and the VSG parameters are set to be the same. As shown in Fig. 2.20, as n goes from 1 to 2, the damping of medium-frequency oscillation decreases distinctly and the system has more than one oscillation frequency in the low frequency band. As n continues to increase, the response curve of the output power changes little, since the low-frequency f_1' resonance peak is very small, and the increase of low-frequency f_1'' and medium frequency f_2 resonance peak is not obvious as Fig. 2.13 shows.

In a word, the power coupling and oscillation characteristic analysis discussed in Sect. 2.4 are in agreement with the above CHIL results, which indicating the introduced $[P\ Q]$–$[\omega\ V]$ "admittance" model reflects the dynamic performance of the system well.

2.4.9 *Interaction Among VSGs*

The interaction between VSGs is verified in this case. Figure 2.21a shows the active power response when a step change is applied to active power reference $\Delta P_1^* = 1$

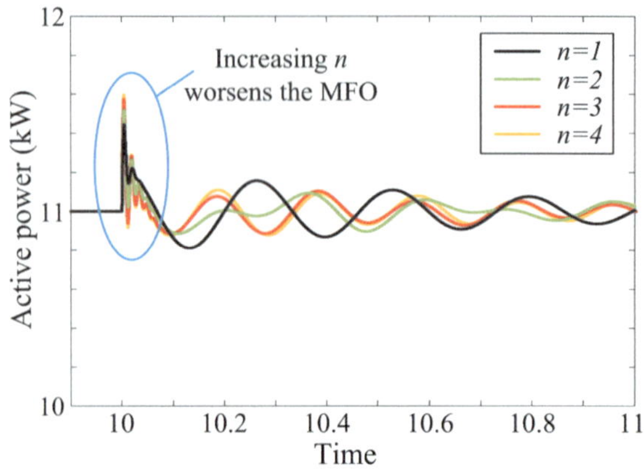

Fig. 2.20 Active power response waveform with different number of parallel VSG under disturbance

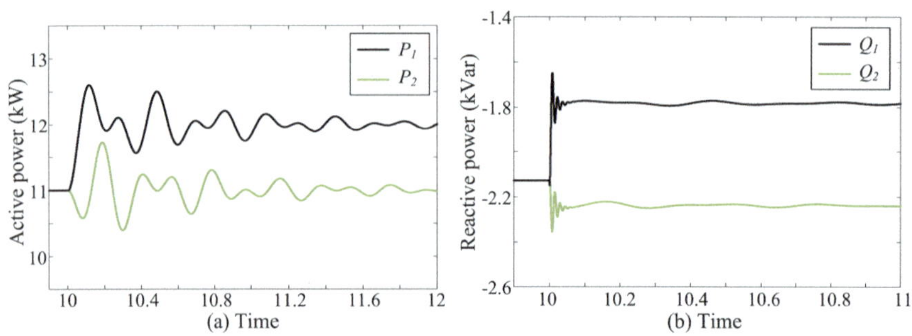

Fig. 2.21 **a** active power response waveforms under active power reference disturbance, and **b** reactive power response waveforms under reactive power reference disturbance

kW at 10 s. It is observed that severe low-frequency oscillations occur in the output active power of the two VSGs. Besides, the oscillation frequency of the output active power ($f_1' = 3\,\text{Hz}$ and $f_1'' = 5\,\text{Hz}$) is obtained by FFT analysis tool. Figure 2.21b shows the reactive power response when a step change is applied to reactive power reference $\Delta Q_1^* = 1$ kVar at 10 s. As seen, the output reactive power of VSGs exhibits severe medium-frequency oscillation ($f_2 = 70\,\text{Hz}$). Thus, the analysis of the interaction between VSGs reflects almost accurately the system oscillation and its frequency. Besides, the reactive power of VSG2 changes in the direction opposite ΔQ_1^* in Fig. 2.21b, since the low-band phase of $\mathbf{F}_{12\_22}(s)$ is 180°. Summing up the above, the correctness of analysis in Sect. 2.4.4, Sect. 2.4 is verified by the CHIL results.

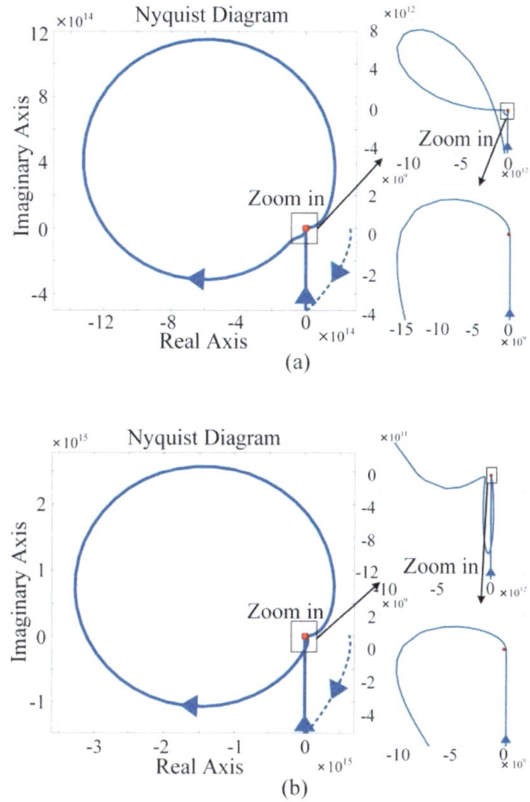

Fig. 2.22 Nyquist plots of: **a** $L_1(s) \times (s+1)^3$, and **b** $L_2(s) \times (s+1)^3$

2.4.10 *Verification of Stability Criterion*

The stability criterion given in Sect. 2.3 is verified in this section. The control parameters are set as follows: $D_p = 300$, $J = 80$, $D_q = 150$ and $K = 1$. In order to avoid Nyquist plots of $L(s)$ passing through the origin, Fig. 2.22a, b draws Nyquist plots of $L_1(s) \times (s+1)^3$ ($n = 1$) and $L_2(s) \times (s+1)^3$ ($n = 2$), respectively. It is observed that the system is stable when $n = 1$, but it becomes unstable when $n = 2$ since it encircles the point (0, 0). Figure 2.23 shows the output current waveform of VSG1. When $t<0.5$s, the single VSG grid-connected system is sTable At t=0.5 s, the second VSG is connected in parallel with the first VSG, and the system becomes unsTable The CHIL results verify the correctness of stability criterion.

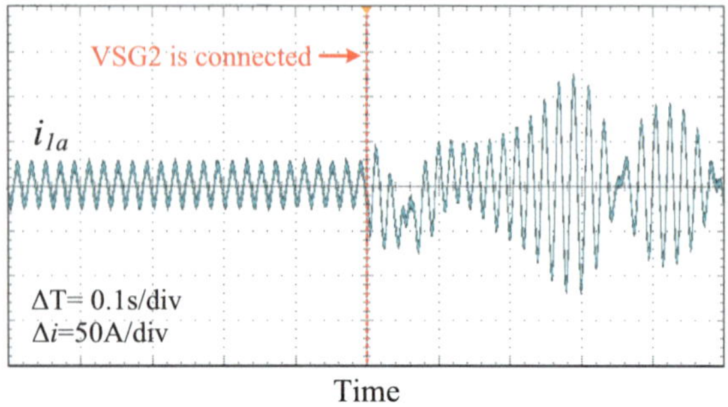

Fig. 2.23 Output current waveform of VSG1

2.5 Conclusion

This chapter establishes a $[P\ Q]$–$[\omega\ V]$ "admittance" model of the multi-VSG grid-connected system. The contributions of the introduced model include: (1) Based on the model, the power and frequency/voltage responses of VSGs under input power reference and grid voltage/frequency disturbances are described. (2) The influence of key parameters on the power dynamic coupling and oscillation characteristics of the multi-VSG system is analyzed. Small damping coefficient D_p leads to low frequency oscillation and small voltage-droop coefficient D_q leads to medium frequency oscillation, which are detrimental to the system stability. (3) With the increase of grid resistance R_g, medium frequency oscillation is suppressed and the low-frequency oscillation is more serious. (4) Based on the model, the stability criterion of multi-VSG grid-connected system is obtained. The accuracy of the established admittance model and the correctness of the oscillation characteristic analysis have been verified by CHIL results.

Appendix

$\mathbf{C}_i$, $\mathbf{D}_i$, $\mathbf{C}_{ai}$, $\mathbf{D}_{ai}$, $\mathbf{C}_g$ and $\mathbf{D}_g$ are 2 × 2 matrices and the elements can be expressed as:

$$C_{i-11} = Z_{li}\frac{X_{fi}E_{i0}V_{p0}\cos\delta_{ip0}+\left(R_{fi}+L_{fi}s\right)E_{i0}V_{p0}\sin\delta_{ip0}}{s}$$
$$C_{i-12} = Z_{li}\left[\left(R_{fi}+L_{fi}s\right)\left(2E_{i0}-V_{p0}\cos\delta_{ip0}\right)+X_{fi}V_{p0}\sin\delta_{ip0}\right]$$

$$C_{i-21} = Z_{li}\frac{X_{fi}E_{i0}V_{p}\sin\delta_{ip0}-\left(R_{fi}+L_{fi}s\right)E_{i0}V_{p}\cos\delta_{ip0}}{s}$$
$$C_{i-22} = Z_{li}\left[2X_{fi}E_{i0}-X_{fi}V_{p0}\cos\delta_{ip0}-\left(R_{fi}+L_{fi}s\right)V_{p0}\sin\delta_{ip0}\right]$$
$$D_{i-11} = C_{i-11};\ D_{i-21} = C_{i-21}$$
$$D_{i-12} = Z_{li}\left[\left(R_{fi}+L_{fi}s\right)E_{i0}\cos\delta_{ip0}-X_{fi}E_{i0}\sin\delta_{ip0}\right]$$
$$D_{i-22} = Z_{li}\left[X_{fi}E_{i0}\cos\delta_{ip0}+\left(R_{fi}+L_{fi}s\right)E_{i0}\sin\delta_{ip0}\right]$$

$$C_{ai-11} = Z_{li}\frac{X_{fi}V_{p0}E_{i0}\cos\delta_{ip0} - \left(R_{fi}+L_{fi}s\right)V_{p0}E_{i0}\sin\delta_{ip0}}{s}$$
$$C_{ai-12} = Z_{li}\left[\left(R_{fi}+L_{fi}s\right)V_{p0}\cos\delta_{ip0} + X_{fi}V_{p0}\sin\delta_{ip0}\right]$$
$$C_{ai-21} = Z_{li}\frac{-X_{fi}V_{p0}E_{i0}\sin\delta_{ip0} - \left(R_{fi}+L_{fi}s\right)V_{p0}E_{i0}\cos\delta_{ip0}}{s}$$

$$C_{ai-22} = Z_{li}\left[X_{fi}V_{p0}\cos\delta_{ip0} - \left(R_{fi}+L_{fi}s\right)V_{p0}\sin\delta_{ip0}\right]$$
$$D_{ai-11} = C_{ai-11};\ D_{ai-21} = C_{ai-21}$$
$$D_{ai-12} = Z_{li}\left[\left(R_{fi}+L_{fi}s\right)\left(2V_{p0} - E_{i0}\cos\delta_{ip0}\right) - X_{fi}E_{i0}\sin\delta_{ip0}\right]$$
$$D_{ai-22} = Z_{li}\left[2X_{fi}V_{p0} + \left(R_{fi}+L_{fi}s\right)E_{i0}\sin\delta_{ip0} - X_{fi}E_{i0}\cos\delta_{ip0}\right]$$

$$C_{g-11} = Z_{lg}\frac{V_{p0}V_{g0}\left(R_g+sL_g\right)\sin\delta_{pg0} + X_gV_{p0}V_{g0}\cos\delta_{pg0}}{s}$$
$$C_{g-21} = Z_{lg}\left[\left(R_g+sL_g\right)\left(2V_{p0} - V_{g0}\cos\delta_{pg0}\right) + X_gV_{g0}\sin\delta_{pg0}\right]$$
$$C_{g-12} = Z_{lg}\frac{V_{p0}V_{g0}X_g\sin\delta_{pg0} - V_{p0}V_{g0}\left(R_g+sL_g\right)\cos\delta_{pg0}}{s}$$

$$C_{g-22} = Z_{lg}\left[2X_gV_{p0} - V_{g0}X_g\cos\delta_{pg0} - V_{g0}\left(R_g+sL_g\right)\sin\delta_{pg0}\right]$$

$$D_{g-11} = C_{g-11};\ D_{g-21} = C_{g-21}$$
$$D_{g-12} = Z_{lg}\left[-\left(R_g+sL_g\right)\cos\delta_{pg0}V_{g0} + X_gV_{p0}\sin\delta_{pg0}\right]$$

$$D_{g-22} = Z_{lg}\left[-V_{p0}X_g\cos\delta_{pg0} - V_{p0}\left(R_g+sL_g\right)\sin\delta_{pg0}\right]$$

where $X_{fi} = \omega_n L_{fi}$, $X_g = \omega_n L_g$, $\delta_{ip0} = \delta_{i0} - \delta_{p0}$, $\delta_{pg0} = \delta_{p0} - \delta_{g0}$ and $Z_{li} = \frac{1.5}{\left(R_{fi}+L_{fi}s\right)^2 + X_{fi}^2}$, $Z_{lg} = \frac{1.5}{\left(R_g+sL_g\right)^2 + X_g^2}$, The variables subscript "0" represent the corresponding steady-state value.

References

1. J. Roldan-Perez, A. Rodriguez-Cabero, M. Prodanovic, Design and analysis of virtual synchronous machines in inductive and resistive weak grids. IEEE Trans. Energy Convers. **34**(4), 1818–1828 (2019)
2. K. Shi, W. Song, H. Ge, P. Xu, Y. Yang, F. Blaabjerg, Transient analysis of microgrids with parallel synchronous generators and virtual synchronous generators. IEEE Trans. Energy Convers. **35**(1), 95–105 (2020)
3. L. Ruttledge, N.W. Miller, J.O. Sullivan, D. Flynn, Frequency response of power systems with variable speed wind turbines. IEEE Trans. Sustain. Energy **3**(4), 683–691 (2012)
4. P. Kundur, Power system stability and control, in *The EPRI Power system Engineering Series*
5. Mohammad Amin, Marta Molinas, Small-signal stability assessment of power electronics based power systems: a discussion of impedance-and eigenvalue-based methods. IEEE Trans. Ind. Appl. **53**(5), 5014–5030 (2017)
6. X. Guo et al., Dynamic phasors-based modeling and stability analysis of droop-controlled inverters for microgrid applications. IEEE Trans. Smart Grid **5**(6), 2980–2987 (2014)
7. J. Sun, Impedance-based stability criterion for grid-connected inverters. IEEE Trans. Power Electron. **26**(11), 3075–3078 (2011)
8. H. Liu, S. Shah, J. Sun, An impedance-based approach to HVDC system stability analysis and control development, in *2014 International Power Electronics Conference, Japan* (2014)

9. L. Hong, W. Shu, J. Wang, R. Mian, Harmonic resonance investigation of a multi-inverter grid-connected system using resonance modal analysis. IEEE Trans. Power Delivery **34**(1), 63–72 (2018)
10. K. Sun, W. Yao, J. Fang, X. Ai, J. Wen, S. Cheng, Impedance modeling and stability analysis of grid-connected DFIG-based wind farm with a VSC-HVDC. IEEE J. Emerg. Select. Top. Power Electron. **8**(2), 1375–1390 (2020)
11. Q. Zhong, G. Weiss, Synchronverters: inverters that mimic synchronous generators. IEEE Trans. Industr. Electron. **58**(4), 1259–1267 (2011)
12. B. Qin, Y. Xu, C. Yuan, J. Jia, A unified method of frequency oscillation characteristic analysis for multi-VSG grid-connected system. IEEE Trans. Power Delivery **37**(1), 279–289 (2022)
13. B. Qin, Y. Xu, C. Yong, J. Jia, The P/ω admittance modeling and power-frequency oscillation analysis of multi-VSGs grid-connected systems. Proc. CSEE **40**(9), 2932–2942 (2020)

Chapter 3
General $[P\ \ Q]$–$[\omega\ V]$ Modeling Method of Hybrid GFM/GFL Multi-VSC Systems

3.1 Classical Converter Synchronous Control Type

According to the different characteristics of the control scheme, voltage source converters (VSCs) are divided into grid-following (GFL) control-based VSCs and grid-forming (GFM) control-based VSCs [1, 2]. The GFL converter is controlled into a controlled current source, and the voltage phase at the common coupling point of the grid-connected converter is measured, which is used to obtain the power or current given. The GFM converter is controlled into a controlled voltage source, and the core idea of the grid converter is to learn from the physical synchronization mechanism of synchronous generator in the power electronic converter, and make the converter possess voltage support and frequency regulation ability.

Thus, this chapter establishes general $[P\ \ Q]$–$[\omega\ V]$ modeling method of hybrid GFM/GFL multi-VSC systems, which is applicable to general fully connected network topology. It can be used as a standard model to locate the weak source of serious low and medium frequency power oscillation under disturbance, and to evaluate the effectiveness of various low and medium frequency oscillation suppression methods. It provides a convenient and universal tool to analyze the low and medium frequency power oscillation and stability.

3.2 $[P\ \ Q]$–$[\omega/V]$ Model of a Single VSC

3.2.1 Modeling of Grid-Forming-Control-Based VSC

VSG control is a classic grid-forming control scheme and its block diagram is illustrated in Fig. 3.1. The $[P\ \ Q]$–$[\omega\ V]$ modeling of the VSG will be introduced in detail. As shown in Fig. 3.1, V_{pi} is the amplitude of the inverter output voltage. δ_{pi} is the output voltage phase. S_i represents the output apparent power of the VSG-i.

X. Hou et al., *Self-organized Operation and Stability Control of Electronized Power Networks*, https://doi.org/10.1007/978-981-96-9067-1_3

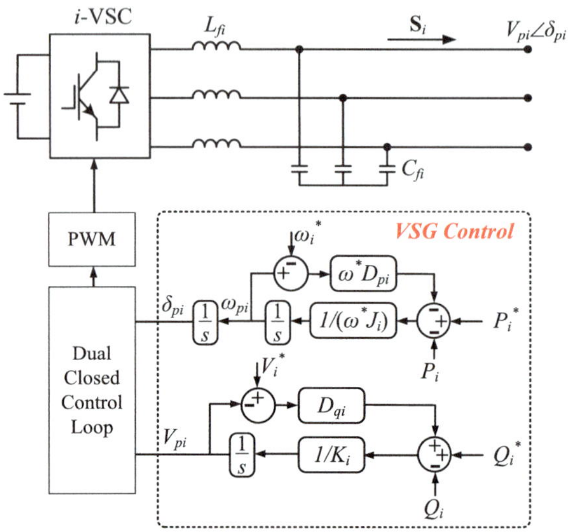

Fig. 3.1 Block diagram of a classic grid-forming control, i.e. VSG control

When establishing the model of the inverter, considering the dual closed loop helps study the high-frequency oscillation of the system. However the order of the whole model is high, which increases the difficulty of calculation. Thus, for simplicity, we assume that the bandwidth of the inner loop is much larger than that of the power control, ignoring the dual closed loop and focusing on studying the low and medium frequency (a few tenths to tens Hz) power oscillation in this chapter. Then, the small signal model of the VSG is expressed as

$$\tilde{\mathbf{v}}_{pi} = \mathbf{G}_{bi}(s)\,\tilde{\mathbf{v}}_i^* - \mathbf{Z}_{bi}(s)\,\tilde{\mathbf{S}}_i \tag{3.1}$$

where $\sim$ represents the small variations in variables, $\tilde{\mathbf{v}}_i^* = [\tilde{\omega}_i^*\ \tilde{V}_i^*]^T$ represents the reference disturbance, $\tilde{\mathbf{v}}_{pi}^* = [\tilde{\omega}_{pi}^*\ \tilde{V}_{pi}^*]^T$ is the output angular frequency and voltage amplitude, $\tilde{\mathbf{S}}_i = [\tilde{P}_i\,\tilde{Q}_i]^T$, $\mathbf{Z}_{bi}(s) = diag\{\frac{1}{J_i\omega^* s + D_{pi}\omega^*}\ \frac{1}{K_i s + D_{qi}}\}$ and $\mathbf{G}_{bi}(s) = diag\{\frac{D_{pi}\omega^*}{J_i\omega^* s + D_{pi}\omega^*}\ \frac{D_{qi}}{K_i s + D_{qi}}\}$.

According to the conventional modeling method [3], the model of the VSC can be expressed as

$$\tilde{\mathbf{u}}_c = \mathbf{G}_s(s)\,\tilde{\mathbf{u}}_r - \mathbf{Z}_s(s)\,\tilde{\mathbf{i}}_g \tag{3.2}$$

Usually, the VSC under grid-forming control is designed to be stable under current-source loads in island mode. Thus, all the transfer functions $\mathbf{G}_s(s)$,$\mathbf{Z}_s(s)$, $\mathbf{G}_{bi}(s)$ and $\mathbf{Z}_{bi}(s)$ are sTable Furthermore, the models of the VSC can be described as a “voltage source” and an “impedance” in series with it as shown in Fig. 3.2. The equivalent circuit shown in Fig. 3.2a is derived from (3.2), and the circuit shown in Fig. 3.2b is derived from (3.1), where a constant power load is considered. Further, the idea of the impedance-based stability analysis can be borrowed [4].

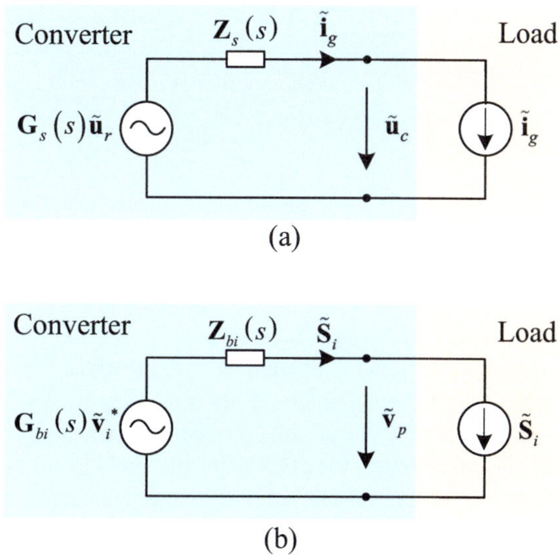

Fig. 3.2 (**a**) Conventional impedance model and (**b**) the introduced model of an off-grid GFM control-based VSC

Besides, the port equivalent [P Q]–[ω V] model of a general RL load is derived. The load inductance dynamic model is expressed as:

$$\begin{cases} L_{load}\frac{di_{id}}{dt} = V_{pid} - i_{id}R_{load} + \omega_n L_{load} i_{iq} \\ L_{load}\frac{di_{iq}}{dt} = V_{piq} - i_{iq}R_{load} - \omega_n L_{load} i_{id} \end{cases} \tag{3.3}$$

where i_{idq} represents output current, ω_n is the steady angular frequency, $V_{pid} = V_{pi}\cos\delta_{in}$, $V_{piq} = V_{pi}\sin\delta_{in}$ and $\delta_{in} = \delta_{pi} - \delta_n$. R_{Load} and L_{Load} are the load resistance and inductance, $X_{Load} = \omega_n L_{Load}$.

From (3.3), the steady-state values of the current i_{idq} are obtained:

$$\begin{cases} i_{id0} = \frac{R_{load}V_{pid0} + X_{load}V_{piq0}}{R_{load}{}^2 + X_{load}{}^2} \\ i_{iq0} = \frac{-X_{load}V_{pid0} + R_{load}V_{piq0}}{R_{load}{}^2 + X_{load}{}^2} \end{cases} \tag{3.4}$$

where $X_{ij} = \omega_n L_{ij}$,$V_{pid0} = V_{pi0}\cos\delta_{in0}$ and $V_{piq0} = V_{pi0}\sin\delta_{in0}$. The subscript "0" stands for the steady-state value.

Linearizing (3.3) around the equilibrium points yields

$$\begin{cases} \tilde{i}_{id} = \frac{(R_{load} + L_{load}s)\tilde{V}_{pid} + X_{load}\tilde{V}_{piq}}{(R_{load} + L_{load}s)^2 + X_{load}{}^2} \\ \tilde{i}_{iq} = \frac{-X_{load}\tilde{V}_{pid} + (R_{load} + L_{load}s)\tilde{V}_{piq}}{(R_{load} + L_{load}s)^2 + X_{load}{}^2} \end{cases} \tag{3.5}$$

where $\tilde{V}_{pid} = \tilde{V}_{pi}\cos\delta_{in0} - V_{pi0}\sin\delta_{in0}\tilde{\delta}_{in}$ and $\tilde{V}_{piq} = \tilde{V}_{pi}\sin\delta_{in0} + V_{pi0}\cos\delta_{in0}$ $\tilde{\delta}_{in}$.

Calculating the load power and linearizing it around the equilibrium points yields

$$\begin{cases} \tilde{P}_i = 1.5\left(\tilde{V}_{pid} i_{id0} + V_{pid0}\tilde{i}_{id} + \tilde{V}_{piq} i_{iq0} + V_{piq0}\tilde{i}_{iq}\right) \\ \tilde{Q}_i = 1.5\left(-\tilde{V}_{pid} i_{iq0} - V_{pid0}\tilde{i}_{iq} + \tilde{V}_{piq} i_{id0} + V_{piq0}\tilde{i}_{id}\right) \end{cases} \tag{3.6}$$

Substituting (3.3)–(3.5) into (3.6), after simplification, the port equivalent $[P\ Q]$–$[\omega\ V]$ model of a general RL load is represented as:

$$\tilde{\mathbf{S}}_i = \mathbf{C}_i\,(s)\,\tilde{\mathbf{v}}_{pi} \tag{3.7}$$

The matrix $\mathbf{C}_i(s)$ is presented in Appendix (3.29). Every element of $\mathbf{C}_i(s)$ does not have the right half-plane poles and $\mathbf{C}_i(s)$ is sTable

The difference is that the conventional impedance-based model chooses voltage and current as state variables, and the introduced model chooses P, Q, ω and V as state variables. Compared with the conventional model, it is conducive and convenient for the introduced model to analyze directly power, frequency, and voltage oscillations under disturbances and related affecting factors without complex coordinate transformation and calculation based on voltage and current information. The introduced modeling method is equally applicable to power converters based on classical droop control, VSG control, as well as other classical GFM controls. The output model is expressed as (3.1), and the matrices $\mathbf{G}_{bi}(s)$ and $\mathbf{Z}_{bi}(s)$ in model based on five classical GFM control methods are shown in Appendix Table 3.2.

3.2.2 Modeling of Grid-Following-Control-Based VSC

In this sub-section, the $[P\ Q]$–$[\omega\ V]$ modeling of the VSC under the grid-following control with inertia emulation will be introduced. And the control block diagram is shown in Fig. 3.3. If $k_{df} = k_{pf} = k_Q = 0$, the control scheme becomes a classic grid-following control scheme. It is noting that under weak network conditions, the design of the PLL is to be more accurate considering the impedance on the grid side [5]. However, the nodal equivalent $[P\ Q]$–$[\omega\ V]$ dynamic characteristics model of one converter is mainly established in this part, and the $[P\ Q]$–$[\omega\ V]$ model of the transmission line is separately established in the Sect. 3.3.

For simplicity, the bandwidth of the inner current loop is assumed to be much larger than that of the power control loop. Then, the effect of the inner current loop is ignored, i.e. $i_{id} = i_{idref}$ and $i_{iq} = i_{iqref}$.

After the dq transformation as shown in Fig. 3.3, the voltage vectors of PCC can be expressed as $\mathbf{V}_{pidq} = V_{pi}e^{j(\delta_{pi}-\delta_{si})}$, where is the output phase angle of the PLL. The error angle $\Delta\delta = \delta_{pi} - \delta_{si}$ is zero when the system reaches a steady state.

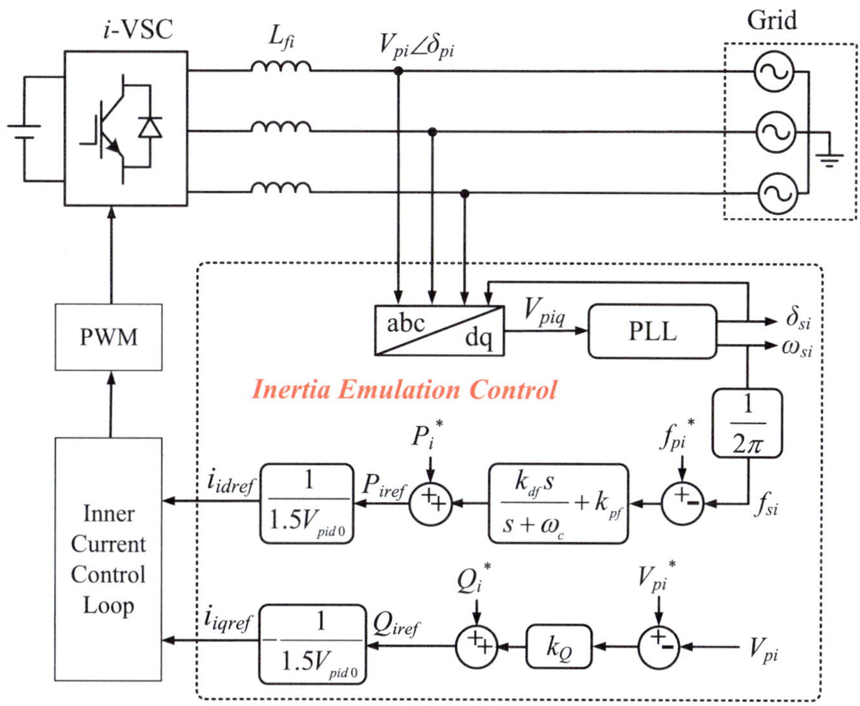

Fig. 3.3 Block diagram of a classic GFL control, i.e. inertia emulation control [2]

The dynamic equations of the PLL in small signal sense are expressed as

$$\begin{cases} \tilde{\delta}_{si} = \frac{1}{s}\tilde{\omega}_{si} \\ \tilde{\omega}_{si} = \left(k_{p,l} + \frac{k_{i,l}}{s}\right) V_{pi0} \left(\tilde{\delta}_{pi} - \tilde{\delta}_{si}\right) \end{cases} \tag{3.8}$$

Further, substituting $\tilde{\delta}_{pi} = \tilde{\omega}_{pi}/s$ into (3.3) yields

$$\tilde{\omega}_{si} = G_p(s)\,\tilde{\omega}_{pi} \tag{3.9}$$

where $G_p(s) = \frac{(k_{p,l}s+k_{i,l})V_{pi0}}{s^2+(k_{p,l}s+k_{i,l})V_{pi0}}$

According to Fig. 3.3, the output current of the GFL converter in small signal sense is expressed as

$$\begin{cases} \tilde{i}_{id} = \frac{1}{1.5V_{pid0}}\left[\tilde{P}_i^* - \left(\frac{k_{df}s}{s+\omega_c} + k_{pf}\right)\tilde{f}_{si}\right] \\ \tilde{i}_{iq} = -\frac{1}{1.5V_{pid0}}\left[\tilde{Q}_i^* - k_Q\tilde{V}_{pi}\right] \end{cases} \tag{3.10}$$

where $V^*_{pid0} = V_{pi0}$. The subscript "0" stands for the steady-state value.

The output power of the GFL converter can be calculated by (3.6). Substituting (3.9) and (3.10) into (3.6) yields

$$\begin{cases} \tilde{P}_i = \tilde{P}^*_i + \frac{P_i{}^*}{V_{pi0}}\tilde{V}_{pi} - G_{11}(s)\,\tilde{\omega}_{pi} \\ \tilde{Q}_i = \tilde{Q}^*_i - \left(k_Q - \frac{Q_i{}^*}{V_{pid0}}\right)\tilde{V}_{pi} + \frac{P_i{}^*(1-G_p(s))}{s}\tilde{\omega}_{pi} \end{cases} \tag{3.11}$$

where $G_{11}(s) = \frac{G_p(s)\left[(k_{df}+k_{pf})s+k_{pf}\omega_c\right]}{2\pi(s+\omega_c)} + \frac{Q^*(1-G_p(s))}{s}$

Rewriting (3.11) in matrix form as

$$\tilde{\mathbf{S}}_i = \mathbf{G}_{ai}(s)\,\tilde{\mathbf{S}}^*_i - \mathbf{Y}_{ai}(s)\,\tilde{\mathbf{v}}_{pi} \tag{3.12}$$

where $\tilde{\mathbf{S}}^*_i = [\tilde{P}^*_i\ \tilde{Q}^*_i]^T$ represents the power reference disturbance, $\mathbf{G}_{ai}(s)$ is a 2×2 unit matrix, and

$$\mathbf{Y}_{ai}(s) = \begin{bmatrix} G_{11}(s) & -\frac{P_i{}^*}{V_{pi0}} \\ -\frac{P_i{}^*(1-G_p(s))}{s} & k_Q - \frac{Q_i{}^*}{V_{pi0}} \end{bmatrix} \tag{3.13}$$

According to the conventional modeling method [6], the model of the VSC under the grid-following control can be represented as

$$\mathbf{i} = \mathbf{G}(s)\,\mathbf{i}_{ref} - \mathbf{Y}_s(s)\,\mathbf{u}_c \tag{3.14}$$

Usually, the VSC under grid-following control is designed to be stable under ideal voltage source in grid-connected mode. Thus, all the transfer functions $\mathbf{G}_s(s)$, $\mathbf{Y}_s(s)$,$\mathbf{G}_{ai}(s)$ and $\mathbf{Y}_{ai}(s)$ are sTable To facilitate the analysis, the model of the VSC described by (3.12) and (3.14) can be described as a "current source" and an "admittance" in parallel with it as shown in Fig. 3.4. Figure 3.4a represents the equivalent circuit under the conventional model and Fig. 3.4b represents the equivalent circuit under the introduced model. In addition, several models of the VSC under the other typical GFL control methods are listed in Appendix Table 3.3.

3.3 Multi-VSC System Model Under General Network

This section establishes a [P Q]–[ω V] model for a multi-VSC system under general network. An arbitrary network structure including n VSC units shown in Fig. 3.5 can be equivalently transformed into a fully meshed network as shown in Fig. 3.6 [5].

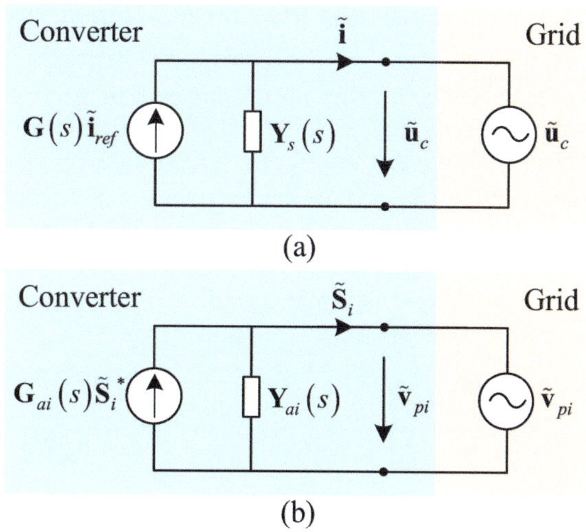

Fig. 3.4 (**a**) Conventional admittance model and (**b**) the introduced model of a grid-following-control-based VSC

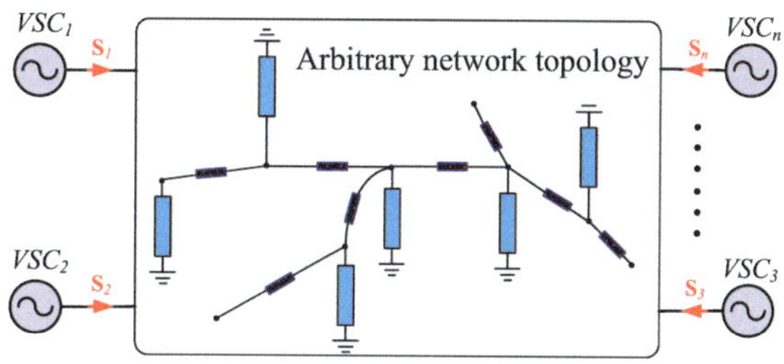

Fig. 3.5 Structure diagram of an n VSC system with arbitrary network topology

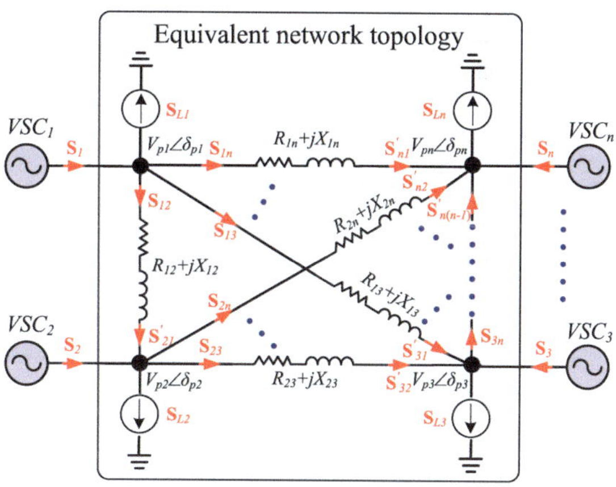

Fig. 3.6 Equivalent fully meshed network of an n VSC system

3.3.1 Modeling of Line Power Transmission

Without loss of generality, the transmission power from node i to node j is presented. Defining when $i < j$, $S_{ij} = [P_{ij}\,Q_{ij}]^T$ represents the apparent power out from node i to node j, and $S_{ji'} = [P'_{ji}\,Q'_{ji}]^T$ represents the apparent power injected to node j from node i. S_{ij} and S'_{ji} are unequal because there exists power loss on the line impedance between node i and node j. $R_{ij} + jX_{ij}$ represents the line impedance between node i and node j.

Aligning the steady phase angle δ_n with axis d, after dq transformation, the inductance dynamic model is expressed as:

$$\begin{cases} L_{ij}\frac{di_{ijd}}{dt} = V_{pid} - V_{pjd} - i_{ijd}R_{ij} + \omega_n L_{ij} i_{ijq} \\ L_{ij}\frac{di_{ijq}}{dt} = V_{piq} - V_{pjq} - i_{ijq}R_{ij} - \omega_n L_{ij} i_{ijd} \end{cases} \tag{3.15}$$

where i_{ijdq} represents current out from node i to node j, ω_n is the steady angular frequency,$V_{pjd} = V_{pj}\cos\delta_{jn}$, $V_{pjq} = V_{pj}\sin\delta_{jn}$ and $\delta_{jn} = \delta_{pj} - \delta_n$.

Based on (3.15) and refer to the derivation of (3.3)–(3.6), the power $\tilde{P}_{vi}$ and $\tilde{Q}_{vi}$ is expressed in matrix form:

$$\tilde{\mathbf{S}}_{ij} = \mathbf{C}_{ij}(s)\,\tilde{\mathbf{v}}_{pi} - \mathbf{D}_{ij}(s)\,\tilde{\mathbf{v}}_{pj} \tag{3.16}$$

where $\tilde{\mathbf{v}}_{pi} = [\tilde{\omega}_{pi}\,\tilde{V}_{pi}]^T$ and $\tilde{\mathbf{v}}_{pj} = [\tilde{\omega}_{pj}\,\tilde{V}_{pj}]^T$.

In a similar way, the power $\tilde{P}'_{ji}$ and $\tilde{Q}'_{ji}$ are expressed as

$$\tilde{\mathbf{S}}'_{ji} = \mathbf{C}_{ji}'(s)\,\tilde{\mathbf{v}}_{pj} - \mathbf{D}'_{ji}(s)\,\tilde{\mathbf{v}}_{pi} \tag{3.17}$$

where $\tilde{\mathbf{S}}'_{ji} = [\tilde{P}'_{ji}\,\tilde{Q}'_{ji}]^T$. The matrices $\mathbf{C}_{ij}(s)$, $\mathbf{C}'_{ji}(s)$, $\mathbf{D}_{ij}(s)$ and $\mathbf{D}'_{ji}(s)$ are presented in Appendix (3.30)–(3.33). Every element of the matrices $\mathbf{C}_{ij}(s)$, $\mathbf{D}_{ij}(s)$, $\mathbf{C}'_{ji}(s)$ and $\mathbf{D}'_{ji}(s)$ do not have the right half-plane poles. Thus, the matrices $\mathbf{C}_{ij}(s)$, $\mathbf{C}'_{ji}(s)$, $\mathbf{D}_{ij}(s)$ and $\mathbf{D}'_{ji}(s)$ are stable, but their inverse matrix is not necessarily sTable It is noted that when there is no physical connection between node i and node j, $\mathbf{C}_{ij}(s) = \mathbf{D}_{ij}(s) = \mathbf{C}'_{ji}(s) = \mathbf{D}'_{ji}(s) = \mathbf{0}_{2\times 2}$.

3.3.2 Modeling of Network Subsystem

According to the established line power transmission model (3.16) and (3.17) and the principle of power conservation, obtaining the power relation equation of node i as

$$\tilde{\mathbf{S}}_i + \sum_{t<i,t\neq i}^{n} \tilde{\mathbf{S}}'_{it} - \sum_{t>i,t\neq i}^{n} \tilde{\mathbf{S}}_{it} - \tilde{\mathbf{S}}_{Li} = 0 \tag{3.18}$$

where the output power $\tilde{\mathbf{S}}_i$ of the VSC-i is always flowing into node i, and we define the load power $\tilde{\mathbf{S}}_{Li}$ is always flowing out from node i. As previously defined, when $t < i$, $\tilde{\mathbf{S}}_{it}'$ represents the power injected to node i from node t. When $t > i$, $\tilde{\mathbf{S}}_{it}$ represents the power out node i.

Combining (3.16)–(3.18), the network subsystem model is acquired in matrix form as

$$\begin{bmatrix} \tilde{\mathbf{S}}_1 \\ \tilde{\mathbf{S}}_2 \\ \vdots \\ \tilde{\mathbf{S}}_n \end{bmatrix} = \mathbf{Y}_{line}(s) \begin{bmatrix} \tilde{\mathbf{v}}_{p1} \\ \tilde{\mathbf{v}}_{p2} \\ \vdots \\ \tilde{\mathbf{v}}_{pn} \end{bmatrix} + \mathbf{I}_{2n\times 2n} \begin{bmatrix} \tilde{\mathbf{S}}_{L1} \\ \tilde{\mathbf{S}}_{L2} \\ \vdots \\ \tilde{\mathbf{S}}_{Ln} \end{bmatrix} \tag{3.19}$$

where $\mathbf{I}_{2n\times 2n}$ represents a 2n × 2n unit matrix. The transfer function matrix $\mathbf{Y}_{line}(s)$ is shown in Appendix (3.34). $\mathbf{Y}_{line}(s)$ is a stable transfer function, since the elements of matrix $\mathbf{Y}_{line}(s)$ are $\mathbf{C}_{ij}(s)$, $\mathbf{D}_{ij}(s)$, $\mathbf{C}'_{ji}(s)$ and $\mathbf{D}'_{ji}(s)$, or their sum. In the conventional modeling method, both the admittance matrix of power network and its inverse are stable since power network is passive. However, in the introduced $[P\ \ Q]$–$[\omega\ \ V]$ modeling method, only the matrix $\mathbf{Y}_{line}(s)$ is stable, while $\mathbf{Y}_{line}^{-1}(s)$ is not necessarily sTable

3.3.3 Modeling of VSC Subsystem

Based on the analysis discussed in Sect. 3.2, the models of the VSCs under grid-forming control and grid-following control are different. Here, all VSCs are divided into two types.

Assuming there are k grid-forming VSCs and the others are grid-following VSCs in the studied system, then according to (3.1), the model of grid-forming-control-based VSCs (1-VSC to k-VSC) can be expressed as

$$\begin{bmatrix} \tilde{\mathbf{v}}_{p1} \\ \tilde{\mathbf{v}}_{p2} \\ \vdots \\ \tilde{\mathbf{v}}_{pk} \end{bmatrix} = \mathbf{Z}_{GM}(s) \begin{bmatrix} \tilde{\mathbf{S}}_1 \\ \tilde{\mathbf{S}}_2 \\ \vdots \\ \tilde{\mathbf{S}}_k \end{bmatrix} + \mathbf{G}_{GM}(s) \begin{bmatrix} \tilde{\mathbf{v}}_1^* \\ \tilde{\mathbf{v}}_2^* \\ \vdots \\ \tilde{\mathbf{v}}_k^* \end{bmatrix} \tag{3.20}$$

where $\mathbf{Z}_{GM}(s) = \begin{bmatrix} -\mathbf{Z}_{b1}(s) & \mathbf{0} & \mathbf{0} \\ \mathbf{0} & \ddots & \mathbf{0} \\ \mathbf{0} & \mathbf{0} & -\mathbf{Z}_{bk}(s) \end{bmatrix}$, $\mathbf{G}_{GM}(s) = \begin{bmatrix} \mathbf{G}_{b1}(s) & \mathbf{0} & \mathbf{0} \\ \mathbf{0} & \ddots & \mathbf{0} \\ \mathbf{0} & \mathbf{0} & \mathbf{G}_{bk}(s) \end{bmatrix}$ and $\mathbf{0}$ represents a 2 × 2 zero matrix.

According to (3.8), the output model of grid-following control-based VSCs ($k+1$-VSC to n-VSC) can be expressed as

$$\begin{bmatrix}\tilde{\mathbf{S}}_{k+1}\\ \tilde{\mathbf{S}}_{k+2}\\ \vdots\\ \tilde{\mathbf{S}}_{n}\end{bmatrix}=\mathbf{G}_{GF}(s)\begin{bmatrix}\tilde{\mathbf{S}}_{k+1}^{*}\\ \tilde{\mathbf{S}}_{k+2}^{*}\\ \vdots\\ \tilde{\mathbf{S}}_{n}^{*}\end{bmatrix}+\mathbf{Y}_{GF}(s)\begin{bmatrix}\tilde{\mathbf{v}}_{p(k+1)}\\ \tilde{\mathbf{v}}_{p(k+2)}\\ \vdots\\ \tilde{\mathbf{v}}_{pn}\end{bmatrix} \tag{3.21}$$

where $\mathbf{Y}_{GF}(s)=\begin{bmatrix}-\mathbf{Y}_{a(k+1)}(s) & \mathbf{0} & \mathbf{0}\\ \mathbf{0} & \ddots & \mathbf{0}\\ \mathbf{0} & \mathbf{0} & -\mathbf{Y}_{an}(s)\end{bmatrix}$, $\mathbf{G}_{GF}(s)=\begin{bmatrix}\mathbf{G}_{a(k+1)}(s) & \mathbf{0} & \mathbf{0}\\ \mathbf{0} & \ddots & \mathbf{0}\\ \mathbf{0} & \mathbf{0} & \mathbf{G}_{an}(s)\end{bmatrix}$.

Assuming that the VSCs with grid-forming control under current-source loads in island mode and the VSCs with grid-following control under ideal voltage source in grid-connected mode are designed to be stable, then all the transfer functions $\mathbf{Z}_{GFM}(s)$, $\mathbf{G}_{GFM}(s)$, $\mathbf{Y}_{GFL}(s)$ and $\mathbf{G}_{GFL}(s)$ are sTable

3.3.4 Modeling of Whole System

The model of the multi-VSC system is divided into the converter device dynamics with control algorithm and the network dynamics with the transmission line as shown in Fig. 3.7, which help better illustrate how the converter interact with the physical network [7].

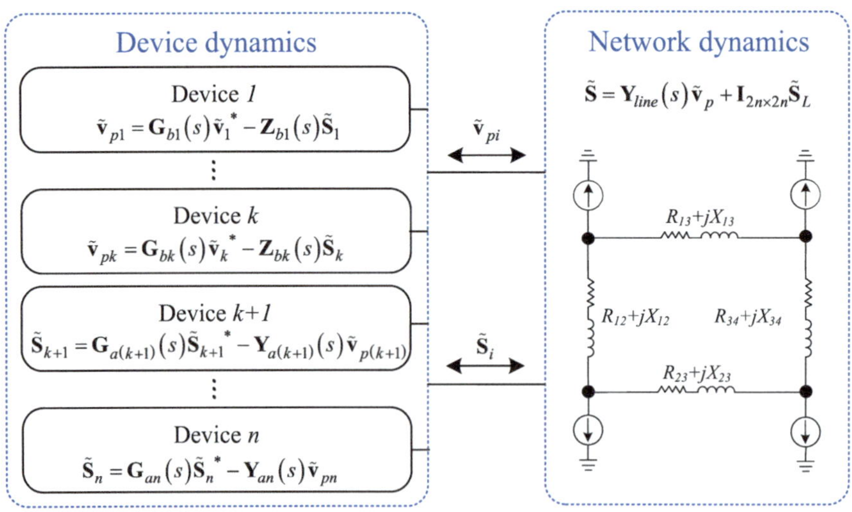

Fig. 3.7 The overall closed-loop diagram of the multi-VSC system

Combining (3.18), (3.20) and (3.21), arranging the state variables of the studied system as $\mathbf{X} = [\tilde{\mathbf{S}}_1\, \tilde{\mathbf{S}}_2 ... \tilde{\mathbf{S}}_n\, \tilde{\mathbf{v}}_{p1}\, \tilde{\mathbf{v}}_{p2} ... \tilde{\mathbf{v}}_{pn}]^T$ and the disturbance variables as $\mathbf{u} = [\tilde{\mathbf{S}}_{L1}\, \tilde{\mathbf{S}}_{L2} ... \tilde{\mathbf{S}}_{Ln}\, \tilde{\mathbf{v}}_1^*\, \tilde{\mathbf{v}}_2^* ... \tilde{\mathbf{v}}_n^*\, \tilde{\mathbf{S}}_{k+1}^*\, \tilde{\mathbf{S}}_{K+2}^* ... \tilde{\mathbf{S}}_n^*]^T$,then the overall closed-loop model of the system can be presented as

$$\mathbf{X} = (\mathbf{A}(s))^{-1}\mathbf{M}(s)\,\mathbf{u} \tag{3.22}$$

where $\mathbf{A}(s) = \begin{bmatrix} \mathbf{I}_{2n\times 2n} & -\mathbf{Y}_{line}(s) \\ \mathbf{A}_{21}(s) & \mathbf{A}_{22}(s) \end{bmatrix}$, $\mathbf{A}_{22}(s) = \begin{bmatrix} \mathbf{I}_{2k\times 2k} & \mathbf{0}_{2k\times 2(n-k)} \\ \mathbf{0}_{2(n-k)\times 2k} & -\mathbf{Y}_{GF}(s) \end{bmatrix}$, $\mathbf{A}_{21}(s) = \begin{bmatrix} -\mathbf{Z}_{GM}(s) & \mathbf{0}_{2k\times 2(n-k)} \\ \mathbf{0}_{2(n-k)\times 2k} & \mathbf{I}_{2(n-k)\times 2(n-k)} \end{bmatrix}$, $\mathbf{I}_{i\times i}$ represents a i × i unit matrix, $\mathbf{I}_{i\times i}$ represents a i × j zero matrix, and $\mathbf{M}(s) = \begin{bmatrix} \mathbf{I}_{2n\times 2n} & \mathbf{0}_{2n\times 2k} & \mathbf{0}_{2n\times 2(n-k)} \\ \mathbf{0}_{2k\times 2n} & \mathbf{G}_{GM}(s) & \mathbf{0}_{2k\times 2(n-k)} \\ \mathbf{0}_{2(n-k)\times 2n} & \mathbf{0}_{2(n-k)\times 2k} & \mathbf{G}_{GF}(s) \end{bmatrix}$

Remark There are many ways to arrange the state variables and combine the subsystem models to form the closed-loop whole system model (3.22). It is worth noting that organizing the closed-loop matrix requires some skills and is not a simple task. If submatrices such as $\mathbf{Y}_{line}^{-1}(s)$ and $\mathbf{Y}_{GFL}^{-1}(s)$, which cannot be guaranteed to be stable, appears in the $\mathbf{A}(s)$ and $\mathbf{M}(s)$, then the number of poles of the right half plane of the submatrix needs to be determined to further judge the whole system stability. In other words, improper combinations will add lots of work to the system stability judgement. Specific theoretical explanations can be seen in Sect. 3.4. When forming $\mathbf{A}(s)$ and $\mathbf{M}(s)$ in this part, only transfer matrices $\mathbf{Y}_{line}(s)$, $\mathbf{Z}_{GFM}(s)$, $\mathbf{G}_{GFM}(s)$ $\mathbf{Y}_{GFL}(s)$ and $\mathbf{G}_{GFL}(s)$ appear that are guaranteed to be sTable In this way, the whole system stability is determined only by whether the Nyquist curve of det($\mathbf{A}(s)$) surrounds the origin, and we do not need to consider the issue of the right half-plane pole, which facilitates system stability analysis.

3.4 Stability and Oscillation Analysis

3.4.1 Stability Analysis

Reconstructing the inverse matrix of $\mathbf{A}(s)$ as a determinant and an adjoint matrix:

$$\mathbf{A}(s)^{-1} = \frac{\text{adj}\,(\mathbf{A}(s))}{\det\,(\mathbf{A}(s))} \tag{3.23}$$

As discussed in the Sect. 3.3, assuming that each GFM and GFL VSCs under ideal test conditions are designed to be stable, then all the transfer functions $\mathbf{Z}_{GFM}(s)$, $\mathbf{G}_{GFM}(s)$ $\mathbf{Y}_{GFL}(s)$ and $\mathbf{G}_{GFL}(s)$ are sTable Besides, when forming $\mathbf{A}(s)$ and $\mathbf{M}(s)$, only stable transfer matrices $\mathbf{Y}_{line}(s)$, $\mathbf{Z}_{GFM}(s)$, $\mathbf{G}_{GFM}(s)$ $\mathbf{Y}_{GFL}(s)$ and $\mathbf{G}_{GFL}(s)$ appear. Thus, $\mathbf{A}(s)$ and $\mathbf{M}(s)$ are stable transfer function matrices. Further, det($\mathbf{A}(s)$)

and adj($\mathbf{A}(s)$) are stable, since $\mathbf{A}(s)$ does not have the right half-plane (RHP) poles (i.e. P=0).

Further, on the grounds of the classical control theory [8], the closed-loop system is stable if and only if

$$Z[\det(\mathbf{A}(s))] \subset \{s \in \mathbb{C} | \Re(s) < 0^-\} \tag{3.24}$$

where Z[det($\mathbf{A}(s)$)] denotes zeros of det($\mathbf{A}(s)$), $\mathbb{C}$ denotes the field of complex numbers, and $\Re(s)$ denotes the real part of s.

Since det($\mathbf{A}(s)$) is analytic on $\mathbb{C}$, except at the poles of $\mathbf{A}(s)$, based on the argument principle (Theorem 1), the stability condition of (3.22) can be equivalently transformed to

$$Z = 0 \Leftrightarrow \frac{1}{2\pi} \Delta_C \arg\left[\det\left(\mathbf{A}\left(s\right)\right)\right]\Big|_{s \in \bar{N}_\infty} = P \tag{3.25}$$

where $\bar{N}_\infty$ is the Nyquist path including the points $+j\infty$ and $-j\infty$, Z and P represent the amount of zeros and poles of det($\mathbf{A}(s)$).

In other words, when and only when the origin is not encircled by the Nyquist diagram of det($\mathbf{A}(s)$), i.e. det($\mathbf{A}(s)$) has no RHP zeros [9], the entire system is sTable

Theorem 1 *Argument Principle [8]: Let C express a positively oriented simple closed contour, and assume that*

(a) a function $f(z)$ is meromorphic in the domain interior to C;
(b) $f(z)$ is analytic and nonzero on C;
(c) Z and P are the amounts of zeros and poles of $f(z)$ inside C, respectively.

Then

$$\frac{1}{2\pi} \Delta_C \arg\left[f\left(z\right)\right] = Z - P \tag{3.26}$$

where $\Delta_C \arg\left[f(z)\right]$ expresses the change in arg $[f(z)]$ after one cycle of z moving along the positive direction of C, which is an integer multiple of 2π.

3.4.2 Oscillation Analysis

The dynamic response of power, frequency and voltage under the action of the reference and load disturbance can be described via (3.22) comprehensively. The frequency domain characteristics of the system reflect the oscillation characteristics of the system to some extent. For instance, the amplitude-frequency gain peaks and their corresponding frequency are associated with the oscillation overshoot/amplitude and oscillation frequency in the time-domain waveform [10]. Thus, the system oscillation can be analyzed by the amplitude-frequency characteristic curves, i.e. bode plots of the transfer function matrix $\mathbf{W}(s) = (\mathbf{A}(s))^{-1}\mathbf{M}(s)$.

3.4.3 Washout-Filter Coupling Feedback Control Method

This chapter proposes a new washout-filter coupling feedback (WFCF) control method to suppress the medium frequency oscillation, and it can be expressed as

$$\begin{cases} \omega_i = \frac{1}{J_i\omega^* s}\left[P_i^* - P_i - D_{pi}\omega^*\left(\omega_i - \omega^*\right)\right] + n_{xi}\frac{s}{s+\omega_c}Q_i \\ KE_i = \frac{1}{s}\left[Q_i^* - Q_i - D_{qi}\left(E_i - E_i^*\right) - m_{xi}\frac{s}{s+\omega_c}P_i\right] \end{cases} \tag{3.27}$$

where m_{xi} and n_{xi} are the active power feedback coefficient and reactive power feedback coefficient, respectively. ω_c represents the washout-filter cut-off frequency. A single inverter system is taken as an example here to illustrate the principle of the introduced method. The small-signal closed-loop model can be represented by Fig. 3.8, where $k_{p_\delta}(s)$, $k_{pE}(s)$, $k_{q\delta}(s)$ and $k_{qE}(s)$ can be derived by analogy with (3.11)–(3.14), and its expression is:

$$\begin{aligned} k_{p\delta}(s) &= \frac{E_{i0}{}^2X_g}{\left(R_g+L_gs\right)^2+X_g{}^2} - \frac{E_{i0}{}^2X_g - E_{i0}V_{g0}\cos\delta_{ig0}X_g - E_{i0}V_{g0}\sin\delta_{ig0}R_g}{R_g{}^2+X_g{}^2} \\ k_{pE}(s) &= \frac{V_{g0}\sin\delta_{og0}X_g + R_gE_{i0} - R_gV_{g0}\cos\delta_{ig0}}{R_g{}^2+X_g{}^2} + \frac{E_{i0}\left(R_g+L_gs\right)}{\left(R_g+L_gs\right)^2+X_g{}^2} \\ k_{q\delta}(s) &= \frac{E_{i0}V_{g0}\sin\delta_{og0}X_g + E_{i0}{}^2R_g - E_{i0}V_{g0}\cos\delta_{ig0}R_g}{R_g{}^2+X_g{}^2} - \frac{E_{i0}{}^2\left(R_g+L_gs\right)}{\left(R_g+L_gs\right)^2+X_g{}^2} \\ k_{qE}(s) &= \frac{E_{i0}X_g - V_{g0}\cos\delta_{ig0}X_g - R_gV_{g0}\sin\delta_{ig0}}{R_g{}^2+X_g{}^2} + \frac{E_{i0}X_g}{\left(R_g+L_gs\right)^2+X_g{}^2} \end{aligned} \tag{3.28}$$

Compared with the conventional VSG method, a washout-filter coupling P-V and Q-ω power feedback are added. Usually, the phase angle difference is small, i.e. $\cos\delta_{ig0} \approx 1$ and $\sin\delta_{ig0} \approx 0$. Ignoring inductance dynamics, $k_{q\delta}(s) = \frac{-E_{i0}V_{g0}R_g}{R_g{}^2+X_g{}^2} < 0$, i.e. reactive power and phase angle are negatively correlated ($\tilde{Q}_i \propto -\tilde{\delta}_i$). Thus, adding reactive power positive feedback to the frequency control loop is beneficial to the system stability. Similarly, $k_{pE}(s) = \frac{2R_gE_{i0}-R_gV_{g0}}{R_g{}^2+X_g{}^2} > 0$, i.e. active power and voltage magnitude are negatively correlated ($\tilde{P}_i \propto -\tilde{E}_i$). Thus, negative active power feedback is added to the voltage control loop. It is worth noting that the introduced WFCF control method does not affect the steady-state equilibrium of the system and provides more control freedom to adjust the system dynamic performance. The $[P\ Q]$–$[\omega\ V]$ model of the introduced method is shown in Table 3.2. Besides, the influence of key control parameters m_{xi} and n_{xi} on the power oscillation of the system is discussed in Sect. 3.5.5.

3.5 Control-Hardware-In-Loop Experiment

Figure 3.9 shows the configuration of the CHIL and a specific structure diagram of a four-VSC off-grid system, where VSC-1 adopts inertia emulation control [2] (GFL control), VSC-2 adopts the generalized droop control method [11], VSC-3 adopts

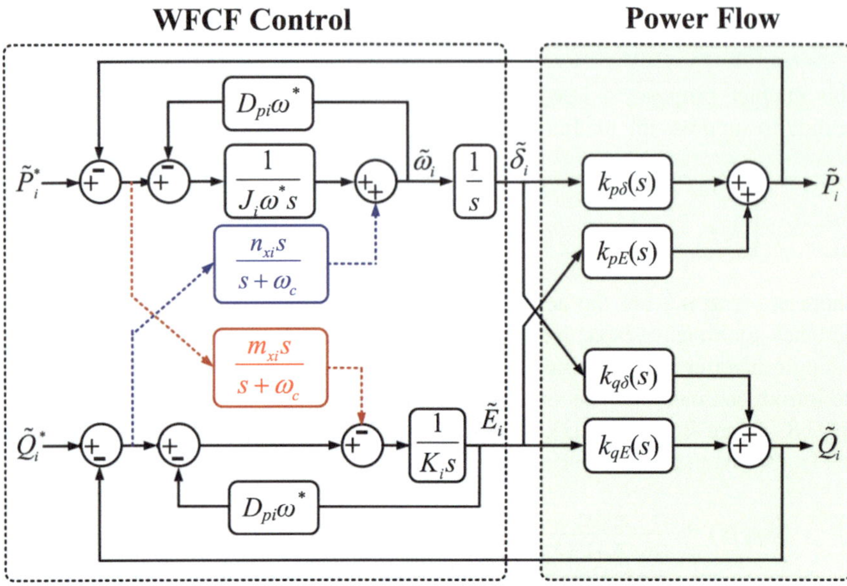

Fig. 3.8 Block diagram of the closed-loop model of a single inverter system

Fig. 3.9 The configuration of the CHIL system and structure diagram of a 4 VSC test system

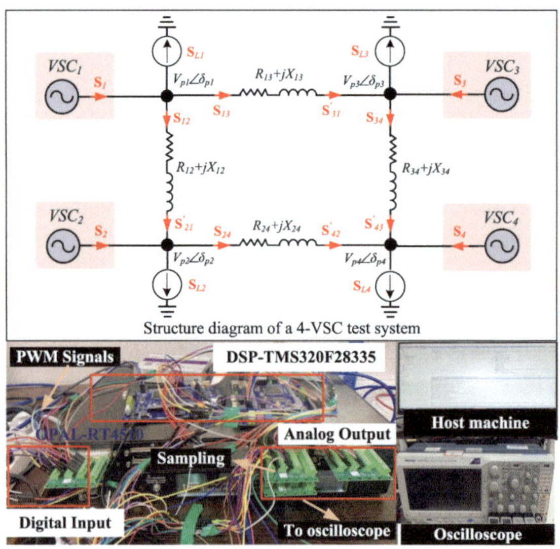

Table 3.1 System parameters

VSC - 1		*VSC - 3*	
Derivative coefficient k_{df}	100	Damping coefficient D_{p3}	$1500/\omega^*$
Filtering coefficient ω_c	30	Inertia coefficient J_3	$300/\omega^*$
Q–V droop controller gain	300	Voltage—droop coefficient D_{q3}	1000
		Integral coefficient K_3	4
VSC - 2		*VSC - 4*	
Secondary control coefficient $G_{CFF}(s)$	0	Damping coefficient D_{p4}	$1500/\omega^*$
Forward channel $G_{PLC}(s) = D_{p1}/(\tau_1 s + 1)$	1500;0.01	Inertia coefficient J_4	$200/\omega^*$
Feedback compensator coefficient $H_{CPF}(s)$	1	Voltage—droop coefficient D_{q4}	1000
Reactive power—droop coefficient k_q	0.001	Integral coefficient K_4	10
Filtering coefficient ω_c	100	Filtering parameter τ_4	0.01
		Damping correction coefficient m_4	0.05
Line impedance		*Load power*	
$R_{12} + jX_{12} = 0.8 + j0.3\,\Omega$	$R_{24} + jX_{24} = 1 + j0.6\,\Omega$	$S_{L1} = P_{L1} + jQ_{L1} = 10\,\text{kW} + j2\,\text{kVar}$	$S_{L2} = P_{L2} + jQ_{L2} = 10\,\text{kW} + j9\,\text{kVar}$
$R_{13} + jX_{13} = 2 + j0.9\,\Omega$	$R_{34} + jX_{34} = 0.3 + j1.6\,\Omega$	$S_{L3} = P_{L3} + jQ_{L3} = 10\,\text{kW} + j9\,\text{kVar}$	$S_{L4} = P_{L4} + jQ_{L4} = 10\,\text{kW} + j9\,\text{kVar}$

the conventional VSG control method, and VSC-4 adopts the decentralized mutual damping control method [3]. OPAL-RT4510 simulator simulates the circuit system and the DSP performs the controllers of VSCs. The control parameters and line impedance parameters are presented in Table 3.1.

3.5.1 *Performance of Single VSC*

The original single VSC controller design is discussed in this case. VSC-3 based on conventional VSG control is taken as an example to show that the oscillation issue does not result from the original VSC controllers' design [12]. According to the deduce in Sect. 3.2, the small signal model of the VSG is expressed as (3.1). The $\mathbf{Z}_{b3}$ and $\mathbf{G}_{b3}$ are diagonal matrices, and the diagonal elements are all first-order inertia elements. The bode diagram for the diagonal elements of $\mathbf{Z}_{bi}$, i.e. $Z_{b3,11}$ and $Z_{b3,22}$

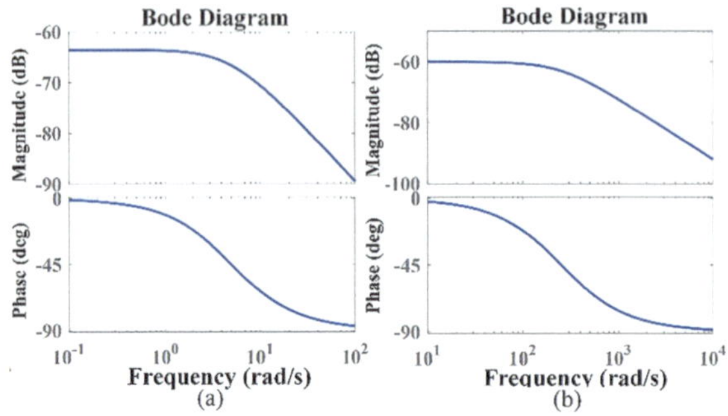

Fig. 3.10 Bode plots of the transfer functions: (**a**) $Z_{b3,11}$ and (**b**) $Z_{b3,22}$

are shown in Fig. 3.10, and the used parameter is shown in Table 3.1. From the bode diagram, it is observed that the original VSC controller design will not result in the oscillation issue. Similarly, other VSCs with parameter in Table 3.1 can be tested in the same way and original single VSC controller design show good performance.

3.5.2 Oscillation Characteristics Analysis and Verification

In this case, the power oscillation characteristics of the 4 VSC system are analyzed by the amplitude-frequency characteristic curve of the transfer function matrix $\mathbf{W}(s)$, and m_4 is set to 0. Figure 3.11a shows the bode plot of the transfer function $\Delta P_i/\Delta P_{L4}$. The peak coordinate value of the low-frequency resonance point of $\Delta P_3/\Delta P_{L4}$ and $\Delta P_4/\Delta P_{L4}$ is marked, and they have a large low-frequency resonance peak value. It indicates that low-frequency power oscillations are prone to occur between VSC-3 and VSC-4, and the oscillation frequency is $f_o = 5$ Hz. Figure 3.11b shows active power waveforms under a step-up load change $\Delta P_{L4} = 10$ kW. As observed, the active power of VSC-3 and VSC-4 show severe low-frequency oscillations. By measurement, the oscillation frequency is 4.9 Hz. It is nearly consistent with what we infer from Fig. 3.11a.

Figure 3.12a shows the bode plot of the transfer functions $\Delta P_i/\Delta E_3^*$. The amplitude-frequency curves of $\Delta P_3/\Delta E_3^*$ and $\Delta P_4/\Delta E_3^*$ have large low-frequency and medium-frequency resonance peak values. It indicates that low-frequency and medium-frequency power oscillations are prone to occur between VSC-3 and VSC-4, and the oscillation frequencies are $f_{o1} = 5$ Hz and $f_{o2} = 53$ Hz. Figure 3.12b shows active power waveforms under $\Delta E_3^* = 10$ V. As observed , the active power of VSC-3 and VSC-4 show severe low-frequency and medium-frequency oscillations. By measurement, the oscillation frequencies are $f_{o1} = 4.9$ Hz and $f_{o2} = 52.5$ Hz. It is nearly consistent with what we infer from Fig. 3.12a. Thus, the oscillation analysis based on the introduced model is consistent with the CHIL results, and it can be used to locate the weak nodes with severe power oscillations.

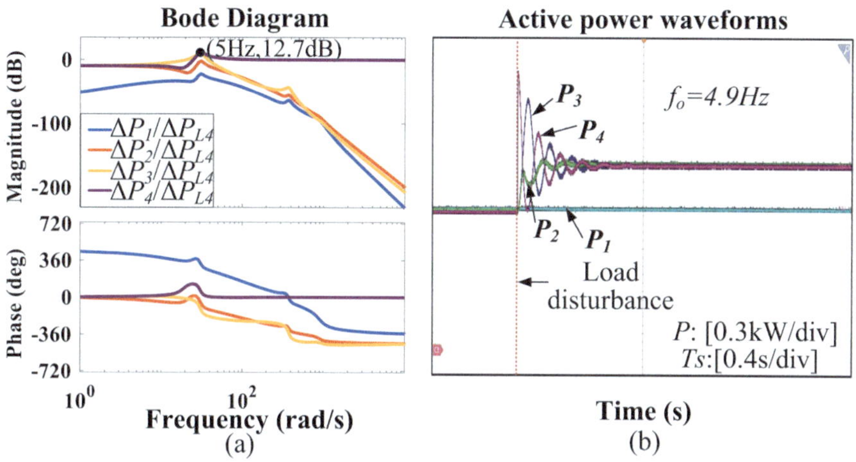

Fig. 3.11 (**a**) Bode plots of the transfer functions $\Delta P_i/\Delta P_{L4}$ and (**b**) active power waveforms under load disturbance ΔP_{L4}

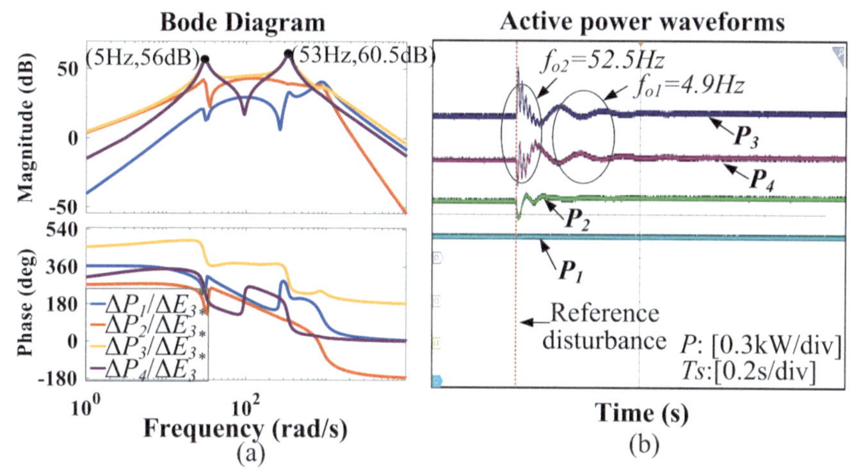

Fig. 3.12 (**a**) Bode plots of the transfer functions $\Delta P_i/\Delta E_3^*$ without WFCF control and (**b**) active power waveforms under voltage reference disturbance ΔE_3^*

3.5.3 Influence of Line Impedance

In this case, the influence of line impedance on the power oscillation characteristics of the system are analyzed. Figure 3.13a shows the bode plot of the transfer function $\Delta P_3/\Delta P_{L4}$ with $L_{34} = 5$, 4 and 2 mH. When increasing the L_{34}, the peak coordinate value of the low-frequency resonance point of $\Delta P_3/\Delta P_{L4}$ is increased from 12.7 to 20 dB, and the frequency of LFO is increased. It indicates that low-frequency

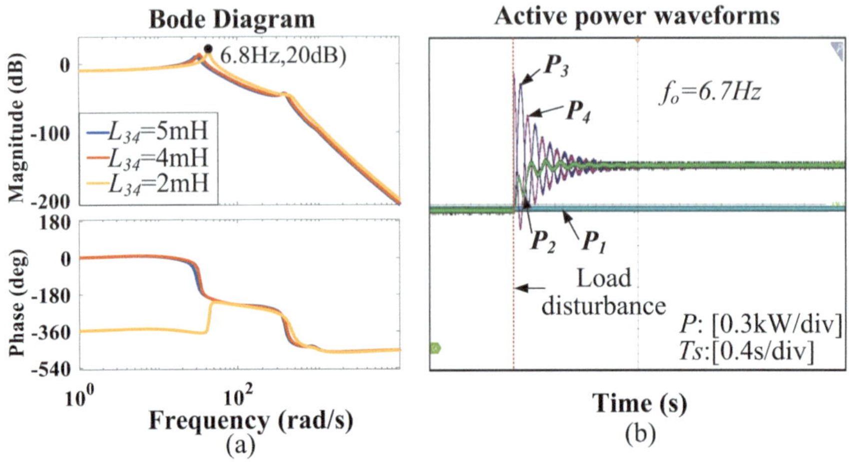

Fig. 3.13 (**a**) Bode plots of the transfer functions $\Delta P_3/\Delta P_{L4}$ with L_{34} = 5, 4 and 2 mH, (**b**) active power waveforms when L_{34} = 2 mH under load disturbance ΔP_{L4}

power oscillations of VSC-3 is more serious with the increase of L_{34}. Figure 3.13b shows active power waveforms under a step-up load change ΔP_{L4} = 10 kW when L_{34} = 2 mH. As observed, the LFO of active power between VSC-3 and VSC-4 is more serious compared with Fig. 3.11b. Besides, by measurement, the oscillation frequency is 6.7 Hz. It is nearly consistent with what we infer from Fig. 3.13a.

3.5.4 Effect Assessment of Oscillation Suppression Method

Many methods were introduced to suppress the power oscillation among multiple VSGs. The established model in this chapter can be used to evaluate the performance of existing power oscillation suppression methods in improving the power and frequency dynamics.

This case takes the decentralized mutual damping method [3] as an example to illustrate the assessment ability of the model on the effect of these methods. Figure 3.14a shows the bode plot of the transfer function $\Delta P_i/\Delta P_{L4}$ where VSC-3 and VSC-4 adopt the decentralized mutual damping method. The parameters are set as $\tau_i = 0.01$ and $m_i = 0.05$. Compared with Fig. 3.9, the low-frequency resonant peaks of $\Delta P_3/\Delta P_{L4}$ and $\Delta P_4/\Delta P_{L4}$ decrease significantly. It indicates that low-frequency power oscillations that occur between VSC-3 and VSC-4 can be effectively suppressed with the method [3]. Figure 3.14b presents the active power waveforms under ΔP_{L4} = 10 kW where the VSC-3 and VSC-4 adopt the mutual damping method. As observed, the decentralized mutual damping method suppressed the power oscillation. Thus, the introduced model can be used to assess the oscillation suppression methods.

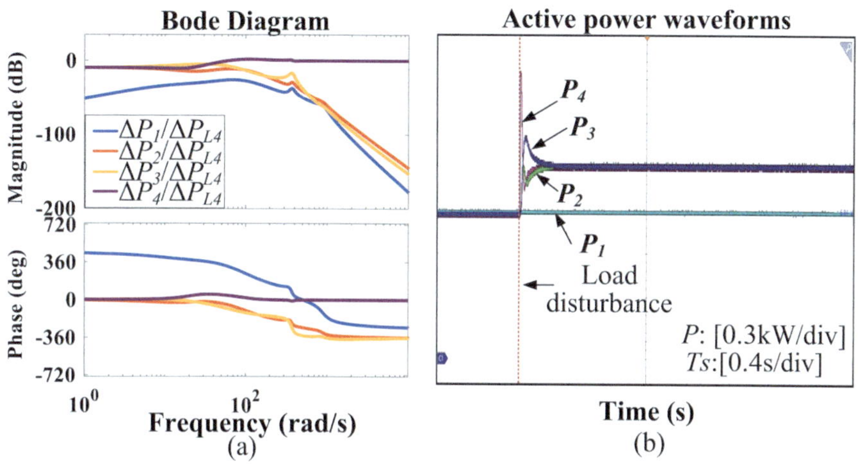

Fig. 3.14 (**a**) Bode plots of the transfer functions $\Delta P_i/\Delta P_{L4}$ and (**b**) active power waveforms under load disturbance ΔP_{L4}

3.5.5 Effectiveness of the WFCF Method

To verify the effectiveness of the introduced WFCF method, this case analyzes the effect of the crucial control coefficient m_x and n_x on the medium-frequency oscillation based on the extended $[P\ Q]$–$[\omega\ V]$ model. As observed in Fig. 3.12, under voltage reference disturbance, VSC-3 and VSC-4 have severe medium-frequency oscillations. Thus, applying the introduced control method to VSC-3 and VSC-4.

Figure 3.15 shows the bode plots of (a) the transfer functions $\Delta P_3/\Delta E_3^*$ with $m_x = 0, 0.3$, and 1 and $n_x = 0$, and (b) the transfer functions $\Delta P_3/\Delta E_3^*$ with $m_x = 0$, $n_x = 0$, 3e−4, 1e−3. It is observed that the active power feedback coefficient m_x and n_x affect the medium-frequency resonance peak of the transfer function, and the larger m_x and n_x are, the smaller the medium-frequency resonance peak value is. It is inferred that increasing m_x and n_x can effectively suppress the medium-frequency oscillation of the system.

Figure 3.16 presents the active power waveforms under voltage reference disturbance $\Delta E_3^* = 10$ when the VSC-3 and VSC-4 adopt the introduced method (a) with $m_x = 1, n_x = 0$ and (b) $m_x = 0, n_x = 1\text{e}{-3}$. Compared with the results in Fig. 3.12, the medium-frequency oscillation can be effectively suppressed with the introduced method. The effectiveness of the introduced control strategy is verified, and it is consistent with the conclusions we draw from the analysis based on the established model.

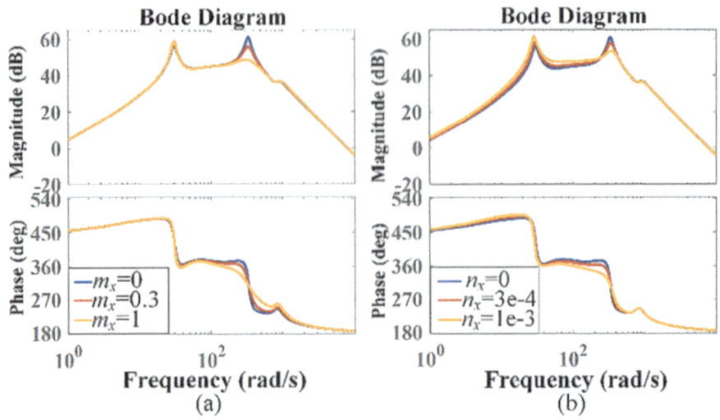

Fig. 3.15 Bode plots of the transfer functions $\Delta P_3/\Delta E_3^*$ with (**a**) different m_x and (**b**) different n_x

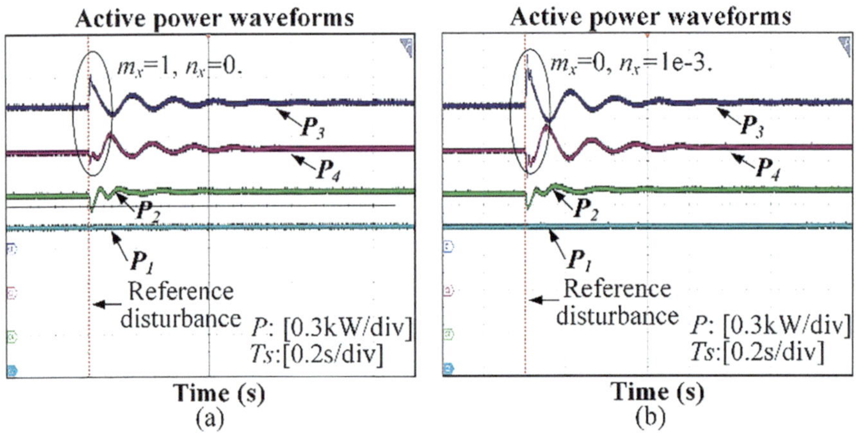

Fig. 3.16 Active power waveforms under voltage reference disturbance ΔE_3^* with (**a**) $m_x = 5$, $n_x = 0$, and (**b**) $m_x = 0$, $n_x = 1\text{e}{-3}$

3.5.6 Comparison with the Virtual Active Damping

In this case, the effect of medium frequency power oscillation suppression under the introduced WFCF method and the virtual active damping in [13] is compared. Since VSC-3 and VSC-4 have severe medium-frequency oscillations, the virtual active damping in [13] is applied to VSC-3 and VSC-4. The coverter model can be expressed as (3.1), where

$$\mathbf{Z_b(s)} = \left\{ \frac{\mathbf{1}}{\mathbf{J_i\omega^*s + D_{pi}\omega^*}}, \mathbf{0}; \frac{\mathbf{G_{filter}G_{ll}}}{\mathbf{J_i\omega^*s + D_{pi}\omega^*}}, \frac{\mathbf{1}}{\mathbf{K_is + D_{qi}}} \right\}$$

and $\mathbf{G_b(s)} = \{\frac{\mathbf{D_{pi}\omega^*}}{\mathbf{J_i\omega^*s+D_{pi}\omega^*}}, \mathbf{0}; \frac{\mathbf{D_{pi}\omega^*G_{filter}G_{ll}}}{\mathbf{J_i\omega^*s+D_{pi}\omega^*}}, \frac{\mathbf{D_{qi}}}{\mathbf{K_is+D_{qi}}}\}$. Figure 3.7a shows the bode plot of the transfer functions $\Delta P_i/\Delta E_3^*$. Compared with Fig. 3.12a, the low-frequency

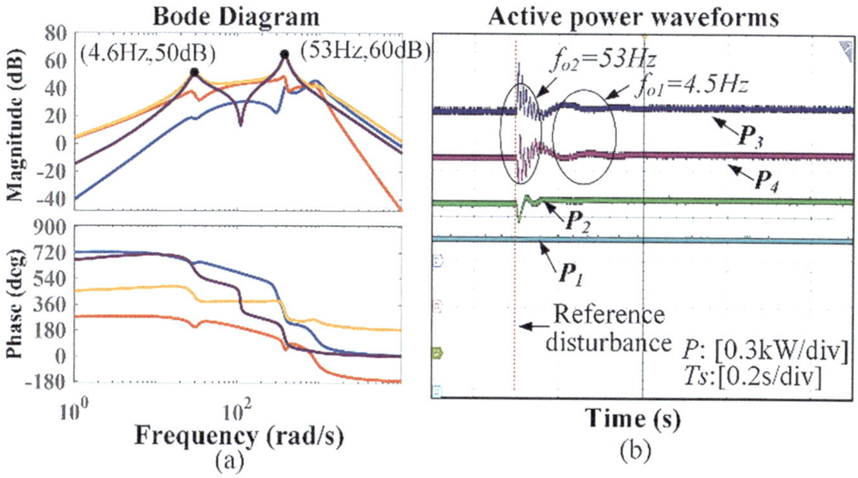

Fig. 3.17 (**a**) Bode plots of the transfer functions $\Delta P_i/\Delta E_3^*$ with method in [13] and (**b**) active power waveforms under voltage reference disturbance ΔE_3^*

resonance peak values is decreased and medium-frequency resonance peak values is increased. Figure 3.17b shows active power waveforms under $\Delta E_3^* = 10\,\text{V}$. It is observed that the method in [13] helps suppress the low-frequency power oscillation, but the medium frequency power oscillation is not suppressed and worsened. Thus, the introduced WFCF method has better effect on suppressing medium frequency power oscillation.

3.5.7 Influence of Control Parameters on Stability

In this case, the voltage-droop coefficient of VSC-3 $D_{q3} = 100$, $K_3 = 1$, and other control parameters are shown in Table 3.1. All the sub-systems are stable under the parameter settings. The VSC-3 and VSC-4 adopt the introduced WFCF method with $n_x = 0.2\text{e}{-3}$. The Nyquist curves of det($\mathbf{A}(s)$) with $m_x = 1$ and $m_x = 0$ are displayed in Fig. 3.18. When $m_x = 1$, the overall system is stable since the point (-0, 0) is not surrounded by the Nyquist curve. When $m_x = 0$, the Nyquist curve surrounds (-0, 0), indicating that the overall system gets unsTable Fig. 3.19 shows the CHIL result of the output current of VSC-4. When m_x changes from 1 to 0, the system becomes unsTable It indicates that appropriately the introduced WFCF method is conducive to system stability. The CHIL result is consistent with the result of the obtained stability criterion.

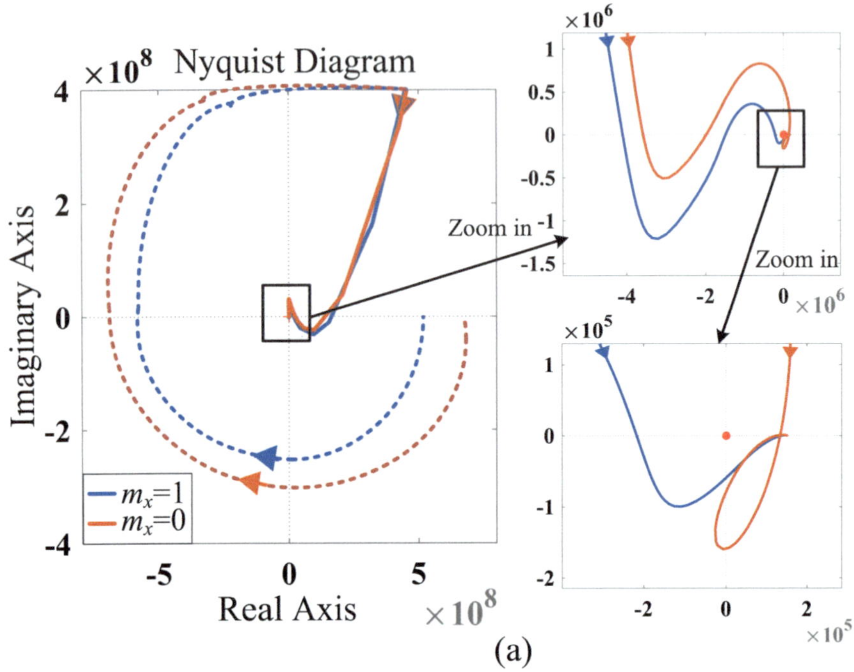

Fig. 3.18 Nyquist loci of det($\mathbf{A}(s)$) with $m_x = 1$ and $m_x = 0$ with the introduced $[P\ Q]$–$[\omega\ V]$ model

Fig. 3.19 CHIL result of the output current of VSC-4

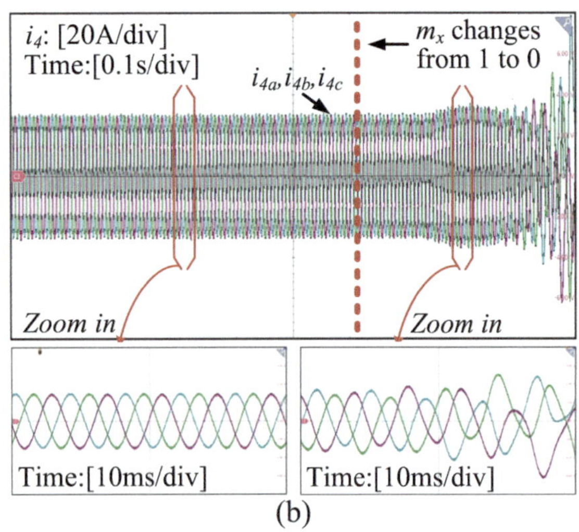

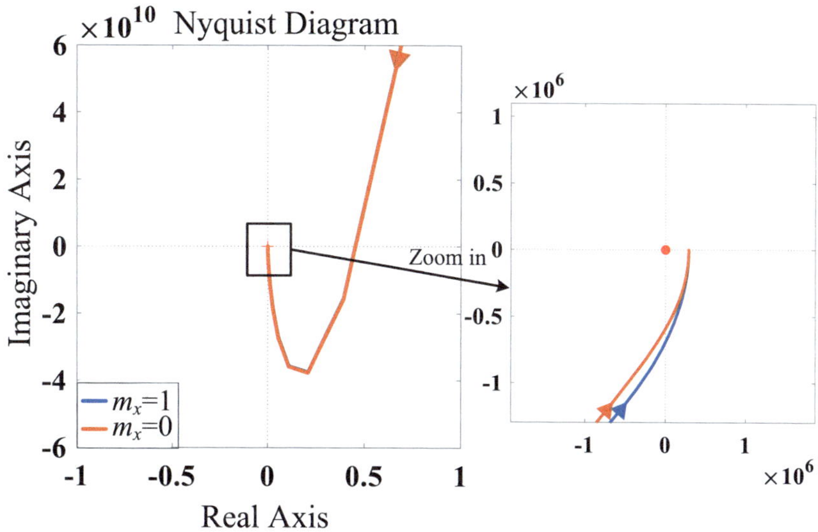

(b)

Fig. 3.20 Nyquist loci of det($\mathbf{A}(s)$) with $m_x = 1$ and $m_x = 0$ with the P-ω model

3.5.8 *Comparison with the P-ω Model*

In this case, the accuracy of the introduced $[P\ Q]$–$[\omega\ V]$ model and P-ω model is compared. The control parameters are set same with the case 3.5.7. The Nyquist curves of det($\mathbf{A}(s)$) with $m_x = 1$ and m_x =0 are displayed in Fig. 3.20. It is observed that when $m_x = 1$ and $m_x = 0$, the overall system is stable since the point (-0, 0) is not surrounded by the Nyquist curve, indicating that the overall system is sTable However, Fig. 3.19 shows the CHIL result of the output current of VSC-4. When m_x changes from 1 to 0, the system becomes unsTable It indicates that the introduced $[P\ Q]$–$[\omega\ V]$ model is more accurate than the P-ω model.

3.6 Conclusion

To analyze the power, voltage and frequency oscillation or instability of a multi-(VSC) system, a general $[P\ Q]$–$[\omega\ V]$ model for a multi-VSC system under arbitrary structure is established. The system allows both grid-forming and grid-following VSCs, which provides a new power-system-based oscillation and stability analysis standard model for hybrid GFM/GFL VSCs system. Based on the $[P\ Q]$–$[\omega\ V]$ model, the weak nodes with severe low-frequency and medium-frequency (a few tenths to tens Hz) power oscillations are located and the effect of the oscillation suppression method is evaluated. Besides, the introduced WFCF method suppresses the medium-frequency oscillation effectively and the system stability is improved as the experimental results show. However, there are still some shortcomings of the general [P Q]–[ω V] model. It is a small signal stability model is based on the well-known system parameters. It needs to recalculate the steady-state equilibrium of the system (including steady-state power angles and voltage amplitude), when the operating point changes. In the future, systems with unknown parameters will also be the focus of our research.

Appendix

In Table 3.2, matrix **I** represents a 2 × 2 unit matrix. Case 1 represents the model of VSC under the classical droop control [14]. Case 2 shows the case under the generalized droop control [11]. Case 3 considers the case under the power decoupling control [15] where $\mathbf{Z}_{b3}(s)$ is no longer a diagonal matrix. Case 4 represents the case under the decentralized mutual damping control method [3]. Case 5 shows the case that the voltage control loop of VSG applies a PSS controller [16].

In Table 3.3, case 1 represents the model of an inverter under the classical DC-link voltage regulation control strategy [17]. Case 2 presents a control method that changes the DC-link capacitor voltage reference according to grid frequency feedback [18]. Case 3 constructed a correction term of DC-link that is in proportion to the frequency which provides inertia support [19].

In the port equivalent [P Q]–[ω V] model of a general R_L load $\mathbf{C}_i$ is 2 × 2 matrix and expressed as (3.29), where $Z_{Load}(s)$=R_{Load} +$L_{Load}s$. In the line power transmission model, $\mathbf{D}_{ij}$,$\mathbf{D}'_{ji}$, $\mathbf{C}_{ij}$ and $\mathbf{C}'_{ji}$ are 2 × 2 matrices and expressed as (3.30-3.33), where $X_{ij} = \omega_n L_{ij}$, $Z_{ij}(s)$=R_{ij} +$L_{ij}s$ and $\delta_{ij0} = \delta_{pi0}$-$\delta_{pj0}$.

Table 3.2 Correlation matrix of the grid-forming control-based model

Case	The correlation matrix of the model
1 [14]	$\mathbf{Z}_{b1}(\mathrm{s}) = \mathrm{diag}\{\frac{1}{D_{pi}}, \frac{1}{D_{qi}}\}, \mathbf{G}_{b1}(\mathrm{s}) = \mathbf{I}$
2 [11]	$\mathbf{Z}_{b2}(\mathrm{s}) = \mathrm{diag}\{\frac{G_{PLC}(s)H_{CPF}(s)}{1+G_{PLC}(s)G_{CFF}(s)}, \frac{k_q\omega_c}{s+\omega_c}\}, \mathbf{G}_{b2}(\mathrm{s}) = \mathbf{I}$
3 [15]	$\mathbf{Z}_{b3}(\mathrm{s}) = \{\frac{k_p\omega_f \sin\delta_0}{s+\omega_f}, -\frac{k_p\omega_f \cos\delta_0}{s+\omega_f}; \frac{k_q\omega_f \cos\delta_0}{s+\omega_f}, \frac{k_q\omega_f \sin\delta_0}{s+\omega_f}\}, \mathbf{G}_{b3}(\mathrm{s}) = \mathbf{I}$
4 [3]	$\mathbf{Z}_{b4}(\mathrm{s}) = \mathrm{diag}\{\frac{1}{(\tau s+1)(J_i\omega^* s+D_{pi}\omega^*)}, \frac{1}{K_i s+D_{qi}}\}$, $\mathbf{G}_{b4}(\mathrm{s}) = \mathrm{diag}\{\frac{D_{pi}\omega^*}{J_i\omega^* s+D_{pi}\omega^*}, \frac{D_{qi}}{K_i s+D_{qi}}\}$
5 [16]	$\mathbf{Z}_{b5}(\mathrm{s}) = \{\frac{1}{J_i\omega^* s+D_{pi}\omega^*}, 0; -\frac{T_{PSS}(s)}{K_i J_i\omega^{*2}+K_i D_{qi}\omega^* s}, \frac{1}{K_i s}\}$, $\mathbf{G}_{b5}(\mathrm{s}) = \{\frac{D_{pi}\omega^*}{J_i\omega^* s+D_{pi}\omega^*}, 0; \frac{J_i\omega^* \cdot T_{PSS}(s)}{K_i J_i\omega^{*2}+K_i D_{qi}\omega^* s}, 0\}$
6	$\mathbf{Z}_{b6}(\mathrm{s}) = \{\frac{1}{J_i\omega^* s+D_{pi}\omega^*}, -\frac{n_{xi} J_i\omega^{*2}}{(J_i\omega^* s+D_{pi}\omega^*)(s+\omega_c)}; \frac{m_{xi} s}{(s+\omega_c)(K_i s+D_{qi})}, \frac{1}{K_i s+D_{qi}}\}$, $\mathbf{G}_{b6}(\mathrm{s}) = \mathrm{diag}\{\frac{D_{pi}\omega^*}{J_i\omega^* s+D_{pi}\omega^*}, \frac{D_{qi}}{K_i s+D_{qi}}\}$

Table 3.3 Correlation matrix of the grid-following control-based model

Case	The correlation matrix of the model
1 [17]	$\mathbf{Y}_{a1}(s)=\begin{bmatrix} \frac{Q^*(1-G_p(s))}{s} & -\frac{P^*}{V_{pi0}} \\ -\frac{P^*(1-G_p(s))}{s} & \frac{k_v s+k_{vi}}{s(\tau_v s+1)}-\frac{Q_i^*}{V_{pi0}} \end{bmatrix}$ $\mathbf{G}_{a1}(s)=\operatorname{diag}\left\{\frac{1.5V_{pi0}(k_p s+k_i)}{CV_{DC0}s^2+1.5V_{pi0}s(k_p s+k_i)}\ 0\right\}$
2 [18]	$\mathbf{Y}_{a2}(s)=\begin{bmatrix} \frac{G_p(s)}{2\pi}\frac{1.5V_{pi0}s(k_p s+k_i)S_{VSC}H_{VSC}}{Nf_0[CV_{DC0}s^2+1.5V_{pi0}s(k_p s+k_i)]}+\frac{Q^*(1-G_p(s))}{s} & -\frac{P^*}{V_{pi0}} \\ -\frac{P^*(1-G_p(s))}{s} & \frac{k_v s+k_{vi}}{s(\tau_v s+1)}-\frac{Q_i^*}{V_{pi0}} \end{bmatrix}$ $\mathbf{G}_{a2}(s)=\operatorname{diag}\left\{\frac{1.5V_{pi0}(k_p s+k_i)}{CV_{DC0}s^2+1.5V_{pi0}s(k_p s+k_i)}\ 0\right\}$
3 [19]	$\mathbf{Y}_{a3}(s)=\begin{bmatrix} \frac{G_p(s)1.5V_{pi0}C_{dc}V_{dc0}s(K_{vp}s+K_{vi})K_{esc}(s)}{2\pi\ C_{dc}V_{dc0}s^2+1.5V_{pi0}s(K_{vp}s+K_{vi})}+\frac{Q^*(1-G_p(s))}{s} & -\frac{P^*}{V_{pi0}} \\ -\frac{P^*(1-G_p(s))}{s} & k_Q-\frac{Q_i^*}{V_{pi0}} \end{bmatrix}$ $\mathbf{G}_{a3}(s)=\operatorname{diag}\left\{\frac{1.5V_{pi0}(K_{vp}s+K_{vi})}{C_{dc}V_{dc0}s^2+1.5V_{pi0}s(K_{vp}s+K_{vi})}\ 1\right\}$

$$\mathbf{C}_i = 1.5\begin{bmatrix} \frac{1}{s}\left[\frac{-X_{load}V_{pi0}^2}{R_{load}^2+X_{load}^2}+\frac{V_{pi0}^2X_{ij}}{Z_{load}(s)^2+X_{ij}^2}\right] & \frac{R_{load}V_{pi0}}{R_{load}^2+X_{load}^2}+\frac{V_{pi0}Z_{load}(s)}{Z_{load}(s)^2+X_{ij}^2} \\ \frac{1}{s}\left[\frac{R_{load}V_{pi0}^2}{R_{load}^2+X_{load}^2}-\frac{Z_{load}(s)V_{pi0}^2}{Z_{load}(s)^2+X_{ij}^2}\right] & \frac{X_{load}V_{pi0}}{R_{load}^2+X_{load}^2}+\frac{V_{pi0}X_{load}}{Z_{load}(s)^2+X_{load}^2} \end{bmatrix} \tag{3.29}$$

$$\mathbf{D}_{ij} = 1.5\begin{bmatrix} \frac{Z_{ij}(s)V_{pj0}V_{pi0}\sin\delta_{ij0}+X_{ij}V_{pj0}V_{pi0}\cos\delta_{ij0}}{s[Z_{ij}(s)^2+X_{ij}^2]} & \frac{Z_{ij}(s)V_{pi0}\cos\delta_{ij0}-X_{ij}V_{pi0}\sin\delta_{ij0}}{Z_{ij}(s)^2+X_{ij}^2} \\ \frac{V_{pi0}V_{pj0}\sin\delta_{ij0}X_{ij}-V_{pi0}Z_{ij}(s)V_{pj0}\cos\delta_{ij0}}{s[Z_{ij}(s)^2+X_{ij}^2]} & \frac{V_{pi0}\cos\delta_{ij0}X_{ij}+V_{pi0}Z_{ij}(s)\sin\delta_{ij0}}{Z_{ij}(s)^2+X_{ij}^2} \end{bmatrix} \tag{3.30}$$

$$\mathbf{D}'_{ij} = 1.5\begin{bmatrix} \frac{Z_{ij}(s)V_{pi0}V_{pj0}\sin\delta_{ij0}-X_{ij}V_{pi0}V_{pj0}\cos\delta_{ij0}}{s[Z_{ij}(s)^2+X_{ij}^2]} & \frac{-Z_{ij}(s)V_{pi0}\cos\delta_{ij0}-X_{ij}V_{pi0}\sin\delta_{ij0}}{Z_{ij}(s)^2+X_{ij}^2} \\ \frac{X_{ij}V_{pi0}V_{pj0}\sin\delta_{ij0}+Z_{ij}(s)V_{pi0}V_{pj0}\cos\delta_{ij0}}{s[Z_{ij}(s)^2+X_{ij}^2]} & \frac{-X_{ij}V_{pi0}\cos\delta_{ij0}+Z_{ij}(s)V_{pi0}\sin\delta_{ij0}}{Z_{ij}(s)^2+X_{ij}^2} \end{bmatrix} \tag{3.31}$$

$$\mathbf{C}_{ij} = 1.5\begin{bmatrix} \frac{1}{s}\left[\frac{X_{ij}V_{pi0}V_{pj0}\cos\delta_{ij0}-X_{ij}V_{pi0}^2+R_{ij}V_{pi0}V_{pj0}\sin\delta_{ij0}}{R_{ij}^2+X_{ij}^2}+\frac{V_{pi0}^2X_{ij}}{Z_{ij}(s)^2+X_{ij}^2}\right] & \frac{R_{ij}V_{pi0}-R_{ij}V_{pj0}\cos\delta_{ij0}+X_{ij}V_{pj0}\sin\delta_{ij0}}{R_{ij}^2+X_{ij}^2}+\frac{V_{pi0}Z_{ij}(s)}{Z_{ij}(s)^2+X_{ij}^2} \\ \frac{1}{s}\left[\frac{R_{ij}V_{pi0}^2-R_{ij}V_{pi0}V_{pj0}\cos\delta_{ij0}+X_{ij}V_{pi0}V_{pj0}\sin\delta_{ij0}}{R_{ij}^2+X_{ij}^2}-\frac{Z_{ij}(s)V_{pi0}^2}{Z_{ij}(s)^2+X_{ij}^2}\right] & \frac{X_{ij}V_{pi0}-X_{ij}V_{pj0}\cos\delta_{ij0}-R_{ij}V_{pj0}\sin\delta_{ij0}}{R_{ij}^2+X_{ij}^2}+\frac{V_{pi0}X_{ij}}{Z_{ij}(s)^2+X_{ij}^2} \end{bmatrix} \tag{3.32}$$

$$\mathbf{C}'_{ji} = 1.5\begin{bmatrix} \frac{1}{s}\left[\frac{R_{ij}V_{pi0}V_{pj0}\sin\delta_{ij0}+X_{ij}V_{pj0}^2-X_{ij}V_{pi0}V_{pj0}\cos\delta_{ij0}}{R_{ij}^2+X_{ij}^2}-\frac{V_{pj0}^2X_{ij}}{Z_{ij}(s)^2+X_{ij}^2}\right] & \frac{R_{ij}V_{pi0}\cos\delta_{ij0}+X_{ij}V_{pi0}\sin\delta_{ij0}-R_{ij}V_{pj0}}{R_{ij}^2+X_{ij}^2}-\frac{V_{pj0}Z_{ij}(s)}{Z_{ij}(s)^2+X_{ij}^2} \\ \frac{1}{s}\left[\frac{R_{ij}V_{pi0}V_{pj0}\cos\delta_{ij0}+X_{ij}V_{pi0}V_{pj0}\sin\delta_{ij0}-R_{ij}V_{pj0}^2}{R_{ij}^2+X_{ij}^2}+\frac{V_{pj0}^2Z_{ij}(s)}{Z_{ij}(s)^2+X_{ij}^2}\right] & \frac{X_{ij}V_{pi0}\cos\delta_{ij0}-R_{ij}V_{pi0}\sin\delta_{ij0}-X_{ij}V_{pj0}}{R_{ij}^2+X_{ij}^2}-\frac{V_{pj0}X_{ij}}{Z_{ij}(s)^2+X_{ij}^2} \end{bmatrix} \tag{3.33}$$

$$\mathbf{Y}_{line}(s) = \begin{bmatrix} \mathbf{C}_{12} + \mathbf{C}_{13} + \cdots + \mathbf{C}_{1n} & -\mathbf{D}_{12} & \cdots & -\mathbf{D}_{1(n-1)} & -\mathbf{D}_{1n} \\ -\mathbf{D}'_{21} & \mathbf{C}'_{21} + \mathbf{C}_{23} + \cdots + \mathbf{C}_{2n} & \cdots & -\mathbf{D}_{2(n-1)} & -\mathbf{D}_{2n} \\ \vdots & \vdots & \ddots & \vdots & \vdots \\ -\mathbf{D}'_{(n-1)1} & -\mathbf{D}'_{(n-1)2} & \cdots & \mathbf{C}'_{(n-1)1} + \mathbf{C}'_{(n-1)2} + \cdots + \mathbf{C}_{(n-1)n} & -\mathbf{D}_{(n-1)n} \\ -\mathbf{D}'_{n1} & -\mathbf{D}'_{n2} & \cdots & -\mathbf{D}'_{n(n-1)} & \mathbf{C}'_{n1} + \mathbf{C}'_{n2} + \cdots + \mathbf{C}'_{n(n-1)} \end{bmatrix} \tag{3.34}$$

References

1. F. Milano, F. Dörfler, G. Hug, D.J. Hill, G. Verbič, Foundations and challenges of low-inertia systems (invited paper), in *2018 Power Systems Computation Conference (PSCC), Dublin, Ireland* (2018), pp. 1–25
2. I. Serban, C. Marinescu, Control strategy of three-phase battery energy storage systems for frequency support in microgrids and with uninterrupted supply of local loads. IEEE Trans. Power Electron. **29**(9), 5010–5020 (2014)
3. S. Fu, Y. Sun, L. Li, Z. Liu, H. Han, M. Su, Power oscillation suppression of multi-VSG grid via decentralized mutual damping control. IEEE Trans. Industr. Electron. **69**(10), 10202–10214 (2022)
4. J. Sun, Impedance-based stability criterion for grid-connected inverters. IEEE Trans. Power Electron. **26**(11), 3075–3078 (2011)
5. M.A.K. Magableh, A.A.A. Radwan, Y.A.-R.I. Mohamed, Dynamic stabilization of a centralized weak grid-tied VSC system considering PV generator dynamics. IEEE Trans. Power Syst. **39**(1), 2209–2226 (2024)
6. J. He, Y.W. Li, D. Bosnjak, B. Harris, Investigation and active damping of multiple resonances in a parallel-inverter-based microgrid. IEEE Trans. Power Electron. **28**(1), 234–246 (2013)
7. L. Huang, H. Xin, F. Dörfler, H∞-control of grid-connected converters: design, objectives and decentralized stability certificates. IEEE Trans. Smart Grid **11**(5), 3805–3816 (2020)
8. G.F. Franklin, J.D. Powell, A. Emami-Naeini, J.D. Powell, *Feedback control of dynamic systems* (Prentice hall, Upper Saddle River, 2002)
9. S. Jiang, G. Konstantinou, Impedance-based stability analysis: nodal admittance or bus admittance. IEEE Trans. Power Syst. **39**(1), 2327–2340 (2024)
10. B. Qin, Y. Xu, C. Yuan, J. Jia, A unified method of frequency oscillation characteristic analysis for multi-VSG grid-connected system. IEEE Trans. Power Delivery **37**(1), 279–289 (2022)
11. X. Meng, J. Liu, Z. Liu, A generalized droop control for grid-supporting inverter based on comparison between traditional droop control and virtual synchronous generator control. IEEE Trans. Power Electron. **34**(6), 5416–5438 (2019)
12. A. Yazdani, R. Iravani, *Voltage-Sourced Converters in Power Systems: Modeling, Control, and Applications* (Wiley, Hoboken, 2010)
13. K. Guo, Y. Qi, J. Yu, D. Frey, Y. Tang, A converter-based power system stabilizer for stability enhancement of droop-controlled islanded microgrids. IEEE Trans. Smart Grid **12**(6), 4616–4626 (2021)
14. J. Liu, Y. Miura, T. Ise, Comparison of dynamic characteristics between virtual synchronous generator and droop control in inverter-based distributed generators. IEEE Trans. Power Electron. **31**(5), 3600–3611 (2016)
15. T. Wu, Z. Liu, J. Liu, S. Wang, Z. You, A unified virtual power decoupling method for droop-controlled parallel inverters in microgrids. IEEE Trans. Power Electron. **31**(8), 5587–5603 (2016)
16. P. Sun, J. Yao, Y. Zhao, X. Fang, J. Cao, Stability assessment and damping optimization control of multiple grid-connected virtual synchronous generators. IEEE Trans. Energy Convers. **36**(4), 3555–3567 (2021)
17. M. Ashabani, Y.A.-R.I. Mohamed, Novel Comprehensive control framework for incorporating VSCs to smart power grids using bidirectional synchronous-VSC. IEEE Trans. Power Syst. **29**(2), 943–957 (2014)

18. J. Zhu, C.D. Booth, G.P. Adam, A.J. Roscoe, C.G. Bright, Inertia emulation control strategy for VSC-HVDC transmission systems. IEEE Trans. Power Syst. **28**(2), 1277–1287 (2013)
19. J. Fang, H. Li, Y. Tang, F. Blaabjerg, Distributed power system virtual inertia implemented by grid-connected power converters. IEEE Trans. Power Electron. **33**(10), 8488–8499 (2018)

Part II
Oscillation Suppression Control

Chapter 4
Revisiting Droop Control and Virtual Impedance Method

4.1 Different Droop Controls Under Different Impedance Types

The basic control of grid-forming converters is droop control that generates angle and amplitude [1, 2] that achieves power sharing and frequency/voltage regulation. Despite its advantages, droop control has several inherent limitations [3]: (1) It is heavily dependent on the line impedance characteristics and requires transmission lines to exhibit inductive behavior; (2) System frequency and voltage are influenced by load power demand, resulting in deviations in voltage frequency and amplitude; and (3) Mismatches in line impedance can lead to inconsistent output voltage amplitudes among micro sources, preventing the achievement of reactive power equalization.

To address the sensitivity of basic droop control to line impedance characteristics, the virtual impedance method is widely utilized to decouple active-reactive power $P - Q$ and compensate reactive-power differences among the various DGs due to the line impedance mismatches. Several studies [4, 5] have introduced a control strategy that combines virtual impedance and droop control. This approach can reshape line impedance characteristics and thereby ensuring the feasibility and applicability of droop control under varying line impedances. However, these studies do not provide a detailed parameter design guideline of virtual impedance for system stability. In remote low-voltage power distribution areas with high line resistance, large virtual inductance is required to meet system line inductance requirements. This can result in excessive voltage drop and reduced transmission capacity. To improve system stability under complex line impedance conditions, droop control within the virtual coordinate transformation framework has been introduced in [6, 7]. However, due to the use of virtual power for droop feedback, accurate power-sharing cannot be achieved. Then, some other power-sharing control methods have been introduced for AC microgrids with low-voltage highly-resistive line characteristics, including active power-amplitude ($P - V$) droop and reactive power-frequency ($Q - \omega$) droop

X. Hou et al., *Self-organized Operation and Stability Control of Electronized Power Networks*, https://doi.org/10.1007/978-981-96-9067-1_4

Table 4.1 Droop relations under different line impedance

Types of line	P–Q relations	Droop laws
L $\theta \approx 90°$	$P \approx V_g V_i \delta / X$ $Q \approx V_g (V_i - V_g)/X$	$\omega = \omega_n - m_p P$ $V_i = V_n - n_q Q$
RL $\theta \approx 45°$	$P - Q \approx \sqrt{2} V_g V_i \delta / Z$ $P + Q \approx \sqrt{2} V_g (V_i - V_g)/Z$	$\omega = \omega_n - m(P - Q)$ $V_i = V_n - n(P + Q)$
R $\theta \approx 0°$	$P \approx V_g (V_i - V_g)/R$ $Q \approx -V_g V_i \delta / R$	$\omega = \omega_n + m_q Q$ $V_i = V_n - n_p P$

control strategies [8, 9], power angle droop [10, 11], intelligent algorithm-based droop control [12], and decoupling-based droop control [13–15]. However, there is no unified droop control strategy that can ensure stable system operation under arbitrary line characteristics, achieve accurate active power sharing among micro-sources, and realize satisfactory dynamic performance. The classical control laws under the above different line impedances are summarized in Table 4.1.

To fill this gap, this section revisits the basic droop control and virtual impedance methods from a new analogy perspective and presents a unified droop control that is suitable for arbitrary transmission line impedances.

4.2 Power Transmission Models Under Different Line Impedances

4.2.1 Equivalent Model of Single-Unit Infinite-Bus

For parallel multi-inverters of AC microgrid, each distributed generation unit can be regarded as an equivalent voltage source node of the system, including voltage-magnitude control and phase-angle-vector control. And the public coupling connection point of a multi-parallel converter can be approximated as an infinity bus when the number of parallel modules is relatively large, thus simplifying the system analysis and design. As shown in Fig. 4.1, the equivalent physical circuit diagram of a distributed power electronic converter unit connected to a common infinity grid-

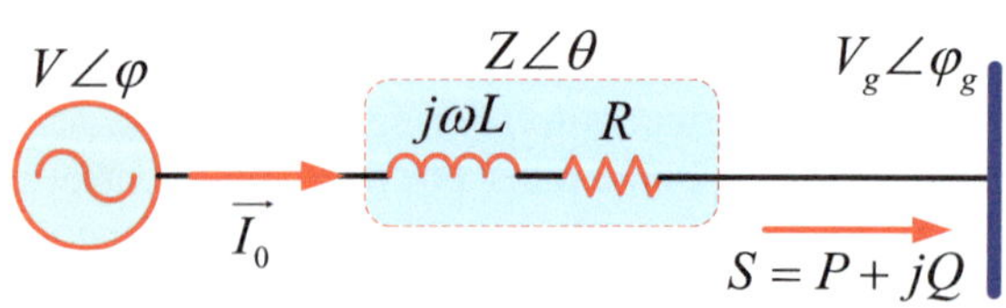

Fig. 4.1 Equivalent physical circuit diagram of the grid-connected unit

connected bus is given. From Fig. 4.1, it can be seen that the output current and apparent power of the distributed power generation unit can be expressed as

$$\vec{I_0} = \frac{V\angle\varphi - V_g\angle\varphi_g}{Z\angle\theta} = \frac{V}{Z}\angle(\varphi - \theta) - \frac{V_g}{Z}\angle(\varphi_g - \theta) \tag{4.1}$$

$$S = P + jQ = \vec{U_g} \cdot \vec{I_0}^* \tag{4.2}$$

In Fig. 4.1, $V\angle\varphi$ and $V_g\angle\varphi_g$ are the output voltage magnitude and vector phase angle of the power conversion unit and the common coupling bus, respectively. $Z\angle\theta$ is the modal magnitude and impedance angle of the line transmission impedance. $\theta = 0$ represents pure resistance and $\theta = \pi/2$ represents pure inductance.

Combining with (4.1) and (4.2), the output active power and reactive power of the distributed grid-connected power conversion unit are expressed as

$$P = \frac{VV_g}{Z}\cos(\theta - \delta) - \frac{V_g^2}{Z}\cos\theta \tag{4.3}$$

$$Q = \frac{VV_g}{Z}\sin(\theta - \delta) - \frac{V_g^2}{Z}\sin\theta \tag{4.4}$$

In (4.3) and (4.5), δ is the voltage phase angle difference (power angle concept) between the grid-connected power conversion unit and the common coupling bus, which can be expressed as

$$\delta = \varphi - \varphi_g = \int (\omega - \omega_g)dt \tag{4.5}$$

To further clearly visualize the coupling relationship of the active/reactive power output from the grid-connected power conversion unit and their variation with voltage amplitude and power angle, the power circle curve for power impact analysis is depicted in Fig. 4.2 from (4.3) and (4.5), where the center of the circle is $(-\frac{V_g^2}{Z}\cos\theta, -\frac{V_g^2}{Z}\sin\theta)$ and the radius is VV_g/Z

4.2.2 *Simplified Power Model Under Different Line Impedances*

It is well known that the output line impedance characteristics of power electronic systems are significantly different at different voltage levels. As shown in Table 4.2, in the high-voltage transmission line, the line impedance characteristics are approximated as highly inductive ($\theta = 72.8°$); in the medium-voltage transmission line, the line impedance characteristics are approximated as complex resistive-inductive

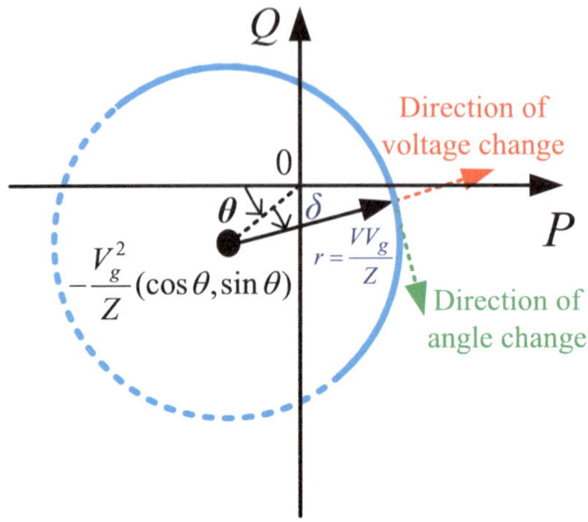

Fig. 4.2 Schematic diagram of the power circle of the grid-connected power electronic conversion unit

Table 4.2 Comparison of line impedance characteristics at different voltage levels

Voltage rating	Resistance	Inductors	Resistance/ inductance ratio	Line impedance
	$r/(\Omega/\text{km})$	$X/(\Omega/\text{km})$	(X/r)	(angle θ)
Low voltage	0.642	0.083	0.13	$\cong 7.5°$
Medium voltage	0.161	0.190	1.18	$\cong 50°$
High voltage	0.060	0.191	3.18	$\cong 72.8°$

Table 4.3 Simplified power model under different line impedance characteristics

Output line impedances	Purely inductive ($\theta = \frac{\pi}{2}$)	Purely resistive ($\theta = 0$)	Equally resistive–inductive ($\theta = \frac{\pi}{4}$)
Active power	$P = \frac{VV_g}{Z}\sin\delta$	$P = \frac{V_g(V\cos\delta - V_g)}{Z}$	$P = \frac{VV_g}{Z}\cos(\frac{\pi}{4} - \delta) - \frac{V_g^2}{Z}\cos\frac{\pi}{4}$
Reactive power	$Q = \frac{V_g(V\cos\delta - V_g)}{Z}$	$Q = -\frac{VV_g}{Z}\sin\delta$	$Q = \frac{VV_g}{Z}\sin(\frac{\pi}{4} - \delta) - \frac{V_g^2}{Z}\sin\frac{\pi}{4}$

impedance ($\theta = 50°$); in the low-voltage transmission line, the line impedance characteristics are approximated as highly resistive ($\theta = 7.5°$). To simplify the subsequent theoretical analysis, the line impedance characteristics at the three voltage levels are replaced by purely inductive, equally resistive-inductive, and purely resistive types, respectively.

Further, from (4.3) and (4.5), the simplified models of active and reactive power under different line impedance characteristics are given in Table 4.3. And a schematic

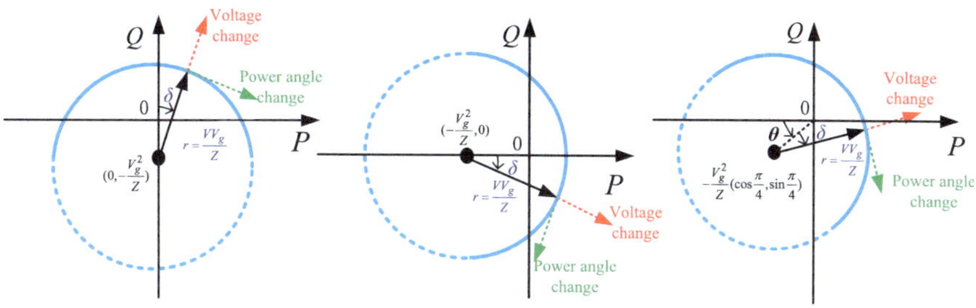

Fig. 4.3 Diagram of power circle unit under different line impedance characteristics

diagram of the power circle under different line impedance characteristics is drawn, as shown in Fig. 4.3. It is derived that under the assumed purely inductive line ($\theta = \pi/2$), the active power output regulation basically relies on the power angle, while the reactive power output regulation mainly relies on the voltage magnitude difference. Under the purely resistive line ($\theta = 0$), the coupling of active and reactive power is just the opposite, and the transmission of active and reactive power relies on the voltage magnitude difference and voltage power angle, respectively. In addition, when the resistance is comparable($\theta = \pi/4$), the transmission of active and reactive power is affected by the voltage power angle and voltage magnitude difference, which makes the coupling mechanism more complicated and increases the difficulty of the controller design. In addition, considering the line transmission capacity under different line impedance characteristics and the feasible power angle domain constraint ($\delta \in [-\pi/2, \pi/2]$), the solid line power circle curves of Fig. 4.3 represent the feasibility interval.

4.3 Analogy of Classical Droop Control and Virtual Impedance

4.3.1 Basic Principle of Classical $P - \omega/Q - V$ Droop Control

For AC microgrids, basic $P - \omega/Q - V$ droop control has become one of the most mainstream decentralized control strategies due to its high reliability, plug-and-play characteristics, and non-communication self-synchronization. The basic principle is to feedback on the locally-collected active and reactive power signals to control the regulated output voltage frequency and amplitude reference, respectively. In addition, the natural essence of conventional droop control is to simulate the natural synchronous generator external output characteristics, so conventional active-frequency droop control has the global frequency autonomous synchronization capability, which can guarantee the exact proportional active power sharing among multiple parallel inverters without any communication condition. Its steady-state expressions are presented as follows

$$\omega_r = \omega^* - mP, \quad 0 < m \leq \frac{\omega_{\max} - \omega_{\min}}{P^*} \tag{4.6}$$

$$V_r = V^* - nQ, \quad 0 < n \leq \frac{V_{\max} - V_{\min}}{Q^*} \tag{4.7}$$

In (4.6) and (4.7), ω_r and V_r are the voltage angular-frequency reference and voltage magnitude reference of the individual inverter; ω^* and V^* are the angular frequency given and voltage magnitude given under rated-condition; m and n represent the active-frequency droop and reactive-voltage droop gain, respectively; P^* and Q^* are the rated active power and reactive power of individual inverter; $\omega_{\max}$ and $\omega_{\min}$ represent the maximum and minimum values of angular frequency allowed for normal operation of the system; $V_{\max}$ and $V_{\min}$represent the maximum and minimum values of voltage allowed for normal operation of the system. $\omega_{\min}$ and $\omega_{\max}$ represent the maximum and minimum values of angular-frequency allowed for normal system operation; The specific droop control curves are shown in Fig. 4.4.

The overall control scheme of a single voltage-source inverter containing droop control is shown in Fig. 4.5. Its physical inverter circuit is mainly composed of a source terminal, power electronic switch, LC filter, and parallel line impedance, and its controller is mainly divided into local signal acquisition, power calculation, droop control loop, double voltage-current tracking closed-loop and SPWM modulation. In particular, the droop control link provides the voltage frequency and amplitude reference for the voltage loop tracking control based on the locally calculated output active power and reactive power signals. In general, the calculation of the average active and reactive power requires first-order low-pass filtering of the instantaneous power, and the control bandwidth is designed to be approximately [2–10] Hz. In addition, for fast and accurate tracking of the outer-loop voltage reference, the voltage loop control is designed to have a medium bandwidth of approximately [400–600] Hz. Such a bandwidth design ensures that the four control links of power, voltage,

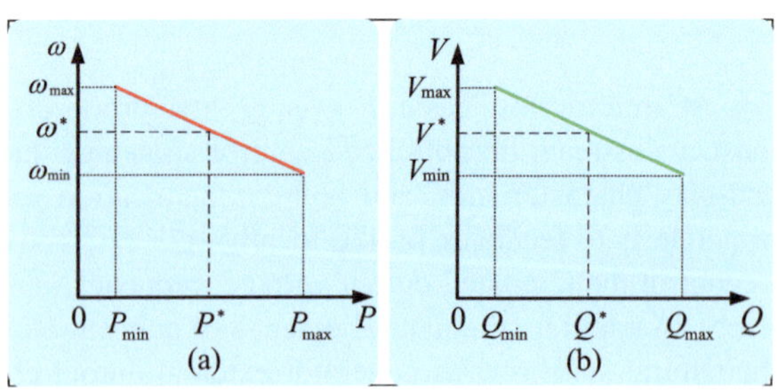

Fig. 4.4 Schematic curves of droop control. (**a**) active-frequency and (**b**) reactive-voltage

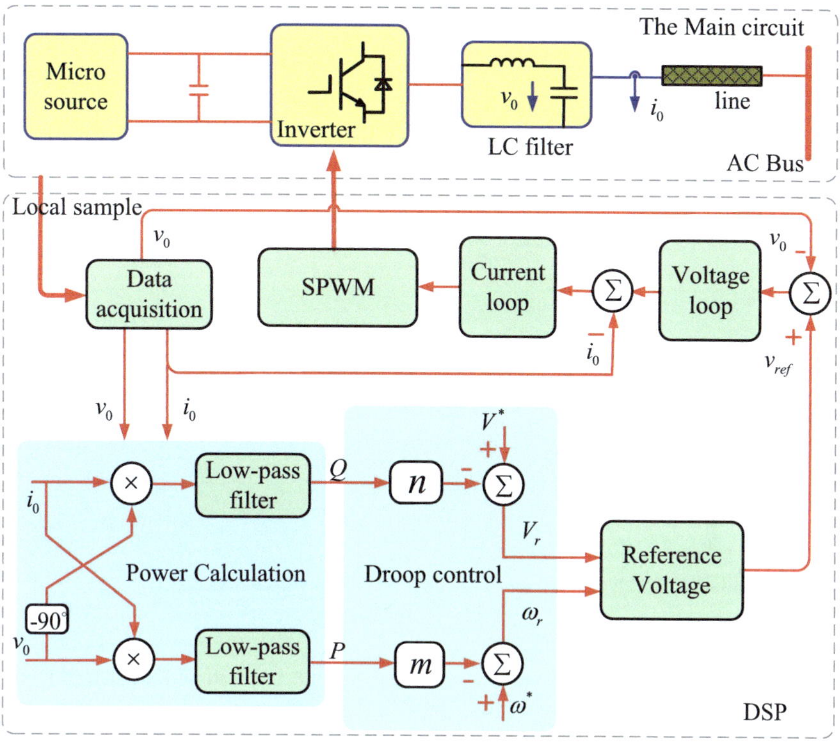

Fig. 4.5 Overall control scheme for a single inverter with basic droop control

current, and modulation loops are effectively separated in time scale, which facilitates the overall design and dynamic response of the inverter controller.

It is worthwhile to note that although conventional droop control forms the basis for decentralized control of parallel inverter systems, it still has some shortcomings of its own: (1) Dependence on highly-inductive output impedance, presence of active-reactive power coupling; (2) Poor dynamic adjustment; (3) Line mismatch leading to uneven reactive power sharing.

4.3.2 Revisiting Control Mechanism of Virtual Impedance

For the intuitive physical concept of virtual impedance, it is introduced to reshape the overall output impedance characteristics of inverters. To solve the problem of line impedance mismatch between multiple parallel inverters, the total output impedance is flexibly adjusted to highly-inductive, highly-resistive, or complex impedance according to the control target requirements. The virtual impedance is analyzed

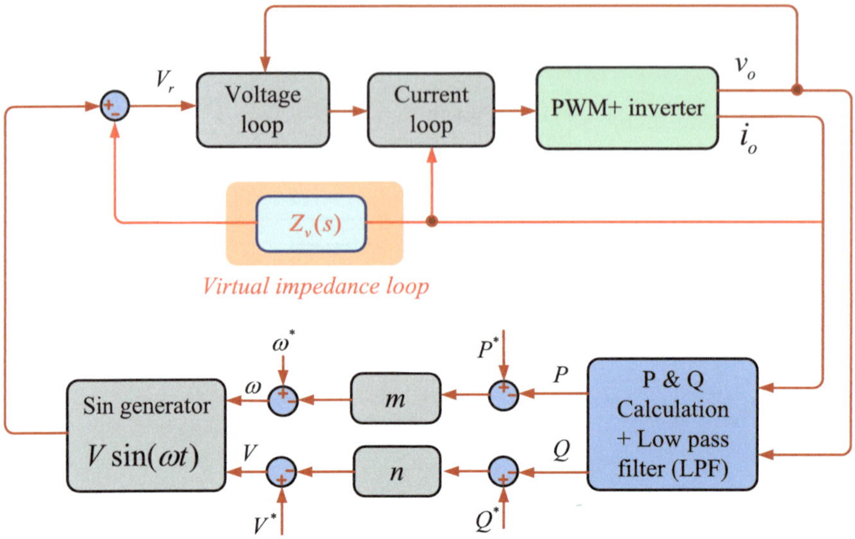

Fig. 4.6 Overall control block of a single inverter including virtual impedance

on the active power-reactive power from the theoretical and practical application perspectives. The influence of virtual impedance on the power dynamic decoupling ability, power oscillation damping ability, tidal current control ability, and system fault current limiting ability is further analyzed.

Figure 4.6 gives the overall control block diagram of a single inverter containing virtual impedance. From it, the mathematical expression of the virtual impedance is given as

$$v_o = v_r - Z_r i_o \tag{4.8}$$

In (4.8), $Z_v = R_v + jX_v$ is the virtual impedance to be shaped; $v_o = V_o\angle\delta_r$ is the actual output voltage of the inverter; represents the actual output current of the inverter; and $v_r = V_r\angle\delta_r$ is the voltage phase reference given by the outer power loop of the droop control.

As seen in Fig. 4.6, the intuitive concept of virtual impedance is to multiply the impedance to be virtualized with the output current vector. And it is subtracted from the voltage loop reference in the expectation of changing the impedance characteristics of the original line. It is worth mentioning that the virtual impedance is mainly used for steady-state sharing and dynamic response performance regulation, without current dynamic filtering and real active losses, compared to the actual physical impedance configured directly in series at the grid-connected inverter.

From the traditional perspective based on intuitive physical concepts, the goal of introducing the virtual impedance concept in a multi-parallel inverter is to reshape the overall equivalent output impedance of the inverter, as shown in detail in Fig. 4.7, where the equivalent output impedance contains both the constructed virtual

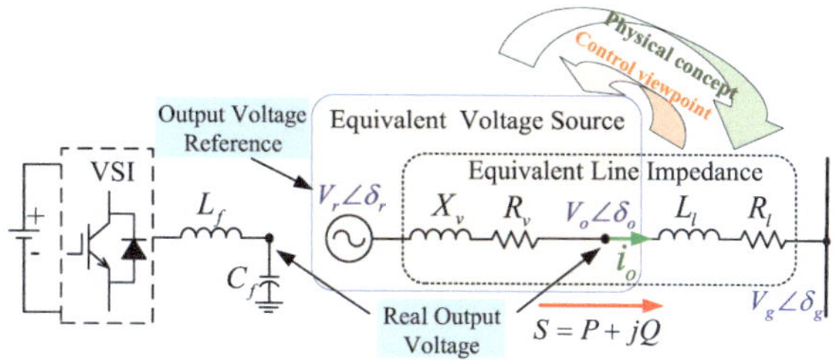

Fig. 4.7 Two perspective comparisons of virtual impedance: intuitive physical concept and real control mechanism

impedance and the actual physical line impedance. However, from the perspective of exploring the real control connotation of virtual impedance, its essential control law is to multiply the virtual impedance with the output current vector and feed it back to the internal reference of the output voltage at the front end of the voltage loop. In other words, the control law of virtual impedance can be studied and analyzed from a new control perspective to investigate the effect of virtual impedance on the actual output node characteristics of the voltage source and try to replace the current feedback with power feedback. In this way, a new outer power-loop control law is constructed to equate the effect of the inner-loop output impedance, providing new ideas for distributed units that want to reshape the output impedance.

From Fig. 4.7, the mathematical relationship between the output power of the inverter and the virtual impedance can be obtained as

$$P + jQ = V_o\angle\delta_o\left(\frac{V_r\angle\delta_r - V_o\angle\delta_o}{R_v + jX_v}\right)^* \tag{4.9}$$

Replacing the output power in (4.9) with the output current in (4.8), the derivation yields the voltage difference and phase angle difference triggered by the output active and reactive power on the virtual impedance, respectively, expressed as

$$\Delta V = V_r - V_o \cong \frac{R_v P + X_v Q}{V_o} \tag{4.10}$$

$$\Delta\delta_v = \delta_r - \delta_o \cong \frac{X_v P - R_v Q}{V_o V_r} \tag{4.11}$$

In (4.10) and (4.11), since the output voltage amplitude is around the rated voltage (within approximately 5% deviation), the steady-state values of V_r and V_o can be replaced by V^* for ease of analysis and expression. Thus, the core control law of the virtual impedance is further expressed as

$$\begin{cases} V_o = V_r - n_r P - n_x Q \\ \delta_o = \delta_r - m_x P + m_r Q \end{cases} \tag{4.12}$$

Rewrite (4.12) under the frequency coordinate framework

$$\begin{cases} V_o = V_r - n_r P - n_x Q \\ \omega_o = \omega_r - m_x \frac{dP}{dt} + m_r \frac{dQ}{dt} \end{cases} \tag{4.13}$$

where,

$$\begin{cases} m_x = \frac{X_v}{V^{*2}} \\ n_x = \frac{X_v}{V^*} \end{cases} ; \quad \begin{cases} m_r = \frac{R_v}{V^{*2}} \\ n_r = \frac{R_v}{V^*} \end{cases} \tag{4.14}$$

From (4.12) and (4.13), it is derived that the virtual impedance control law is a combination of (4.13) power angle droop and (4.12) voltage droop control, especially finding the influence law of virtual inductor X_v and virtual resistance R_v on the phase angle and voltage amplitude regulation, respectively. Therefore, the m_x , m_r , n_x , n_r parameters can be designed flexibly to replace the role of virtual impedance, and have higher parameter freedom than virtual impedance.

In addition, the establishment of this equivalent relationship, in (4.13) power angle droop control, increasing the $\delta - P$ droop coefficient m_x is essentially an equivalent increase in virtual inductance, which in turn compensates for the overall output impedance at the active control level. Thus, it can improve active power sharing accuracy. Meanwhile, in the voltage droop control of (4.12), increasing the $V - Q$ droop coefficient n_x is essentially an equivalent increase in virtual inductance, which in turn compensates the overall output impedance at the reactive power control level and can improve the reactive power sharing accuracy, and also illustrates the root cause of the inability to achieve accurate sharing of voltage-reactive power droop. Moreover, in the system stability analysis, it can also be explained from the perspective of virtual inductance that a moderate increase in the virtual inductance equivalent parameter m_x can appropriately improve the system dynamic response and damping power oscillation capability, but an excessive droop factor m_x is equivalent to increasing the impedance of the system, which will affect the power transmission capability of the inverter and lead to an excessive steady-state power angle of the system, which is not conducive to system stability.

In a multi-parallel inverter AC microgrid, the virtual inductor-dominated approach is often used to satisfy the condition that the conventional droop is dominantly inductive by (4.12) and (4.13), and the control law of the virtual inductor ($R_v = 0$) is given here separately as follows.

$$\delta_o = \delta_r - m_x P \tag{4.15}$$

$$V_o = V_r - n_x Q \tag{4.16}$$

4.3.3 Analogy Between Droop Control and Virtual Impedance

In order to transfer the control law of the virtual inductor from under the phase angle coordinate to under the frequency coordinate, (4.15) can be calculated by taking the differential, and the virtual inductor is then equated to a new frequency droop.

$$\omega_o = \omega_r - m_x \frac{dP}{dt} \tag{4.17}$$

In (4.17), ω_r is the nominal reference value of the angular frequency. It follows that the virtual inductor can be understood as an active differential type of frequency droop.

Further, when the multi-parallel inverter contains both conventional frequency droop and virtual inductance control, a combined droop control method is presented by combining (4.6) and (4.17), and (4.7) and (14.16).

$$\omega_o = \omega^* - mP - m_x \frac{dP}{dt} \tag{4.18}$$

$$V_o = V^* - (n + n_x)Q \tag{4.19}$$

Equation (4.18) is a frequency droop control with proportional-differential (PD) feedback power control. In addition, from (4.19), the reactive-voltage droop gain n_x from the virtual inductor equivalent is superimposed on the droop coefficient of the $Q - V$ control, confirming the homology and equivalence of using virtual inductor control and increasing the reactive droop gain analyzed above. Among them, m_x mainly regulates the dynamic performance of the system, and n_x dominates the stability of the system. By setting the m_x and n_x parameters reasonably instead of the virtual inductance, the combined droop control is universally applicable under any highly-inductive, inductance-resistance, and resistance-inductance line characteristics. At the same time, it also provides a new idea for mapping the droop feedback control law to the physical concept of virtual impedance.

4.3.4 United Multi-degree-of-Freedom Droop Control

The combined droop control in (4.18) and (4.19) extends the general applicability of conventional droop control under complex line impedance and ensures autonomous frequency synchronization and accurate active power sharing among multiple inverters. Inspired by this, in order to simultaneously take into account the basic control objectives of steady-state sharing performance and dynamic response, as well as the universal applicability under complex line impedance, this section further presents a united multi-degree-of-freedom decentralized synchronous control method.

$$\begin{bmatrix} \omega \\ V \end{bmatrix} = \begin{bmatrix} \omega^* \\ V^* \end{bmatrix} - \begin{bmatrix} H_{P\delta}(s) & H_{Q\delta}(s) \\ H_{PV}(s) & H_{QV}(s) \end{bmatrix} \begin{bmatrix} P \\ Q \end{bmatrix} \tag{4.20}$$

where $H_{ij}(s)$ is the proportional-integral-derivative (PID) operator, expressed as

$$H_{ij}(s) = m_{ij} + \frac{m'_{ij}}{s} + m''_{ij}s; \quad (i = P, Q; j = \delta, V) \tag{4.21}$$

Compared with the combined droop control in (4.18) and (4.19), the unified multi-degree-of-freedom control framework in (4.20) has two significant improvements: (1) one is to use the proportional-integral-differential (PID) operator instead of the traditional proportional operator; the proportional operator can flexibly adjust the steady-state operating effect; the integral operator can be applied to the dual-mode switching operation of parallel and off-grid, and the differential operator can flexibly adjust the system dynamic damping; (2) The second is the use of feedback active-reactive coupling-matrix, which can realize the dynamic decoupling of active and reactive power through flexible design parameters and greatly improve the system stability.

The multi-degree-of-freedom means that the united droop control has multi-functions with systematic consideration of steady-state power-sharing performance, dynamic performance, and universal applicability of complex line impedance. For the unified multi-degree-of-freedom droop control framework, the parameters can be flexibly selected according to different control objectives to facilitate comprehensive and optimal design, which is illustrated by the following examples.

(1) With regard to the proportional operator, the priority is given to the steady-state active power sharing while ensuring the synchronization of the autonomous frequency of the system, which is generally chosen from $m_{Q\delta} = 0$, and if the priority is given to the exact reactive power sharing, which is generally chosen from $m_{P\delta} = 0$.
(2) Regarding the integration operator, in order to adapt to dual-mode operation with and without the grid, $H_{ij}(s)$ generally contains the integration operation ($m'_{ij} \neq 0$) under grid-connected operation in order to self-adapt to abnormal grid changes, but under off-grid operation, the integration operator is generally zero ($m'_{ij} = 0$).
(3) With respect to the differential operator, the active or reactive power oscillations can be effectively damped by the flexible selection of appropriately sized differential operators and , respectively.
(4) Conventional droop control applied under highly-inductive line impedance can be seen as a special case of a united control framework with a non-zero main diagonal and a zero anti-diagonal feedback coupling matrix.

(5) When the main diagonal is all zero and the opposite diagonal is not zero, it can be applied to resistive lines for reactive-frequency control and active-voltage control.
(6) The influence law of $3 * 4 = 12$ parameters in the feedback coupling matrix can be studied by the participation factor and sensitivity analysis methods, which can provide guidance for the application of decentralized control of parallel-type inverter networks in engineering practice.

4.4 Stability and Mechanism Analysis Under Different Line Impedances

4.4.1 Stability Analysis

In order to investigate the influence of control parameters on system stability and dynamic response in the unified multi-degree-of-freedom droop control framework of (4.20), this section focuses on the small-signal stability analysis of the power outer-loop control dynamics of the grid-connected inverter system. The steady-state operating point of the system is denoted by capital letters as (V, V_g , δ), and the small-signal disturbance terms are denoted as $(\tilde{V}, \tilde{\delta}, \tilde{\omega}, \tilde{\omega}_g, \tilde{\omega}^*, \tilde{V}^*)$.

First, the power transfer characteristics of (4.3) and (4.5) are linearized as follows.

$$\begin{bmatrix} \tilde{P} \\ \tilde{Q} \end{bmatrix} = \mathbf{G} \begin{bmatrix} \tilde{\delta} \\ \tilde{V} \end{bmatrix} \tag{4.22}$$

In (4.22) , $\mathbf{G}$ is the gain matrix

$$\mathbf{G} = \begin{bmatrix} k_{p\delta} & k_{pV} \\ k_{q\delta} & k_{qV} \end{bmatrix} \tag{4.23}$$

$$\begin{cases} k_{p\delta} = \frac{\partial P}{\partial \delta} = \frac{VV_g}{Z} \sin(\theta - \delta)\,; k_{pV} = \frac{\partial P}{\partial V} = \frac{V_g}{Z} \cos(\theta - \delta) \\ k_{q\delta} = \frac{\partial Q}{\partial \delta} = -\frac{VV_g}{Z} \cos(\theta - \delta)\,; k_{qV} = \frac{\partial Q}{\partial V} = \frac{V_g}{Z} \sin(\theta - \delta) \end{cases} \tag{4.24}$$

Second, considering that the active and reactive power averages in Fig. 4.5 are obtained after first-order low-pass filtering, the linearized model of the unified multi-degree-of-freedom droop control of (4.20) can be obtained as

$$\begin{bmatrix} \tilde{\omega} \\ \tilde{V} \end{bmatrix} = \begin{bmatrix} \tilde{\omega}^* \\ \tilde{V}^* \end{bmatrix} - \frac{1}{\tau s + 1} \begin{bmatrix} H_{P\delta}(s) & H_{Q\delta}(s) \\ H_{PV}(s) & H_{QV}(s) \end{bmatrix} \begin{bmatrix} \tilde{P} \\ \tilde{Q} \end{bmatrix} \tag{4.25}$$

Transforming the frequency coordinate system in (4.25) to the power-angle coordinate system, it is further reduced to

$$\begin{bmatrix} \tilde{\delta} \\ \tilde{V} \end{bmatrix} = \begin{bmatrix} \tilde{\delta}^* \\ \tilde{V}^* \end{bmatrix} - \mathbf{H}(s) \begin{bmatrix} \tilde{P} \\ \tilde{Q} \end{bmatrix} \tag{4.26}$$

In (4.26) , $\mathbf{H}(s)$ is the feedback matrix that

$$\mathbf{H}(s) = \begin{bmatrix} \frac{H_{P\delta}(s)}{s(\tau s+1)} & \frac{H_{Q\delta}(s)}{s(\tau s+1)} \\ \frac{H_{PV}(s)}{\tau s+1} & \frac{H_{QV}(s)}{\tau s+1} \end{bmatrix} \tag{4.27}$$

Coupling (4.22) , (4.23) and (4.26) , (4.27) , the system closed-loop transfer function can be obtained as

$$\Phi(s) = \frac{\mathbf{G}}{\mathbf{I} + \mathbf{GH}(s)} \tag{4.28}$$

In summary, the influence of control parameters or physical parameters on the stability and dynamic response of the system can be derived by analyzing the trend of the dominant pole of the closed-loop transfer function of (4.28) with the variation of parameters.

To verify the general applicability of united droop control under different line impedance characteristics, root trajectory tests are conducted for systems with highly-inductive, equally resistive-inductive, and highly-resistive line characteristics, respectively. Thus the effects of active and reactive droop coefficients and power filtering constants on system stability are investigated. The specific system parameters are shown in Table 4.4. The specific root trajectory result is shown in Fig. 4.8.

(1) Under the three-types line impedance characteristics, the active droop coefficient m has a relatively large impact on the dynamic response of the system, when m is too small, the system is over-damped state; when m is increased moderately, a pair of conjugate dominant poles λ_1 and λ_2 will appear, in order to ensure a better dynamic response of the system, the conjugate poles are generally selected as the imaginary part of the value than the real part of the value is located in the range of 0.707 to 1.414; if m is increased all the time, it will appear under-damped If

Table 4.4 Physical and control parameters of the inverter

Variables	Value	Variables	Value
Rated active/reactive power	6 kW/ 6 kVar	$Q - V$ droop factor n	$n \in [0, 0.01]$
Rated voltage reference	311 V/ 50 Hz	$P - \omega$ droop factor m	$m \in (0, 0.002]$
Allowable voltage range	[280 V, 340 V]	First-order filter	$\tau \in [0.01, 0.1]$
Allowable frequency range	[49 Hz, 51 Hz]	Line impedance	$\lvert Z \rvert = 2, \theta \in [0, \pi/2]$

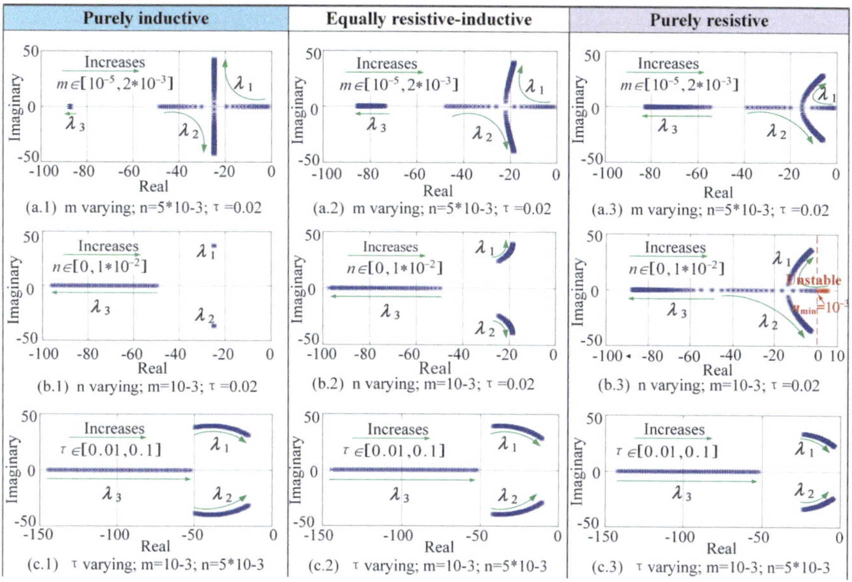

Fig. 4.8 Root trajectory analysis under three line impedances (**a**) m variation; (**b**) n variation; (**c**) filtering constant τ variation

m is increased all the time, the system will be underdamped, which is not good for the dynamic response.

(2) In the highly-inductive and equally resistive-inductive line impedance characteristics, the reactive droop coefficient n has almost no effect on the system stability and dynamic response. However, in the highly-resistive line impedance characteristics, the system stability requires that n cannot be too small, only when it is greater than 0.001 begins to gradually stabilize, according to the above (4.14) in the virtual inductance and droop coefficient of quantitative control relationship, can be equated to the virtual inductance $X_v = nV^* = 0.3\ \Omega$, confirming the necessity of virtual inductance under highly-resistive line characteristic.

(3) Under the three-type line impedance characteristics, the power filtering constant τ varies within a certain range [0.01, 0.1] with negligible effect on the system stability, but too large a filtering constant, which is equivalent to a delay in the power calculation time, is not conducive to system stability.

The above analysis confirms the feasibility of united droop control under different impedance characteristics, especially for the system under highly-resistive characteristics, which can be designed from two perspectives of mutual mapping of virtual inductance and droop coefficient, and the reactive droop coefficient n is reasonably selected to ensure system stability.

4.4.2 Mechanism Analysis and Turning Guidelines

The above analysis shows that the active power droop coefficient m and the filtering time constant τ mainly affect the transient response of the system, and their effects on stability can be approximately ignored within the normal parameter range. In addition, the reactive droop coefficient n has basically no effect on the stability in the case of purely-inductive and purely-resistive lines, but in the case of purely resistive, the reactive droop coefficient n is critical and directly related to the stable operation of the system. In order to further verify the reasonableness of the above root trajectory analysis, this section will analyze the reason from the perspective of the small signal control block diagram.

According to the small-signal analysis, the overall small-signal block diagram models of the stand-alone system in the purely-inductive, equally resistive-inductive, and purely-resistive lines are shown in Fig. 4.9a.1–a.3, respectively. As seen, the strength of the internal coupling effect of the active-reactive control loop varies for different line impedance characteristics: in the purely-inductive case, there is basically no coupling; in the complex resistive-sensitive line and the purely resistive case, the coupling effect is stronger. In order to clearly express the decomposed active and reactive control in the three cases, the small-signal simplified block diagrams are shown in Figs. 4.9b.1–b.3 and c.1–c.3, respectively.

In the small-signal block diagram of active-power frequency control, there are two pathways of direct and indirect coupling effects of the power angle on the forward pathway gain of the output active power, mathematically expressed as

$$k_{p\delta\_sum} = k_{p\delta\_dir} + k_{p\delta\_indir} \tag{4.29}$$

where

$$\begin{cases} k_{p\delta\_dir} = \left(\frac{\partial P}{\partial \delta}\right)_{dir} = \frac{VV_g}{Z}\sin(\theta - \delta) \\ k_{p\delta\_indir} = \left(\frac{\partial P}{\partial V}\right)_{dir} \times \left(\frac{\partial V}{\partial Q}\right)_{dir} \times \left(\frac{\partial Q}{\partial \delta}\right)_{dir} = \frac{n}{\tau s+1}\frac{VV_g^2}{Z^2}\cos^2(\theta - \delta) \end{cases} \tag{4.30}$$

In the block diagram of reactive-power voltage magnitude control, there are two pathways of direct and indirect coupling effects of voltage on the gain of the forward pathway of output reactive power, mathematically expressed as

$$k_{qV\_sum} = k_{qV\_dir} + k_{qV\_indir} \tag{4.31}$$

where

$$\begin{cases} k_{qV\_dir} = \left(\frac{\partial Q}{\partial V}\right)_{dir} = \frac{V_g}{Z}\sin(\theta - \delta) \\ k_{qV\_indir} = \left(\frac{\partial Q}{\partial \delta}\right)_{dir} \times \left(\frac{\partial \delta}{\partial P}\right)_{dir} \times \left(\frac{\partial P}{\partial V}\right)_{dir} = \frac{m}{s(\tau s+1)}\frac{VV_g^2}{Z^2}\cos^2(\theta - \delta) \end{cases} \tag{4.32}$$

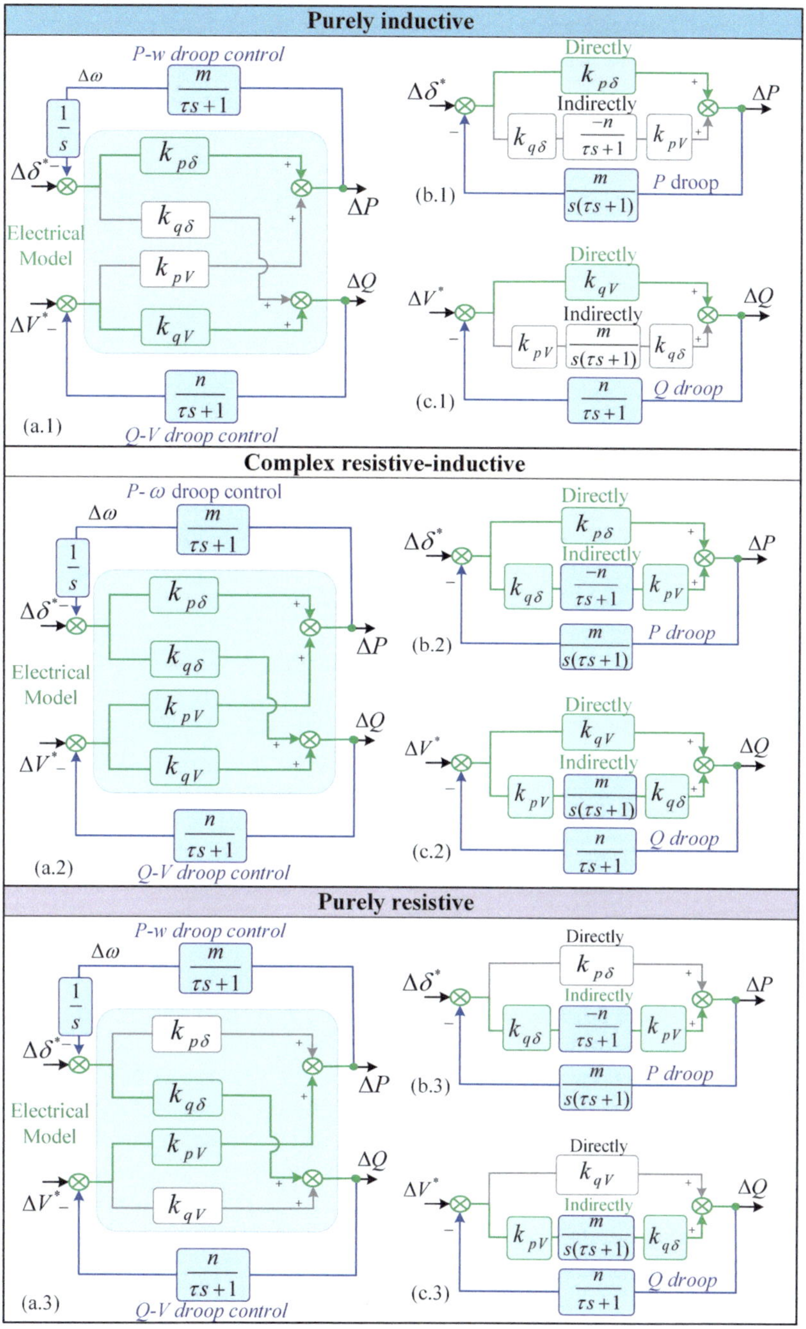

Fig. 4.9 Control block diagram under different line impedances. (**a**) overall small-signal block diagram; (**b**) decomposed active control; (**c**) decomposed reactive control

Table 4.5 Power coupling characteristics based on different line impedance conditions

Output line impedance	Purely inductive ($\theta = \frac{\pi}{2}$)	Purely resistive ($\theta = 0$)	Equally resistive–inductive ($\theta = \frac{\pi}{4}$)
$k_{p\delta\_\mathrm{dir}}$	$\frac{VV_g}{z} \cos\delta$	$-\frac{VV_g}{Z} \sin\delta$	$\frac{VV_g}{Z} \sin\left(\frac{\pi}{4} - \delta\right)$
$k_{p\delta\_\mathrm{indir}}$	$n\frac{VV_g^2}{Z^2} \sin^2\delta$	$n\frac{VV_g^2}{Z^2} \cos^2\delta$	$n\frac{VV_g^2}{Z^2} \cos^2\left(\frac{\pi}{4} - \delta\right)$
$k_{p\delta\_\mathrm{dir}}/k_{p\delta\_\mathrm{indir}}$	$\frac{Z}{nV_g\delta^2} \gg 1$	$\cong \frac{Z^*\delta}{nV_g}$	$Z\tan\left(\frac{\pi}{4} - \delta\right)/nV_g\cos\left(\frac{\pi}{4} - \delta\right)$
k_{qV_dir}	$\frac{V_g}{Z} \cos\delta$	$-\frac{V_g}{Z} \sin\delta$	$\frac{V_g}{Z} \sin\left(\frac{\pi}{4} - \delta\right)$
k_{qV_indir}	$m\frac{VV_g^2}{Z^2} \sin^2\delta$	$m\frac{VV_g^2}{Z^2} \cos^2\delta$	$m\frac{VV_g^2}{Z^2} \cos^2\left(\frac{\pi}{4} - \delta\right)$
$k_{qV\_\mathrm{dir}}:k_{qV\_\mathrm{indir}}$	$\frac{Z}{mVV_g\delta^2} \gg 1$	$\cong \frac{Z^*\delta}{mVV_g}$	$Z\tan\left(\frac{\pi}{4} - \delta\right)/nVV_g\cos\left(\frac{\pi}{4} - \delta\right)$

As seen in (4.30) and (4.32) , the forward path gain is positive due to power indirect coupling regardless of the characteristics of the line parameters, enhancing the ability of the system power angle to regulate active power ($\delta - P$) and voltage to regulate reactive power ($V - Q$). Thus, indirect coupling, although it may lead to coupling of the transient response between active and reactive power of the system, inherently facilitates the stability of the system, which in turn ensures the applicability of conventional droop at various line impedances.

To briefly analyze the weights of direct and indirect coupling at different line impedances. Under the conditions of neglecting the system dynamic characteristics (neglecting the power filtering characteristics) and the small power angle of the system operation, the power coupling characteristics of the micro-source under different line impedance conditions are given in Table 4.5. Among them, since the range of power angle of the microgrid generally operates in the interval $[-\pi/6, \pi/6]$, it can be considered as follows: $\delta \cong \tan\delta \cong \sin\delta$, $\cos\delta \cong 1$.

Under the highly-inductive line, the direct coupling of both active and reactive power control is much greater than the effect of indirect coupling, so the system has a good decoupling capability.

In the case of equally resistive-inductive lines, the direct and indirect coupling of micro-sources are positive and in the same order of magnitude, both of which are conducive to the ability of power angle regulation of active power and voltage regulation of reactive power to ensure that the system is always sTable

Under the purely-resistive low-voltage microgrid, the active and reactive power control in the steady-state power angle is positive, and the influence factor of its direct coupling is negative, which is not conducive to the stability of the system. However, with the moderate increase of the adjustable parameter n, the indirect coupling will gradually dominate, and the system will tend to be sTable The critical point of the system stability is the sum of the two forward channels is a positive number, i.e.

$$k_{p\delta\_sum} = k_{p\delta\_dir} + k_{p\delta\_indir} > 0 \tag{4.33}$$

Combining (4.30) and Table 4.5, the key conditions for system stability in the case where the line is purely-resistive are

$$n > \frac{Z \tan \delta}{V_g \cos \delta} \cong \frac{Z \times \delta}{V_g} \tag{4.34}$$

In the microgrid, since the voltage of the common bus is approximately equal to the rated voltage V^* within the allowed deviation from the rated voltage, V_g in (4.30) can be approximated by V^* instead. In addition, the power angle range of the microgrid generally operates in the interval $[-\pi/6, \pi/6]$, which can be considered as: $\delta \cong \tan \delta \cong \sin \delta$,$\cos \delta \cong 1$. Thus, a more intuitive form of (4.31) is

$$n > \frac{Z}{2V^*} \tag{4.35}$$

Considering the voltage quality of the microgrid, the design principle of the reactive power droop factor of low-voltage resistive microgrid is

$$\frac{Z}{2V^*} < n \leq \frac{V_{\max} - V_{\min}}{Q^*} \tag{4.36}$$

4.5 Experiments of United Droop Control Under Different Impedances

In order to verify the effectiveness and feasibility of the introduced united droop control under different line impedance characteristics, an experimental platform of a three-inverter parallel system was built with the physical main circuit and parameters as shown in Fig. 4.10, which mainly contains the DC power supply, inverter, filter, line impedance, and load. The control scheme of three inverters is presented in Fig. 4.11. The rated AC reference voltage and frequency of the system is 312 V/ 50 Hz. The DC-side voltage and virtual impedance of three inverters are 400 V and 0.2, respectively.

According to the different line impedance characteristics in Table 1, this section verifies the operation of the united droop control under three line characteristics: highly-inductive (impedance angle $\theta = 0.404\pi$), complex impedance (impedance angle $\theta = 0.278\pi$) and highly-resistive (impedance angle $\theta = 0.042\pi$), respectively, and the specific line impedance parameters and control parameters of the three groups are shown in Table 4.6. According to Fig. 10, only the RC load (1.5 kW, −0.75 kVar) was connected to the system in the initial stage of the experiment; to test the dynamic response of the control method under load switching, a new RL load (1.5 kW, +1.5 kVar) was cut in after t = 1 s.

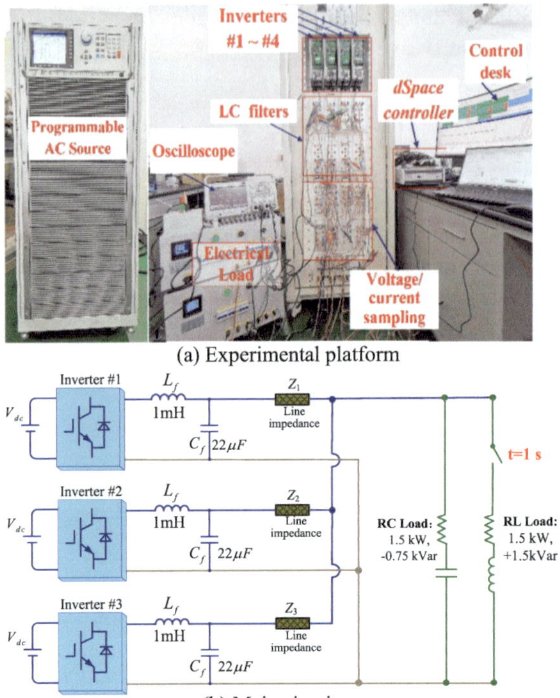

(a) Experimental platform

(b) Main circuits

Fig. 4.10 Prototype setup of the experiment

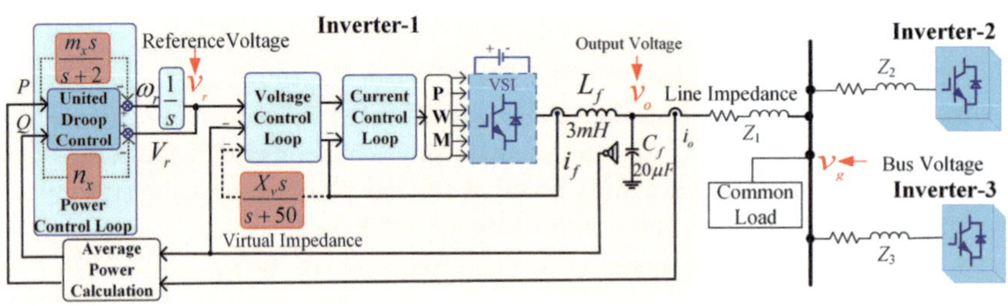

Fig. 4.11 United control scheme of the three-inverter parallel system

4.5.1 Under Highly Inductive Line Impedance

The experimental results under highly inductive line impedance are given in Fig. 4.12. The four experimental waveforms in Fig. 4.12a are the output currents $i_1 \sim i_3$ and common bus voltages u_{bus} of the three inverters. It is seen that the system has high-quality output voltage and current. From Fig. 4.12b, d, it is concluded that the three inverters can achieve accurate active power sharing and frequency autonomous synchronization before and after load switching. From Fig. 4.12c, e, it is concluded

Table 4.6 Physical line parameters and control parameters of the experiment

Line characteristics	Line impedance parameters	Control parameters
(1) Under highly-inductive	$Z_1 = 0.3 + j0.9\Omega$; $Z_2 = 0.25 + j0.75\Omega$; $Z_3 = 0.2 + j0.6\Omega$	$m = 2 \times 10^{-4}, n = 1 \times 10^{-3}$
(2) Under equally resistive-inductive	$Z_1 = 0.6 + j0.7\Omega$; $Z_2 = 0.5 + j0.6\Omega$; $Z_3 = 0.4 + j0.5\Omega$	$m = 2 \times 10^{-4}, n = 4 \times 10^{-3}$
(3) Under highly-resistive	$Z_1 = 1.0 + j0.12\Omega$; $Z_2 = 0.8 + j0.1\Omega$; $Z_3 = 0.6 + j0.07\Omega$	$m = 2 \times 10^{-4}, n = 6 \times 10^{-3}$

that the global voltage inconsistency is caused by the mismatch of the output line impedance values of the three inverters, so the accurate reactive power sharing cannot be guaranteed, and the reactive power sharing can be ensured by adjusting the reactive droop gain within the acceptable deviation range. For the dynamic response in switching load, the system regains a new steady state after about 5 fundamental frequency cycles with good dynamic response capability.

4.5.2 *Under Complex Resistive-Inductive Line Impedance*

Figure 4.13 gives the experimental results under the resistive-inductive equivalent line characteristics. Figure 4.13a shows the output currents i_1–i_3 and the common bus voltage u_{bus} of the three inverters. From Fig. 4.13b, it is concluded that the active power of the three inverters has been in the sharing state before and after load switching, and has a good dynamic response. And the reactive power can only be roughly equalized, and the comparison with the results under highly-inductive line impedance reveals that the results are basically similar except for the different precision of steady-state reactive power sharing. And the increased part of line resistive components basically does not change the dynamic and steady-state performance of the system.

4.5.3 *Under Highly Resistive Line Impedance*

The experimental results of united droop control under highly-resistive line characteristics are given in Fig. 4.14. Figure 4.14a shows the output currents i_1–i_3 and common bus voltages u_{bus} for the three inverters. Compared with the cases under highly-inductive and equally resistive-inductive line impedance, the $Q-V$ droop gain n must be a larger value to strengthen the indirect coupling of Fig. 4.9. The value n is

Fig. 4.12 Experimental results under highly inductive line characteristics

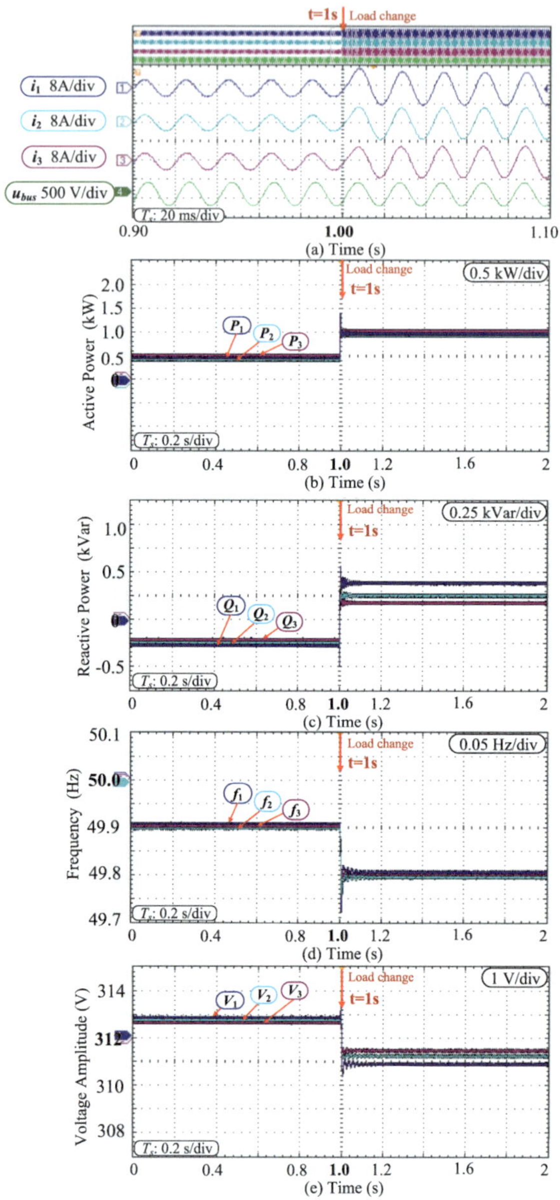

Fig. 4.13 Experimental results under the equally resistive-inductive line characteristics

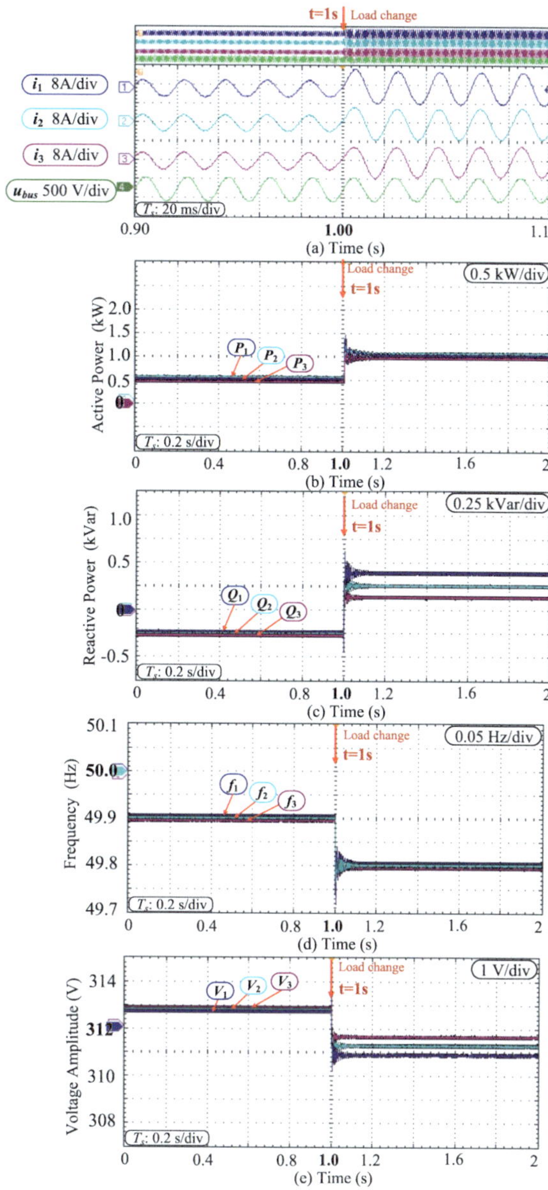

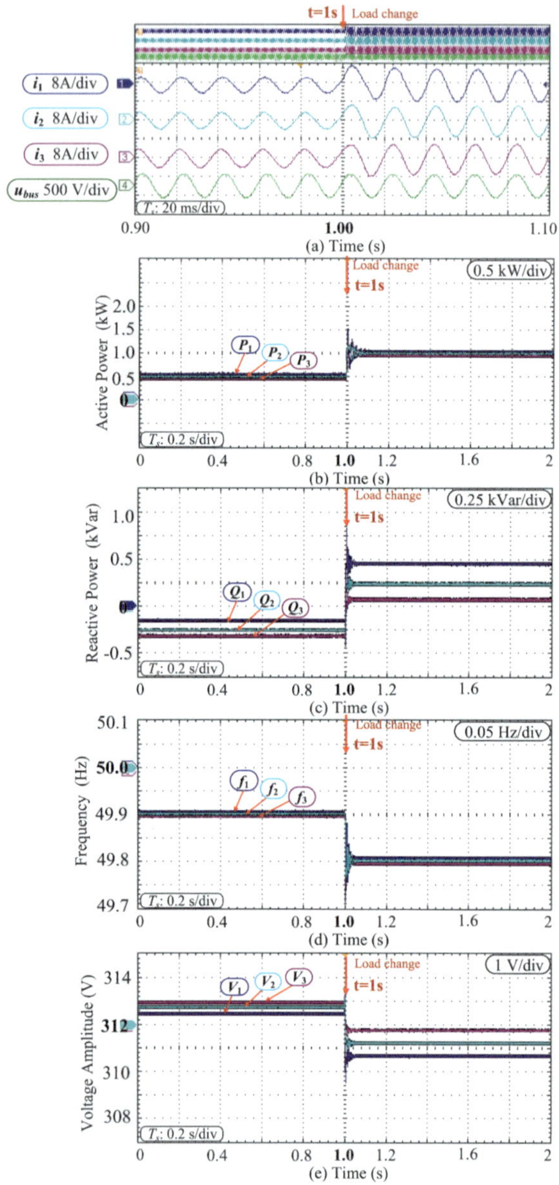

Fig. 4.14 Experimental results under highly-resistive line characteristics

given in Table 4.6 according to the design guideline (4.36). From Fig. 4.14b, it is concluded that the active power of the three inverters before and after load switching has been in the state of equal division with a good dynamic response, and from Fig. 4.14c, it is concluded that the reactive power has not reached the exact equal division due to the line impedance mismatch. Comparing with the results under highly-inductive and equally resistive-inductive line impedance, it is found that the active power and frequency synchronization results are basically similar except for the different precision of steady-state reactive power sharing, which achieves the universal applicability of united droop control under highly-inductive, equally resistive-inductive, and highly-resistive line impedances ensures the frequency synchronization while giving priority to the accurate active power sharing, and ensures the uniformity of the control method in different applications.

4.6 Conclusion

This section revisits the basic droop control and virtual impedance methods from a new analogy perspective and presents a unified droop control that is suitable for arbitrary transmission line impedances. The coupling relationship between active power and reactive power under different line impedance characteristics is revealed. And a unified droop control is introduced that takes into account steady-state power-sharing performance, dynamic performance, and universal applicability of complex line impedance, ensuring universal applicability under arbitrary line characteristics. Because the introduced method can achieve autonomous frequency synchronization and accurate active power sharing simultaneously in a decentralized manner, it has the advantages of communication-cost saving and increased system reliability. Finally, the experiment results verify the feasibility of the introduced unified droop control under the characteristics of highly-inductive, equally resistive-inductive, and highly-resistive line impedances. For future work this revisiting study can inspire researchers to enhance the decentralized droop methods for multiple parallel inverters.

References

1. R.H. Lasseter, P. Paigi, Microgrid: a conceptual solution, in *2004 IEEE 35th Annual Power Electronics Specialists Conference* (2004), vol. 6, pp. 4285–4290
2. Y. Liu, Y. Hu, N. Chen, J. Wang, X. Shao, Z. Pei, C. Liu, Study on power optimization method and coordinated control strategy of optical storage microgrid. J. Electr. Eng. **17**(1), 22–30 (2022)
3. H. Han, X. Hou, J. Yang, J. Wu, M. Su, J.M. Guerrero, Review of power sharing control strategies for islanding operation of AC microgrids. IEEE Trans. Smart Grid **7**(1), 200–215 (2016)
4. J.M. Guerrero, L.G. De Vicuna, J. Matas, M. Castilla, J. Miret, A wireless controller to enhance dynamic performance of parallel inverters in distributed generation systems. IEEE Trans. Power Electron. **19**(5), 1205–1213 (2004)

5. Y.A.R.I. Mohamed, E.F. El-Saadany, Adaptive decentralized droop controller to preserve power sharing stability of paralleled inverters in distributed generation microgrids. IEEE Trans. Power Electron. **23**(6), 2806–2816 (2008)
6. K. De Brabandere, B. Bolsens, J. Van den Keybus, A. Woyte, J. Driesen, R. Belmans, A voltage and frequency droop control method for parallel inverters. IEEE Trans. Power Electron. **22**(4), 1107–1115 (2007)
7. Y. Li, Y.W. Li, Power management of inverter interfaced autonomous microgrid based on virtual frequency-voltage frame. IEEE Trans. Smart Grid **2**(1), 30–40 (2011)
8. X. Yu, A.M. Khambadkone, H. Wang, S.T.S. Terence, Control of parallel-connected power converters for low-voltage microgrid—Part I: a hybrid control architecture. IEEE Trans. Power Electron. **25**(12), 2962–2970 (2010)
9. A. Tuladhar, H. Jin, T. Unger, K. Mauch, Control of parallel inverters in distributed AC power systems with consideration of line impedance effect. IEEE Trans. Ind. Appl. **36**(1), 131–138 (2000)
10. R. Majumder, G. Ledwich, A. Ghosh, S. Chakrabarti, F. Zare, Droop control of converter-interfaced microsources in rural distributed generation. IEEE Trans. Power Deliv. **25**(4), 2768–2778 (2010)
11. R. Majumder, B. Chaudhuri, A. Ghosh, R. Majumder, G. Ledwich, F. Zare, Improvement of stability and load sharing in an autonomous microgrid using supplementary droop control loop. IEEE Trans. Power Syst. **25**(2), 796–808 (2009)
12. H. Bevrani, S. Shokoohi, An intelligent droop control for simultaneous voltage and frequency regulation in islanded microgrids. IEEE Trans. Smart Grid **4**(3), 1505–1513 (2013)
13. Y. Guan, J.M. Guerrero, X. Zhao, J. Vasquez, X. Guo, A new way of controlling parallel-connected inverters by using synchronous-reference-frame virtual impedance loop-Part I: Control principle. IEEE Trans. Power Electron. **31**(6), 4576–4593 (2016)
14. H. Shi, K. Sun, X. Hou, Y. Li, H. Jiang, Equilibrium mechanism between dc voltage and ac frequency for ac-dc interlinking converters. iEnergy, **1**(3), 279-284 (2022)
15. T. Wu, Z. Liu, J. Liu, S. Wang, Z. You, A unified virtual power decoupling method for droop-controlled parallel inverters in microgrids. IEEE Trans. Power Electron. **31**(8), 5587–5603 (2015)

Chapter 5
Power Coupling Index (PCI)-Oriented Stability Analysis

5.1 $[P\ \ Q]$–$[\omega\ \ V]$ Interactive Control Frawework of Microgrid

There are two kinds of classical droop control methods under different line impedances. One is $P - \omega$ droop control under highly inductive line impedance to achieve active power sharing. The other is $Q - \omega$ droop control under highly resistive line impedance to achieve reactive power sharing. The form and function of them are not uniform. In this section, we try to unify an inverter control framework that is enabled to work under any type of network impedance, and then give a comprehensive stability analysis and control design based on the $[PQ] - [\omega V]$ interactive framework.

Figure 5.1 shows the basic control block diagram of a grid-forming converter including the virtual impedance, where the bandwidth of the dual closed loop is assumed to be much larger than that of the power control. Thus, ignoring the dual closed loop, and the virtual impedance can be expressed as

$$\widetilde{\mathbf{E}}_{o,dq} = \widetilde{\mathbf{V}}_{r,dq} - \widetilde{\mathbf{i}}_{o,dq} Z_v(s) \tag{5.1}$$

where $\mathbf{E}_{o,dq}$, $\mathbf{V}_{dq}$ and $\mathbf{i}_{o,dq}$ represents the vectors of output voltage, voltage reference and output current in dq synchronous frame, ˜ is the small variations in variables and $\hat{\mathbf{x}}_{dq} = \hat{x}_d + j\hat{x}_q$. Aligning the steady phase angle δ_n with axis d, after dq transformation, $E_{o,d} = E_o \cos\delta_{on}$, $E_{o,q} = E_o \sin\delta_{on}$,$V_{r,d} = V_r \cos\delta_{rn}$ and $V_{r,q} = V_r \sin\delta_{rn}$. $\delta_{on} = \delta_o - \delta_n$ and $\delta_{rn} = \delta_r - \delta_n$.$\delta_r$ and δ_o are the angles of voltage reference and output voltage, respectively. Virtual resistance and steady-state inductance $Z_v = R_v + jX_v$ are taken as a typical example to derive the equivalent power characteristics of virtual impedance.

The phasor diagram of voltage and current is drew as shown in Fig. 5.2, where $\delta_{ro} = \delta_r - \delta_o$. The output power can be expressed as

X. Hou et al., *Self-organized Operation and Stability Control of Electronized Power Networks*, https://doi.org/10.1007/978-981-96-9067-1_5

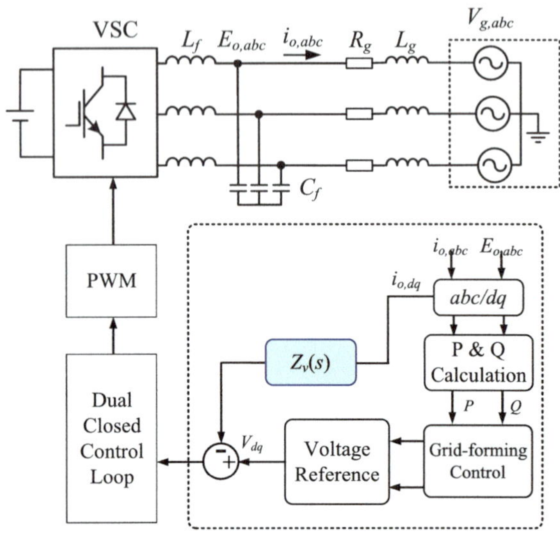

Fig. 5.1 Basic control block diagram including virtual impedance

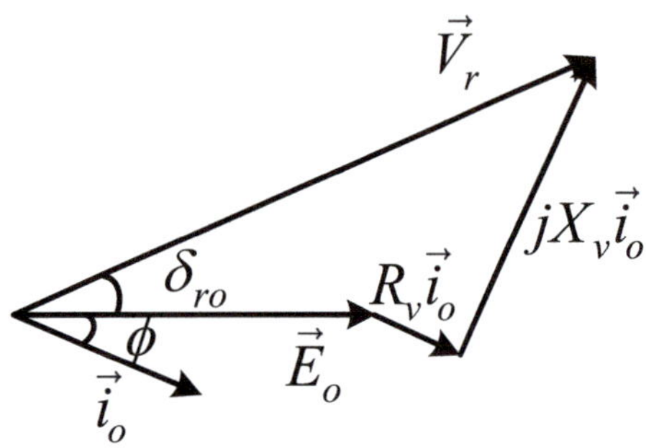

Fig. 5.2 The phasor diagram of voltage and current

$$\begin{cases} P = \frac{1.5}{{R_v}^2 + {X_v}^2}\left(R_v {V_r}^2 - R_v V_r E_o \cos\delta_{ro} + X_v V_r E_o \sin\delta_{ro}\right) \\ Q = \frac{1.5}{{R_v}^2 + {X_v}^2}\left(X_v {V_r}^2 - X_v V_r E_o \cos\delta_{ro} - R_v V_r E_o \sin\delta_{ro}\right) \end{cases} \tag{5.2}$$

Linearizing the output power around equilibrium points yields

$$\begin{cases} \tilde{P} = \frac{1.5}{{R_v}^2 + {X_v}^2}\left[X_v E_{o0} V_{r0}\left(\tilde{\delta}_r - \tilde{\delta}_o\right) - R_v E_{o0} \tilde{E}_o + R_v\left(2V_{r0} - E_{o0}\right)\tilde{V}_r\right] \\ \tilde{Q} = \frac{1.5}{{R_v}^2 + {X_v}^2}\left[-R_v E_{o0} V_{r0}\left(\tilde{\delta}_r - \tilde{\delta}_o\right) + X_v\left(2V_{r0} - E_{o0}\right)\tilde{V}_r - X V_{r0} \tilde{E}_o\right] \end{cases} \tag{5.3}$$

where the subscript suffix "$_o$" stands for the steady-state value. The phase difference between nodes is usually small, assuming $\sin\delta_{or0} \approx 0$ and $\cos\delta_{or0} \approx 1$, where $\delta_{or0} = \delta_{o0} - \delta_{r0}$, i.e. the steady-state angle difference of output voltage and reference voltage.

In addition, the voltage amplitude sag is assumed to be small, i.e. $E_{o0} \approx V_{r0}$. The following relationship can be obtained

$$\begin{cases} \tilde{V}_r - \tilde{E}_o = \frac{2}{3} \cdot \frac{R_v \tilde{P} + X_v \tilde{Q}}{V_{r0}} \\ \tilde{\delta}_r - \tilde{\delta}_o = \frac{2}{3} \cdot \frac{X_v \tilde{P} - R_v \tilde{Q}}{E_{o0} V_{r0}} \end{cases} \tag{5.4}$$

Further, replacing E_{o0} and V_{r0} with voltage reference V^* since the voltage amplitude is limited to the allowable range. Thus, the virtual inductance control can be equivalent to

$$\begin{cases} \delta_o = \delta_r - \frac{X_v}{V^{*2}} P \\ E_o = V_r - \frac{X_v}{V^*} Q \end{cases} \tag{5.5}$$

The classical virtual resistance control can be equivalent to

$$\begin{cases} \delta_o = \delta_r + \frac{R_v}{V^{*2}} Q \\ E_o = V_r - \frac{R_v}{V^*} P \end{cases} \tag{5.6}$$

Similarly, any form of virtual impedance $Z_v(s)$ can be represented based on (5.5) and (5.6). To get rid of the dependence on line impedance, the main and common method is the virtual impedance technique on the basis of classic droop control, which has been verified by the existing research work [1–3, 15]. Therefore, combining (5.5) and (5.6), the unified interactive control framework the control method under the$[PQ] - [\omega V]$ framework of parallel inverter can be expressed as

$$\begin{cases} \omega = \omega^* - m\,(P - P^*) \\ \delta_o = \int \omega\, dt - m_d P + m_x Q \\ E_o = V^* - n_d\,(Q - Q^*) - n_x P \end{cases} \tag{5.7}$$

where m, n_d and m_d represent the $P - \omega$, $Q - V$ and the $P - \delta$ droop coefficient, respectively. Importantly, n_x and m_x represent the coupling $P - V$ droop coefficient and the coupling $Q - \delta$ feedback coefficient, respectively. ω^*, P^* and Q^* are the nominal angular frequency, reference of active power and reactive power.

Further, the correlation diagram among real impedance, virtual impedance and $[PQ] - [\omega V]$ interactive control framework is concluded as Fig. 5.3 shows. All help suppress oscillations, increasing damping and broadening the system stability region. Besides, real resistance consumes power and real inductance helps smooth and filter output current. Virtual impedance control shapes transfer function impedance $Z_v(s)$ not limited to R and L, which the power feedback control can do. Moreover, $[PQ] - [\omega V]$ interactive control framework possesses more independent control freedom.

It is worth noting that the unified $[PQ] - [\omega V]$ interactive control framework is inherently a universal droop control type. It makes the converter adapt to any type of line impedance and ensures frequency synchronization between multiple inverters

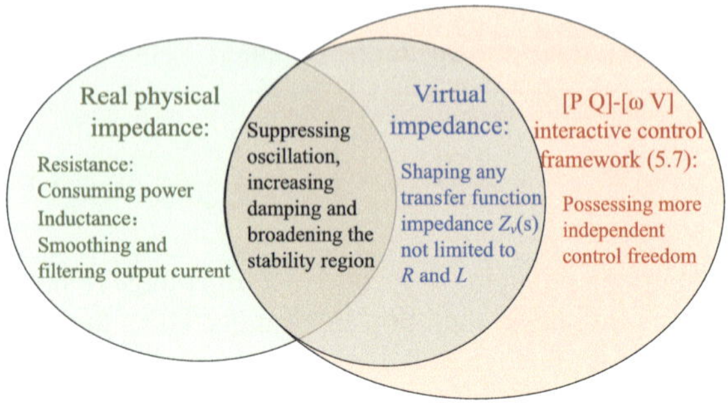

Fig. 5.3 The correlation diagram among real impedance, virtual impedance and $[PQ]-[\omega V]$ interactive control framework

without affecting the accurate active power sharing. The parameter m_d, n_d, m_x, and n_x can also be designed freely and flexibly to provide more independent freedom for the system, which would be discussed in next section.

5.2 Power Coupling Index(PCI)-Oriented Stability Analysis

In this section, the system small signal model is established and a generalized characterization index is constructed to characterize the degree of power coupling. Further, power coupling index (PCI)-oriented stability under different line impedances is analyzed.

5.2.1 *Small Signal Model*

According to Fig. 5.4, the line inductance dynamic model in the small signal is expressed as:

$$\begin{cases} L_g \frac{d\tilde{i}_{o,d}}{dt} = \tilde{E}_{o,d} - \tilde{V}_{gd} - \tilde{i}_{o,d}R_g + \omega_n L_g \tilde{i}_{o,q} \\ L_g \frac{d\tilde{i}_{o,q}}{dt} = \tilde{E}_{o,q} - \tilde{V}_{gq} - \tilde{i}_{o,q}R_g - \omega_n L_g \tilde{i}_{o,d} \end{cases} \tag{5.8}$$

where R_g and L_g represent line resistance and line inductance, $V_{gd} = V_g \cos\delta_{gn}$, $V_{gq} = V_g \sin\delta_{gn}$, and V_g represent grid voltage amplitude.

Linearizing the output power around equilibrium points yields

$$\begin{cases} \tilde{P} = 1.5\left(\tilde{E}_{o,d} i_{o,d0} + E_{o,d0}\tilde{i}_{o,d} + \tilde{E}_{o,q} i_{o,q0} + E_{o,q0}\tilde{i}_{o,q}\right) \\ \tilde{Q} = 1.5\left(-\tilde{E}_{o,d} i_{o,q0} - E_{o,d0}\tilde{i}_{o,q} + \tilde{E}_{o,q} i_{o,d0} + E_{o,q0}\tilde{i}_{o,d}\right) \end{cases} \tag{5.9}$$

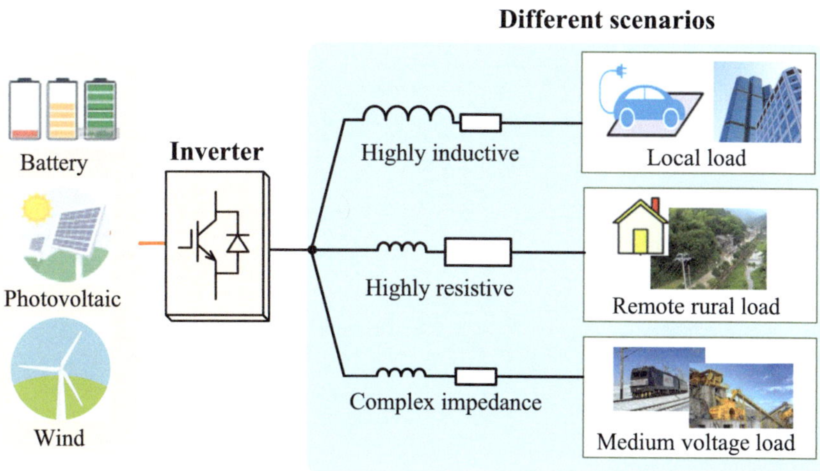

Fig. 5.4 Schematic diagram of scenarios under different types of line impedances

Calculating the output power combining (5.8) and (5.9) yields

$$\begin{aligned} \tilde{P} &= k_{p\delta}(s)\,\tilde{\delta}_o + k_{pE}(s)\,\tilde{E}_o \\ \tilde{Q} &= k_{q\delta}(s)\,\tilde{\delta}_o + k_{qE}(s)\,\tilde{E}_o \end{aligned} \tag{5.10}$$

where

$$\begin{cases} k_{p\delta}(s) = \dfrac{E_{o0}{}^2 X_g}{(R_g+L_g s)^2+X_g{}^2} - \dfrac{E_{o0}{}^2 X_g - E_{o0} V_{g0}\cos\delta_{og0} X_g - E_{o0} V_{g0}\sin\delta_{og0} R_g}{R_g{}^2+X_g{}^2} \\ k_{pE}(s) = \dfrac{V_{g0}\sin\delta_{og0} X_g + R_g E_{o0} - R_g V_{g0}\cos\delta_{og0}}{R_g{}^2+X_g{}^2} + \dfrac{E_{o0}(R_g+L_g s)}{(R_g+L_g s)^2+X_g{}^2} \\ k_{q\delta}(s) = \dfrac{E_{o0} V_{p0}\sin\delta_{og0} X_g + E_{o0}{}^2 R_g - E_{o0} V_{g0}\cos\delta_{og0} R_g}{R_g{}^2+X_g{}^2} - \dfrac{E_{o0}{}^2(R_g+L_g s)}{(R_g+L_g s)^2+X_g{}^2} \\ k_{qE}(s) = \dfrac{E_{o0} X_g - V_{g0}\cos\delta_{og0} X_g - R_g V_{g0}\sin\delta_{og0}}{R_g{}^2+X_g{}^2} + \dfrac{E_{o0} X_g}{(R_g+L_g s)^2+X_g{}^2} \end{cases} \tag{5.11}$$

It is observed that $k_{p\delta}(s)$, $k_{pE}(s)$, $k_{q\delta}(s)$, and $k_{qE}(s)$ have the same conjugate poles:

$$s_{1,2} = -\frac{R_g}{L_g} \pm j\omega_0 \tag{5.12}$$

where the ω_0 represents the rated frequency. Under highly inductive impedance, the resistance is small. The poles $s_{1,2}$ are near the virtual axis, and there is a resonance point at ω_0 [4]. It is studied that a pair of conjugate open-loop poles will be brought in, which possibly cause potential instability [5, 6, 16].

Combining (5.7) and (5.10), the closed-loop small-signal block diagram of the system can be expressed as Fig. 5.5. Rewriting the introduced method (5.7) and the power flow (5.10) in matrix form as:

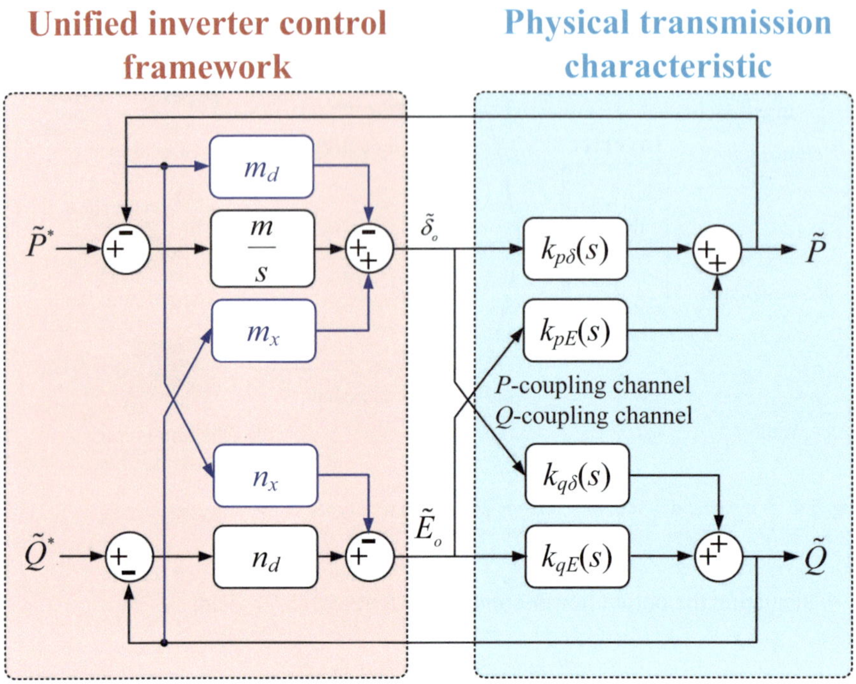

Fig. 5.5 The closed-loop small-signal block diagram of the system

$$\begin{cases} \tilde{\mathbf{x}} = \mathbf{G}_{t1}(s)\,\tilde{\mathbf{S}}^* - \mathbf{G}_{t2}(s)\,\tilde{\mathbf{S}} \\ \tilde{\mathbf{S}} = \mathbf{G}_f(s)\,\tilde{\mathbf{x}} \end{cases} \tag{5.13}$$

where $\widetilde{\mathbf{x}} = [\widetilde{\delta}_o\ \widetilde{\mathbf{E}}_o]^T$, $\widetilde{\mathbf{S}}^* = [\widetilde{P}^*\widetilde{Q}^*]^T$, $\widetilde{\mathbf{S}} = [\widetilde{P}\widetilde{Q}]^T$, $\mathbf{G}_{t1}(s) = diag\{m/s,\ n_d\}$, $\mathbf{G}_{t2}(s) = \begin{bmatrix} \frac{m}{s} + m_d & -m_x \\ n_x & n_d \end{bmatrix}$ and $\mathbf{G}_f(s) = \begin{bmatrix} k_{p\delta}(s) & k_{pE}(s) \\ k_{q\delta}(s) & k_{qE}(s) \end{bmatrix}$

Further, by combining (5.12) and (5.13), the closed-loop system equation is obtained as

$$\tilde{\mathbf{S}} = \mathbf{W}(s)\,\tilde{\mathbf{S}}^* \tag{5.14}$$

where$\mathbf{W}(s) = \left[\mathbf{I} + \mathbf{G}_f(s)\,\mathbf{G}_{t2}(s)\right]^{-1}\mathbf{G}_f(s)\,\mathbf{G}_{t1}(s)$, and I represents a 2× 2 unit matrix. $\mathbf{G_f}(s)$ is stable, and the parameter design easily ensures that$\mathbf{G}_{t1}$ and $\mathbf{G}_{t2}$ are sTable Further, adj$\left[\mathbf{I} + \mathbf{G}_f(s)\,\mathbf{G}_{t2}(s)\right]$ is stable, where 'adj' means adjoint matrix. Thus, the stability of the system depends of the determinant det$\left[\mathbf{I} + \mathbf{G}_f(s)\,\mathbf{G}_{t2}(s)\right]$. If it doesn't have the right-plane zeros or its nyquist diagram doesn't surround the origin, then the system is sTable

5.2.2 Generalized Power Coupling Characterization Index

Relative Gain Array (RGA) analysis is used to acquire a thorough grasp of coupling properties [7]. In a multivariable system, defining c_i as a controlled variable and mj as an operational variable. The relative gain Λ_{ij} of channels m_j to c_i can be represented as

$$\Lambda_{ij} = \frac{w_{ij}}{r_{ij}} = \left.\frac{\partial c_i}{\partial m_j}\right|_{\substack{m_k=const \\ (k\neq j)}} \Big/ \left.\frac{\partial c_i}{\partial m_j}\right|_{\substack{c_k=const \\ (k\neq i)}} \tag{5.15}$$

where w_{ij} is the static channel amplification coefficient of the $m_j - c_i$ channel when all other channels are disconnected, which is called the first amplification coefficient. r_{ij} is the static channel amplification coefficient of $m_j - c_i$ channel when all other channels are closed, which is called the second amplification coefficient.

Ordering w_{ij} in subscript order forms the first amplification coefficient matrix $\mathbf{W}(s)$, which is derived from (8)-(14) and $\mathbf{W}(s) = \left[\mathbf{I} + \mathbf{G}_f(s)\,\mathbf{G}_{t2}(s)\right]^{-1}\mathbf{G}_f(s)$ $\mathbf{G}_{t1}(s)$. The diagonal elements $w_{11}(s)$ and $w_{22}(s)$ stand for the closed-loop dynamic response of active and reactive power when the active and reactive power reference fluctuate. Besides, the degree of cross-coupling between reactive and active power is expressed by $w_{12}(s)$ and $w_{21}(s)$. The matrix $\mathbf{W}(s)$ is a frequency-domain matrix, which can quantitatively evaluate the power coupling levels of different frequency bands.

With the first amplification coefficient matrix $\mathbf{W}(s)$, the RGA matrix can be obtained as

$$\Lambda = \mathbf{W} \otimes \left(\mathbf{W}^{-1}\right)^T \tag{5.16}$$

where $\otimes$ means the multiplication of the same position elements of two matrices. It is worth noting that Λ is independent of the unit of power, and thus it gives a unified dimensionless generalized power coupling index (PCI).

The rule of power coupling analysis using RGA can be summarized as follows [8]:

(1) When $\Lambda_{11} > 1$, the two channels are coupled. As Λ_{11} continues to grow, the P and Q cannot be controlled independently.
(2) When $0 \leq \Lambda_{11} 1 \leq 1$, the closer Λ_{11} is to 1, the weaker the interaction between the two channels is, and the lower the PCI is. On the contrary, the closer Λ_{11} is to 0, the stronger the interaction between two channels is, and the higher the PCI is.

Therefore, when designing controller parameters, Λ_{11} should be as close to 1 as possible. The effect of the key control parameters m, m_d, m_x, n_d andn_x on power coupling under three-type line impedance are studied. Λ_{11} with different (a) m, (b) n_d, (c) m_d, (d) n_x and (e) m_x under L_g=4 mH and $R_g = 0.08\ \Omega$, under L_g=2 mH and $R_g = 0.5\ \Omega$, and under L_g=0.01 mH and $R_g = 2\ \Omega$ are shown in Fig. 5.6, Fig. 5.7 and Fig. 5.8, respectively.

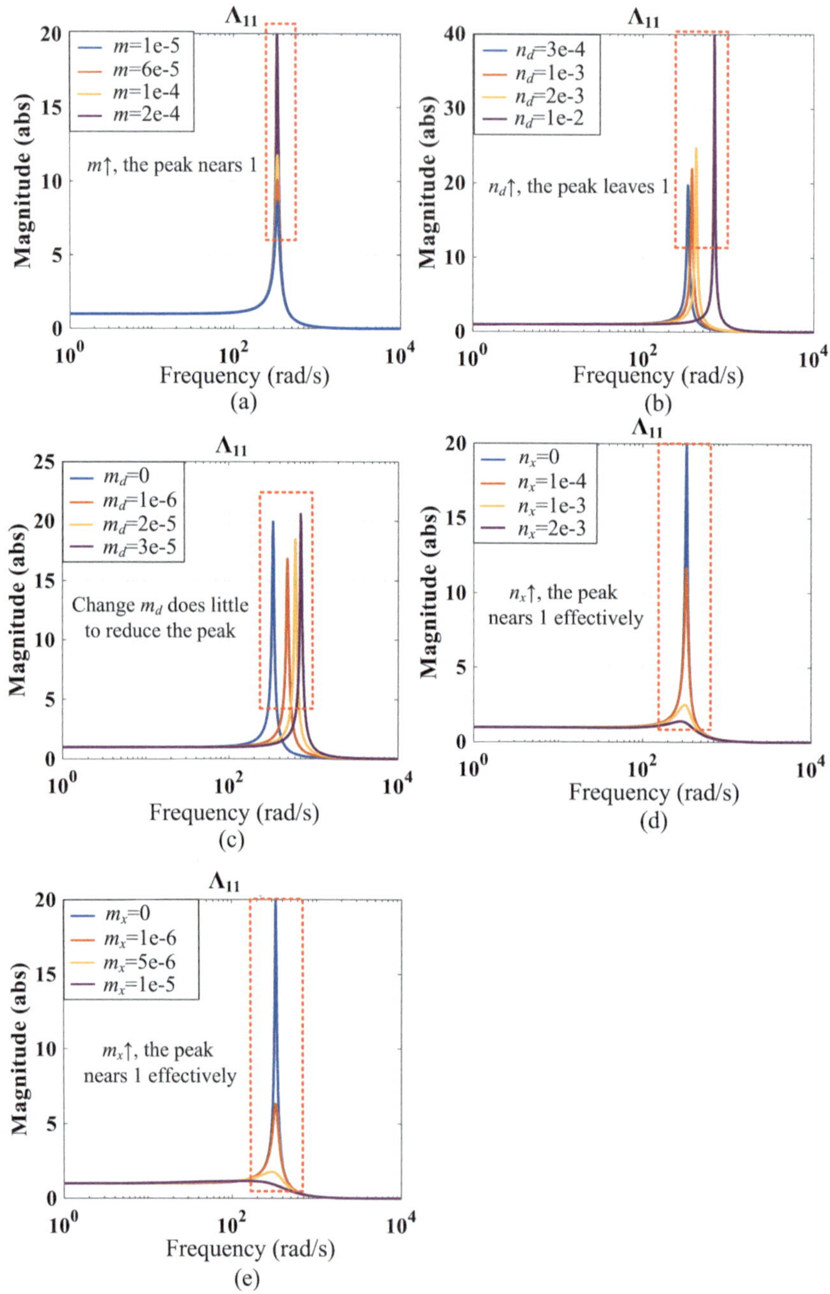

Fig. 5.6 Λ_{11} with different **a** m, **b** n_d, **c** m_d, (**d**)n_x and **e** m_x under $L_g = 4$ mH and $R_g = 0.08\ \Omega$

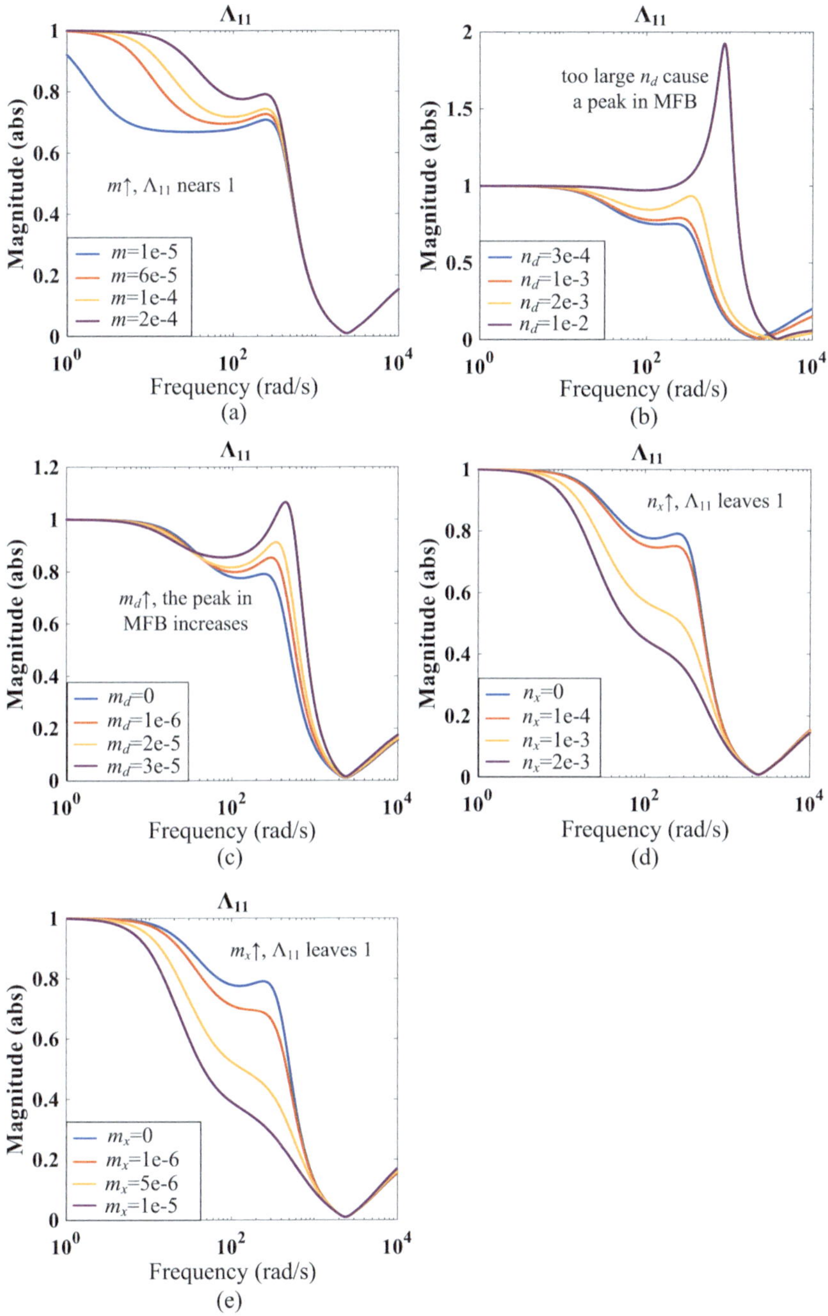

Fig. 5.7 Λ_{11} with different **a** m, **b** n_d, **c** m_d, **d** n_x and **e** m_x under $L_g = 2$ mH and $R_g = 0.5\ \Omega$

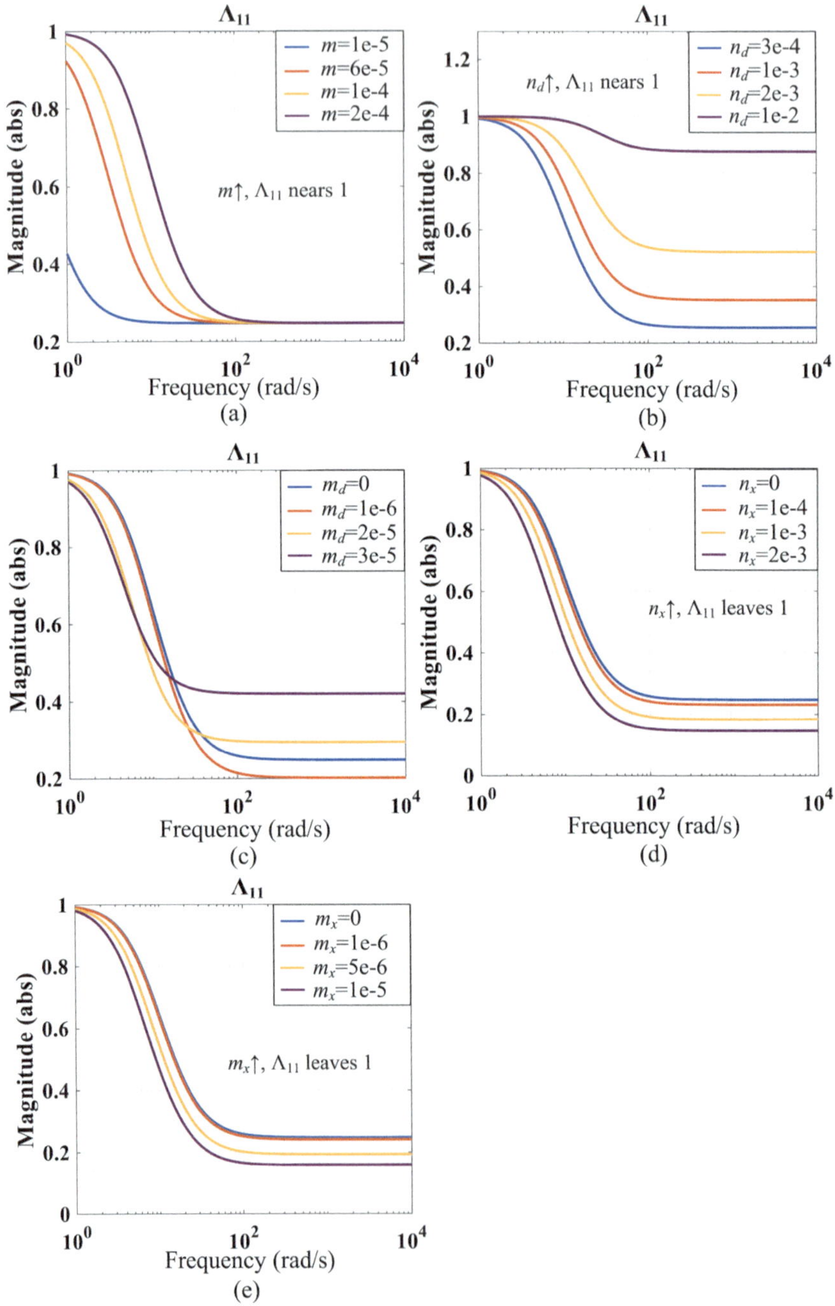

Fig. 5.8 Λ_{11} with different **a** m, **b** n_d, **c** m_d, **d** n_x and **e** m_x under $L_g = 0.01$ mH and $R_g = 2\ \Omega$

When the line impedance is highly inductive, ignoring inductance dynamics, active and reactive power are nearly decoupled which has been discussed in detail in [9, 10]. Here, the influence of the line inductance dynamics and the control parameters on power coupling is discussed. When discussing the impact of parameters m_d, n_x and m_x on decoupling performance, the values of droop coefficients are $m = 2e - 4$ and $n_d = 3e - 4$. In Fig. 5.6a, b, it is observed that decreasing m and n_d can weaken the power coupling in the medium frequency band (MFB), where medium frequency indicates tens to hundreds of Hz. When nd decreases to a certain degree, Λ_{11} is almost unchanged and is much larger than 1 in the MFB. As shown in Fig. 5.6c, it is almost useless to change m_d to promote Λ_{11} closer to 1 in the MFB. When m_d keeps increasing, Λ_{11} keeps away from 1. As Fig. 5.6d, e shows, increasing n_x and m_x can effectively suppress the peak value of the MFB, so as to keep Λ_{11} near 1. Therefore, when the line impedance is highly inductive, it is found that different from the traditional $P - \omega$ control, the appropriate design of coupling power feedback coefficient (n_x and m_x) is more conducive to decreasing the degree of power coupling in MFB.

When the line impedance is complex resistive-inductive in Fig. 5.7a, it is observed that increasing m can reduce the power coupling degree in the low frequency band (LFB), where low frequency indicates a few tenths to several Hz. In Fig. 5.7b, when n_d is too large, the Λ_{11} is close to 1 in the LFB but Λ_{11} has a peak value exceeding 1 in the MFB. As n_d continues to decrease, the peak value is suppressed and the power coupling degree increases slightly. In Fig. 5.7c, appropriately increasing m_d is conducive to decreasing the degree of power coupling, but when m_d is too large, it will increase the peak value of Λ_{11} in the MFB. In Fig. 5.7d, e, increasing n_x and m_x make Λ_{11} slightly away from 1 in the LFB. Even though increasing them can effectively suppress the peak value of the MFB.

When the line impedance is highly resistive, as shown in Fig. 5.8, it is observed that power coupling is serious in the LFB and MFB when m, n_d, and m_d are small. Increasing m, n_d, and m_d can effectively pushes Λ_{11} closer to 1, and the effect of increasing n_d to help decreasing the degree of power coupling is the best.

In conclusion, the main frequency bands of decreasing the degree of power coupling are different under three types of line impedances. When the line impedance is highly inductive, the main contradiction concerned about is the power coupling in the MFB, and increasing m_x and n_x help decrease the power coupling in the MFB. However, when the line impedance is highly resistive, the main contradiction is the power coupling in the LFB and the MFB, and increasing m and n_d will reduce the power coupling of the two frequency bands.

5.2.3 *Small-Signal Stability Analysise*

To confirm the broad applicability of the unified inverter control framework under various line impedance, root trajectories are carried out for systems with highly-inductive, equally resistive-inductive, and highly-resistive line impedances. The effect of the control parameters m, m_d, m_x, n_d and n_x on system stability are studied. The root locus of $\det(\mathbf{I}+\mathbf{G}_f(s)\,\mathbf{G}_t(s))^{-1}$ with (a) $m \in [1\times10^{-5}, 1\times10^{-3}]$, (b) $n_d \in [1\times10^{-4}, 1\times10^{-2}]$, (c) $m_d \in [0, 1\times10^{-4}]$, (d) $n_x \in [0, 2\times10^{-3}]$, (e) $m_x \in [0, 1\times10^{-5}]$ under $L_g = 4$ mH and $R_g = 0.08\ \Omega$, under $L_g = 2$ mH and $R_g = 0.5\ \Omega$, and under $L_g = 0.01$ mH and $R_g = 2\ \Omega$ are shown in Figs. 5.9, 5.10 and 5.11. The following features can be drawn

(1) Considering the inductance dynamics, there are a pair of medium-frequency (MF) roots $\lambda_{1,2}$ and a root λ_3 mainly affecting the system stability and dynamic response. Under the highly inductive line impedance, the MF roots $\lambda_{1,2}$ which are easily unsTable With the decrease of line inductance and the increase of the line resistance, the MF roots $\lambda_{1,2}$ are further away from the imaginary axis and are more sTable Under the highly resistive line impedance, the root λ_3 is close to the imaginary axis that may increase the system adjusting time.
(2) Under the highly inductive and equally resistive-inductive line impedance, decreasing m and increasing m_d help stabilize the MF roots $\lambda_{1,2}$ but also makes the root λ_3 substantially close to the origin, which slows down the system response speed. Thus, it is not the best choice to improve the stability of the roots $\lambda_{1,2}$. With the introduced method, increasing coupling feedback coefficient m_x and n_x, the MF roots $\lambda_{1,2}$ are significantly far away from the imaginary axis, which increases the system damping and inhibits the MF oscillation and is conducive to the stability of the roots $\lambda_{1,2}$. Moreover, the sensitivity of root λ_3 to parameters m_x and n_x is obviously lower than that of the parameters m and m_d.
(3) Under three-types line impedance, increasing n_d moves $\lambda_{1,2}$ away from the real axis, i.e. small n_d is beneficial to damping the MF roots $\lambda_{1,2}$. When the line impedance is highly resistive, the adverse effect on MF roots $\lambda_{1,2}$ is negligible since it is far away from the imaginary axis and decays rapidly. In this case, m and n_d should be large, since large m and n_d move λ_3 away from the imaginary axis that speeds up the system dynamic response. Besides, in this case even if n_d is set to 0, the system can still be stable by the weak coupling between P and ω ($\tilde{P} \approx E_{o0}V_{g0}\sin\delta_{og0}R_g\tilde{\delta}_o/Z_g^2$) where $\sin\delta_{og0}$ is small). However, the dynamic response is slow and the adjusting time is too long.

Figure 5.12 summarizes the parameters control design according to the different degrees of power coupling in different frequency bands and system stability, where $(\lambda_{11}-1)$ represents the degree of λ_{11} deviating from 1. The closer $(\lambda_{11}-1)$ is to 0, the weaker the power coupling. It can be summarized as:

(1) $-\eta_{set} < (\lambda_{11}-1) < \eta_{set}$ represents the low degree of power coupling, which the parameters should be designed to satisfy, where η_{set} is the desired degree of power coupling.
(2) $(\lambda_{11}-1) > \eta_{set}$ and $(\lambda_{11}-1) < \eta_{set}$ represent high power coupling. However, these two cases have inverse control design. When $(\lambda_{11}-1) > \eta_{set}$ in the MFB, increasing n_x and m_x, and decreasing n_d help decrease the degree of power coupling in the MFB and are conducive to system stability. When $(\lambda_{11}-1) < \eta_{set}$, decreasing n_x and m_x, and increasing n_d help decrease the degree of power coupling and help speed up the adjusting time.

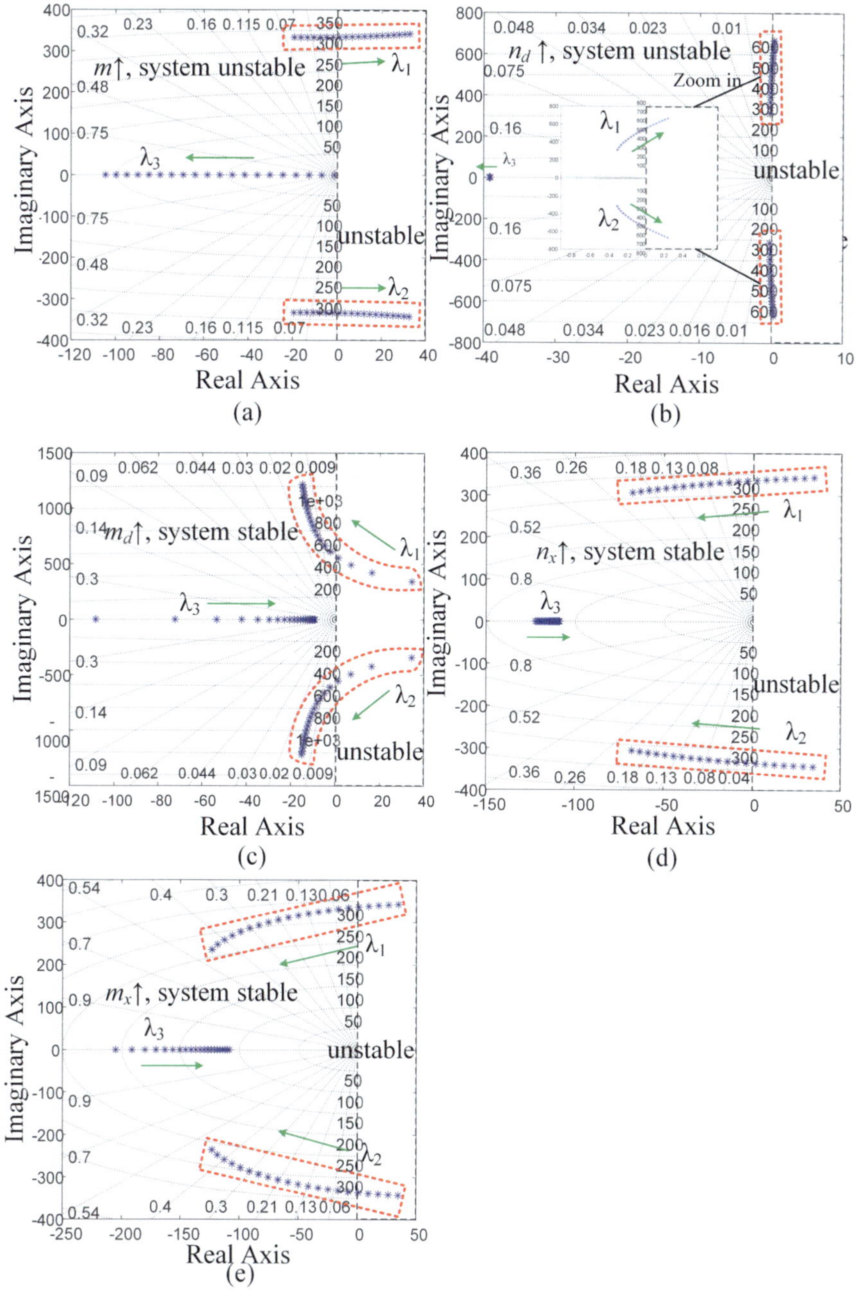

Fig. 5.9 Case 1 (under highly inductive impedance): root locus with **a** $m \in [1 \times 10^{-5}, 1 \times 10^{-3}]$, **b** $n_d \in [1 \times 10^{-4}, 1 \times 10^{-2}]$, **c** $m_d \in [0, 1 \times 10^{-4}]$, **d** $n_x \in [0, 2 \times 10^{-3}]$, **e** $m_x \in [0, 1 \times 10^{-5}]$

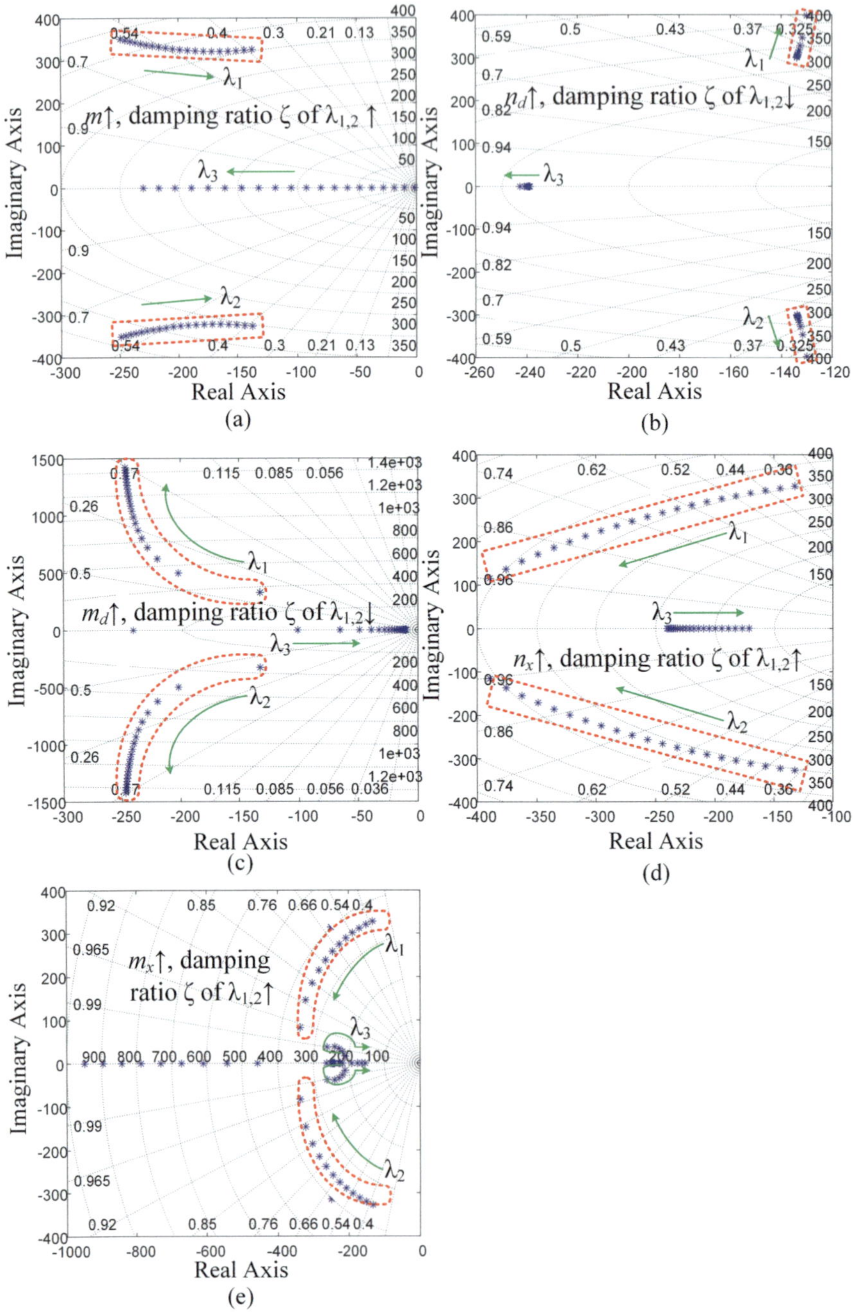

Fig. 5.10 Case 2 (under complex resistive-inductive impedance): root locus with **a** $m \in [1 \times 10^{-5}, 1 \times 10^{-3}]$, **b** $n_d \in [1 \times 10^{-4}, 1 \times 10^{-2}]$, **c** $m_d \in [0, 1 \times 10^{-4}]$, **d** $n_x \in [0, 2 \times 10^{-3}]$, **e** $m_x \in [0, 1 \times 10^{-5}]$

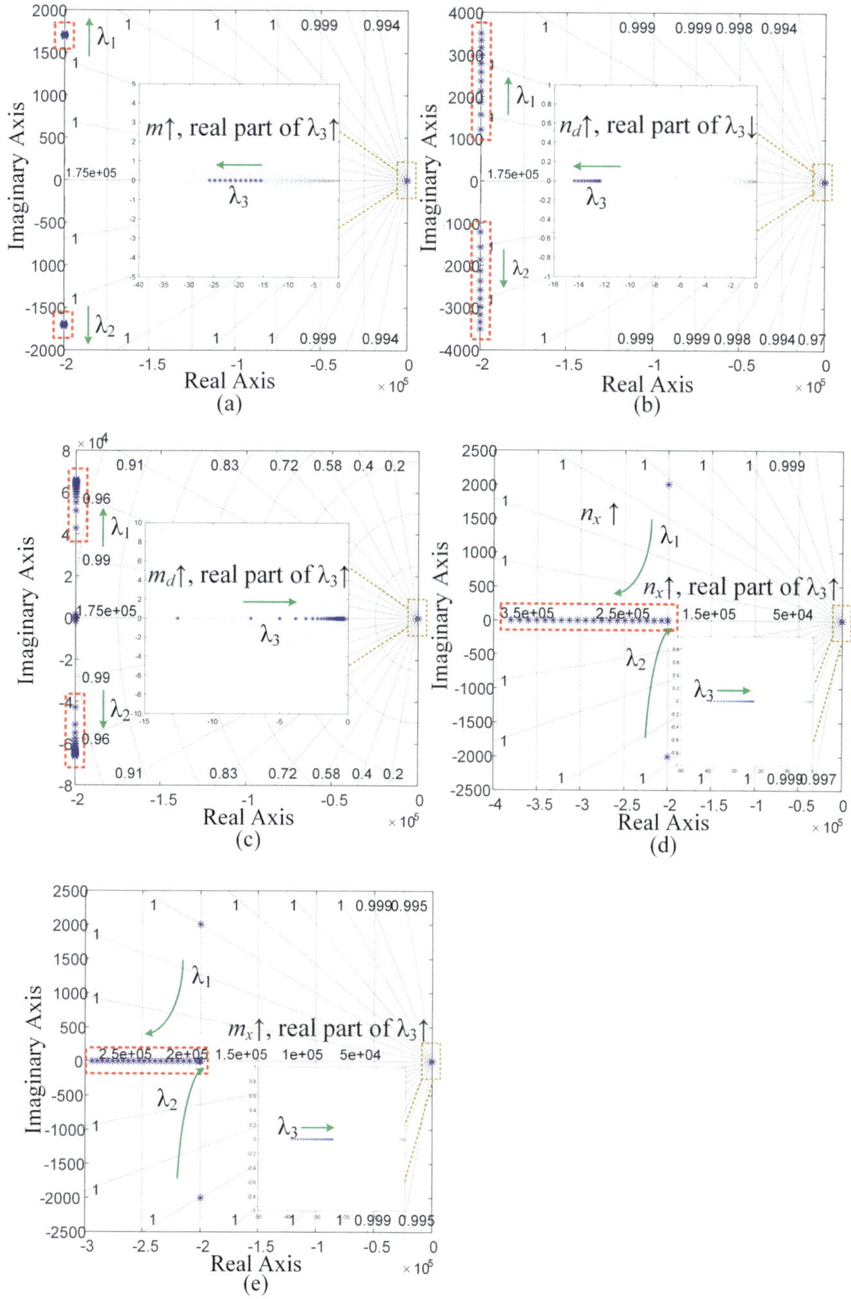

Fig. 5.11 Case 3 (under highly resistive impedance): Root locus with **a** $m \in [1 \times 10^{-5}, 1 \times 10^{-3}]$, **b** $n_d \in [1 \times 10^{-4}, 1 \times 10^{-2}]$, **c** $m_d \in [0, 1 \times 10^{-4}]$, **d** $n_x \in [0, 2 \times 10^{-3}]$, **e** $m_x \in [0, 1 \times 10^{-5}]$

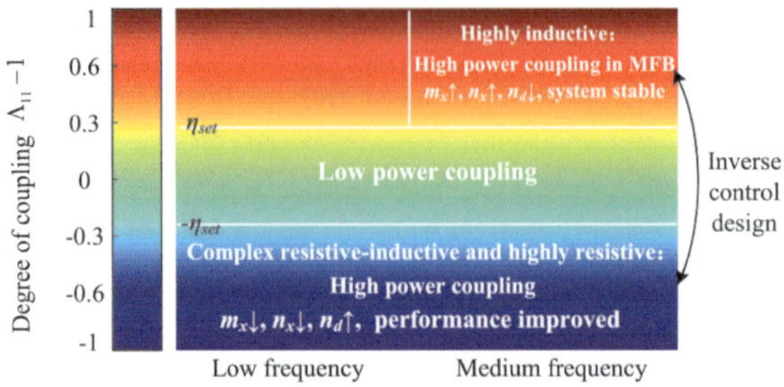

Fig. 5.12 Control parameter design guideline under different degree of power coupling ($\lambda_{11} - 1$) closer to 0 indicates a weaker power coupling)

5.2.4 Large-Signal Stability Analysis

This part conducts a comparative analysis of large-signal stability between the introduced method and traditional GFM (grid-forming) methods, i.e. VSG control [11, 12] under complex line impedance. The parameters of the introduced method are designed according to the design rules under different line X/R ratio that are concluded in the above part. Phase portrait analysis [13, 14] is implemented here. The phase portraits of traditional grid-forming control methods and the introduced method with highly inductive, complex resistive-inductive and highly resistive line impedance are listed in Figs. 5.13, 5.14 and 5.15, respectively. It is observed that the introduced method which follows the parameter design rule also help enhance large-signal stability under grid voltage sags.

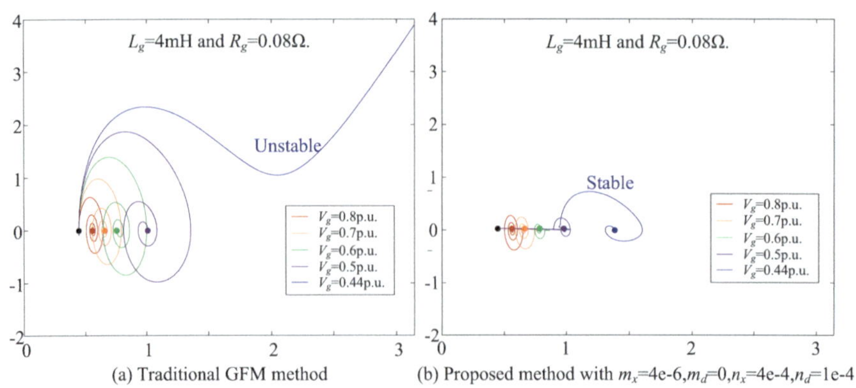

Fig. 5.13 Phase portraits under different grid voltage sags with $L_g = 4$ mH and $R_g = 0.08\ \Omega$: **a** Traditional GFM method **b** introduced method with $m_x = 4 \times 10^{-6}$, $m_d = 0$, $n_x = 4 \times 10^{-4}$

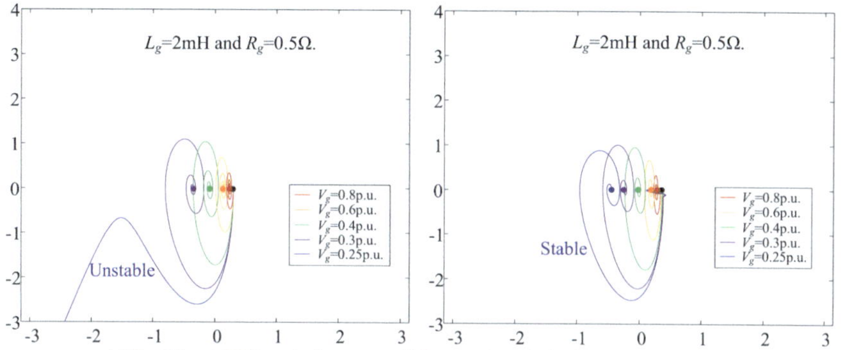

(a) Traditional GFM method (b) Proposed method with m_x=2e-6,m_d=5e-6,n_x=2e-4,n_d=3e-4

Fig. 5.14 Phase portraits under different grid voltage sags with $L_g = 2$ mH and $R_g = 0.5\ \Omega$: **a** Traditional GFM method **b** introduced method with $m_x = 2 \times 10^{-6}$, $m_d = 5 \times 10^{-6}$, $n_x = 2 \times 10^{-4}$

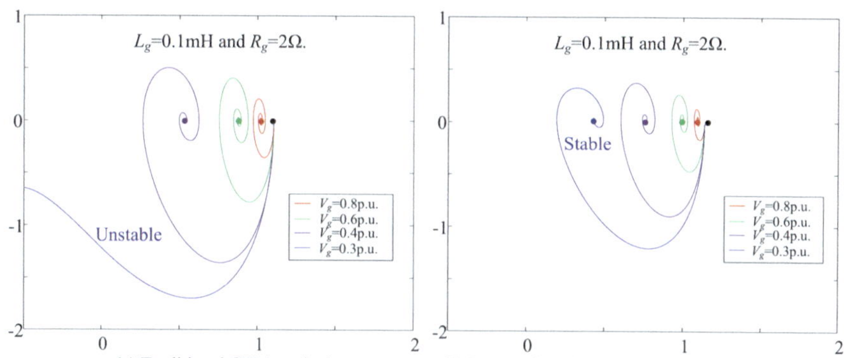

(a) Traditional GFM method (b) Proposed method with n_x=m_x=0,m_d=1e-5,n_d=1e-3

Fig. 5.15 Phase portraits under different grid voltage sags with $L_g = 0.1$ mH and $R_g = 2\ \Omega$: **a** Traditional GFM method **b** introduced method with $m_x = 1 \times 10^{-6}$, $m_d = 5 \times 10^{-6}$, $n_x = 1 \times 10^{-4}$

5.2.5 Parameters Design Guidelines

Improper parameter design may lead to poor dynamic performance of the system. Combining the above analysis, the design rules of the parameters can be summarized as follows:

(1) Under highly inductive line impedance, the medium frequency oscillation is serious, and may even lead to system instability. Meanwhile, the power coupling degree is strong in the medium frequency band and PCI λ_{11} is much greater than 1. Therefore, n_x and m_x should be large, which can effectively reduce the power coupling degree in the MFB, and suppress the medium frequency oscillation.

When an additional voltage drop is not desired, setting $n_x = 0$. Besides, the effect on the pole λ_3 is small that does not slow down the response.

(2) Under highly resistive line impedance, the poles $\lambda_{1,2}$ are far away from the imaginary axis, and its damping ratio is large. However, the root λ_3 is close to the imaginary axis that may increase the adjusting time, and the power coupling degree in the LFB is very strong. Therefore, n_x and m_x can be set to 0. m_d and n_d should be large, which help reduce the degree of power coupling, and its value can be referred to (5.5).

(3) Under complex resistive-inductive line impedance, the damping ratio of conjugate poles $\lambda_{1,2}$ and power coupling in the low-frequency band should be considered synthetically. n_x and m_x should be small, and m_d and n_d should be increased appropriately to reduce the degree of power coupling and speed up the adjusting time. To comprehensively consider the overall systematic dynamic performance, such as suppressing medium frequency oscillation, decreasing the degree of power coupling and reducing adjusting time, the optimization problem of control design can be constructed from the PCI-oriented perspective:

$$\min\ J = \left(\max\left|1 - \Lambda_{11}\left(j2\pi f\right)\right|_{f \le f_{mo}}\right)^2 \tag{5.17}$$

Subject to:

$$0 < m \le \frac{\omega_{\max} - \omega_{\min}}{P_{\max} - P_{\min}} \tag{5.18}$$

$$0 < n_d\left(Q_{\max} - Q_{\min}\right) + n_x P_{\max} \le V_{\max} - V_{\min} \tag{5.19}$$

$$\sigma \le \sigma^* \tag{5.20}$$

$$\zeta \ge \zeta^* \tag{5.21}$$

$$\max\left|1 - \Lambda_{11}\left(j2\pi f\right)\right|_{f \le f_{mo}} \le \eta_{set} \tag{5.22}$$

where $\omega_{\min}$ and $\omega_{\max}$ represents the lower and upper bounds of the allowable range of system angular frequency. ζ and σ are the damping ratio and the real part of the eigenvalues that are calculated by (5.14). ζ^* and σ^* are the set minimum damping ratio and real part that determines the damping margin and system response speed. f_{mo} is the medium frequency of the oscillation. Constraints of the steady-state frequency sag (5.18), voltage quality (5.19), system stability margin (5.20) and (5.21) are comprehensively considered. Besides, the maximum degree of power coupling is limited below η_{set} as (5.22) shows, which can be calculated by (5.16).

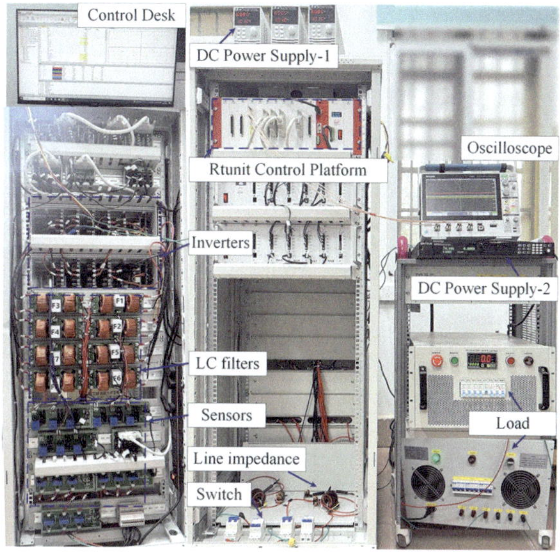

Fig. 5.16 The experimental setups of the two parallel system

Table 5.1 Experimental parameters

Control parameter	
AC voltage reference V^*/ω^*	156 V/314 rad/s
DC-side voltage	180 V
PWM frequency f_{PWM}	10 kHz
Sampling time T_s	0.1 ms
Filter inductance/capacitor L_f/C_f	1.8 mH/25 μF
Low-pass filter cut-off frequency ω_c	30*2π rad/s
Case 1: highly inductive line impedance	$L_1 = 1.5$ mH, $L_2 = 2.2$ mH
Case 2: complex R - L line impedance	$L_1 = 0.7$ mH, $R_1 = 0.4\ \Omega$
	$L_2 = 0.96$ mH, $R_2 = 0.2\ \Omega$
Case 3: highly resistive line impedance	$R_1 = 0.5\ \Omega$, $R_2 = 0.7\ \Omega$

5.3 Case Study

5.3.1 Hardware Experiment

The hardware experiment results are displayed to test the adaptability and effectiveness of the unified inverter control framework under various line impedances (highly inductive, complex impedance and highly resistive), and the experimental setup is shown in Fig. 5.16. It includes two inverters, two DC power supplies, a rtunit controller, and resistance loads. The rtunit platform accomplishes programming tasks based on Matlab/Simulink. The experimental parameters are listed in Table 5.1.

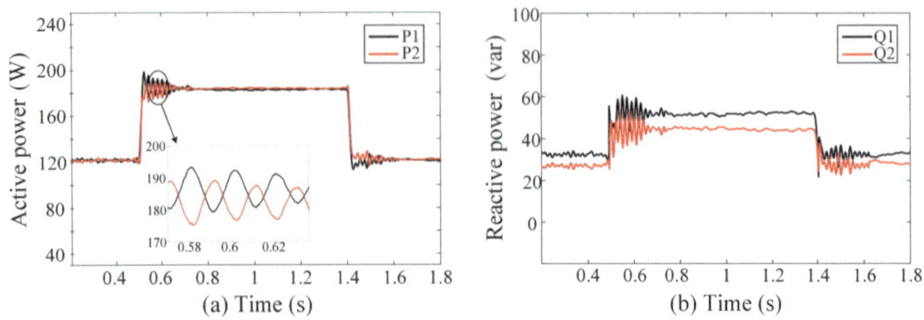

Fig. 5.17 Waveforms under conventional droop control, i.e. $m_d = 0$, $m_x = 0$, $n_x = 0$: (**a**) active power, and (**b**) reactive power

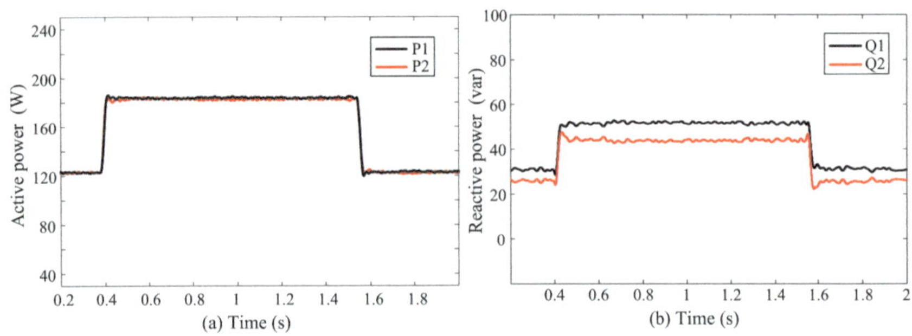

Fig. 5.18 Waveforms under introduced method: (**a**) active power, and (**b**) reactive power

(1) Under highly inductive line impedance

Figures 5.17 and 5.18 shows the power of two inverters under load disturbance with conventional droop control, i.e. m_d=0, m_x=0, $n_x = 0$, and the unified control framework, when the line impedance is highly inductive. The control parameters of VSC_s under the unified control frame in this case are set as $m_1 = 1 \times 10^{-3}$, $n_{d1} = 8 \times 10^{-4}, m_{d1} = 0, m_{x1} = 4 \times 10^{-4}, n_{x1} = 1.1 \times 10^{-3}$, and $m_2 = 1 \times 10^{-3}$, $n_{d2} = 5 \times 10^{-4}$, $m_{d2} = 0$, $m_{x2} = 4 \times 10^{-4}$, $n_{x2} = 2 \times 10^{-3}$, according to the optimal parameters design guidelines (5.17). As Fig. 5.17 shows, at 0.5 s, a load disturbance is introduced. It can be seen that with the conventional $P - \omega$ droop control, the system occurs serious medium frequency oscillation. After measurement, the oscillation frequency of the system is 47.2 Hz. It is observed from Fig. 5.18 that the medium frequency oscillation is suppressed effectively with proper design of m_x and n_x, which proves the correctness of root locus analysis. Compared to the virtual resistance method, there is more freedom of the introduced method. Besides, frequency self-synchronizing is realized and precise active power sharing between two inverters is guaranteed. The output voltage and current of two inverters are shown in Fig. 5.19a. Figure 5.19b shows the output voltage of VSC_1 with conventional droop control when n_d increases to 0.01. It is seen that a large n_d makes the system eventually lose its stability when the line impedance is highly inductive.

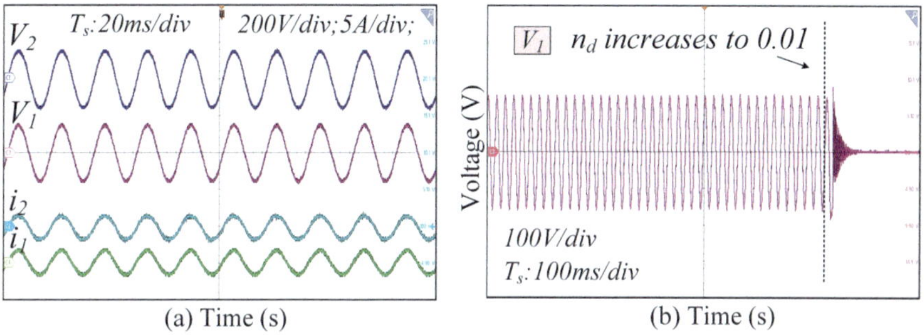

Fig. 5.19 (**a**) Output voltage and current of two inverters, and (**b**) the output voltage of VSC1 with droop control when nd increases to 0.01

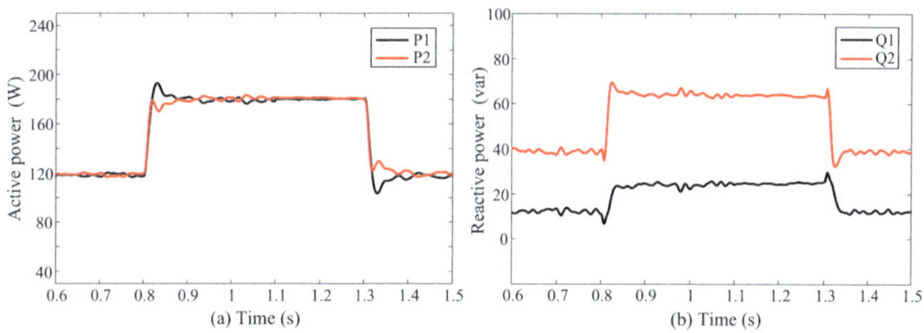

Fig. 5.20 Waveforms under conventional droop control, i.e. $m_d = 0, m_x = 0, n_x = 0$: (**a**) active power, and (**b**) reactive power

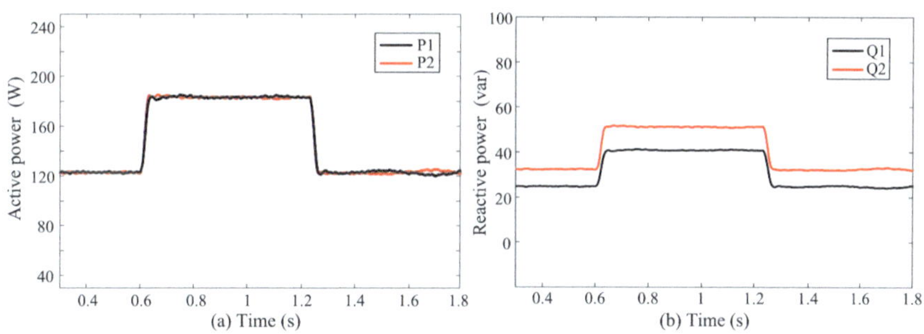

Fig. 5.21 Waveforms under introduced method: **a** active power, and **b** reactive power

(2) Under complex resistive-inductive line impedance

Figures 5.20 and 5.21 shows the power of two inverters under load disturbance with conventional droop control, i.e. $m_d = 0, m_x = 0, n_x = 0$, and the unified control framework, when the line impedance is complex resistive-inductive. The parameters of VSC_s in this case are set as $m_1 = 1 \times 10^{-3}$, $m_{d1} = 1.5 \times 10^{-5}$, $n_{d1} = 0.0151$,

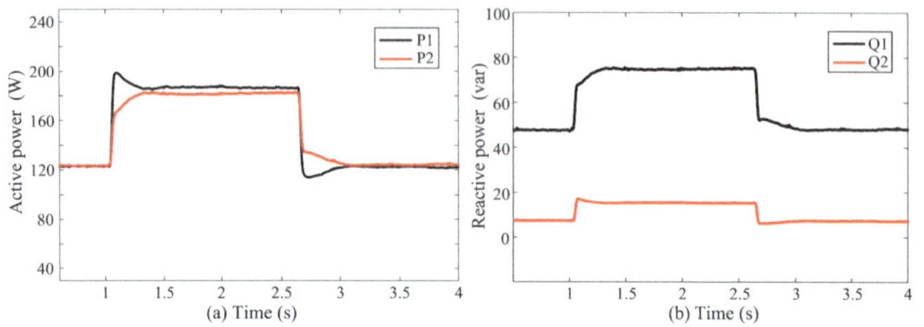

Fig. 5.22 Waveforms under conventional droop control, i.e. $m_d = 0$, $m_x = 0$, $n_x = 0$: **a** active power, and **b** reactive power

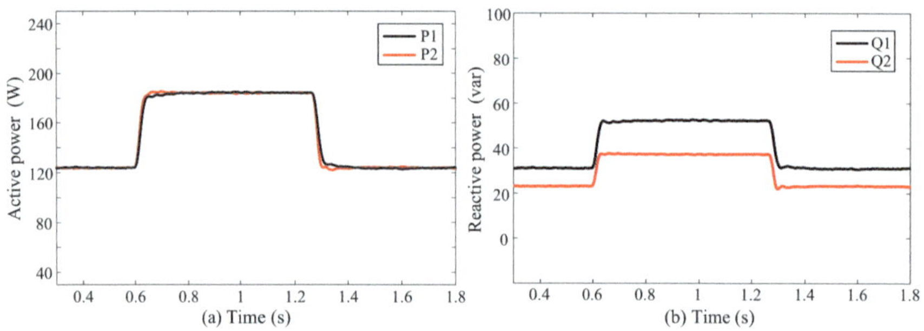

Fig. 5.23 Waveforms under introduced method: **a** active power, and **b** reactive power

$m_{x1} = 0$, $n_{x1} = 1 \times 10^{-4}$, and $m_2 = 1 \times 10^{-3}$, $n_{d2} = 0.02$, $m_{d2} = 0$, $m_{x2} = 1 \times 10^{-5}$, $n_{x2} = 6 \times 10^{-4}$. At 0.6 s, a load disturbance is introduced. It is observed that active power sharing is ensured before and after load disturbance and the power dynamic response is fine during the load disturbance with the introduced control framework. The experimental waveform results are not much different from what happens when the line impedance is highly inductive. To ensure better decoupling of reactive and active power in the case of complex impedance, the value of m_d is appropriately increased. Besides, the dynamic and steady-state performance of the system is little affected by the increase of the line resistance.

(3) Under highly resistive line impedance

The output active power with the conventional droop control, i.e. $m_d = 0$, $m_x = 0$, $n_x = 0$, and the unified control framework under load disturbance is shown in Figs. 5.22 and 5.23, respectively. The parameters of VSC_s in this case are set as $m_1 = 1 \times 10^{-3}$, $m_{d1} = 5 \times 10^{-5}$, $n_{d1} = 0.0283$, $m_{x1} = 0$, $n_{x1} = 0$, and $m_2 = 1 \times 10^{-3}$, $n_{d2} = 0.0351$, $m_{d2} = 6 \times 10^{-5}$, $m_{x2} = 0$, $n_{x2} = 0$. As Fig. 5.19 shows, at 1 s, a load disturbance is introduced. The P and δ (or Q and V) are coupled weakly when the line impedance is highly resistive. Thus, it results in too long adjusting time and poor control performance under conventional $P - \omega$ control with inappropriate parameters.

Table 5.2 Parameters design guideline

Line characteristic	Parameter design guideline
Case 1: highly inductive	▹ Large n_d harms system stability
	▹ Increasing m_x/n_x suppress MF power oscillation
Case 2: complex resistive - inductive	▹ Increasing m_d and n_d reduces power coupling degree
Case 3: highly resistive	▹ Small n_d results in too long adjusting time
	▹ Increasing m_d/n_d reduces power coupling degree and speeds up adjusting time

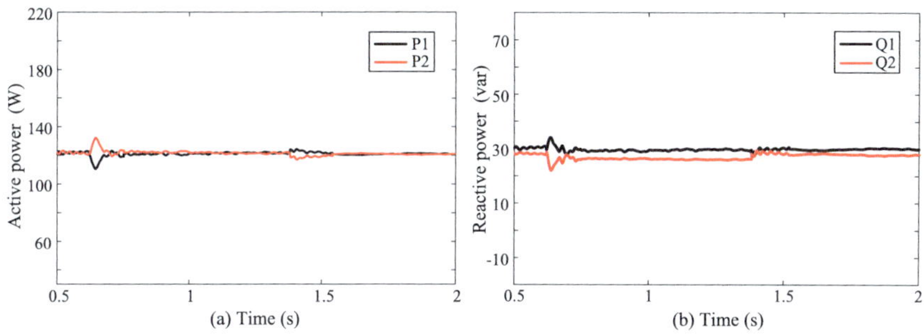

Fig. 5.24 Waveforms under introduced method when the X/R changes: **a** active power, and **b** reactive power

To strengthen the connection between P and ω, n_d and m_d is appropriately increased. From Fig. 5.23, it is observed that the active power of the two inverters before and after load disturbance has been evenly divided with fine dynamic response speed. Besides, the consequences of frequency synchronization are essentially identical to those consequences obtained under highly inductive and equally resistive-inductive line impedance.

Based on the experimental results of the above three cases, it is concluded that $[PQ] - [\omega V]$ interactive control framework ensures universal applicability under highly inductive, complex resistive-inductive, and highly resistive line impedances. It also ensures frequency synchronization and accurate active power sharing. Besides, the parameter control design under the three line impedances are summarized in Table 5.2.

(4) Applicability when the ratio of X/R changes

To verify the practicality of the introduced unified control framework, the waveforms of output active and reactive power with the unified control framework when the X/R of the line impedance changes is shown in Fig. 5.24. The parameters of VSC_s in this case are set as $m_1 = 1 \times 10^{-3}$, $m_{d1} = 0.11 \times 10^{-5}$, $n_{d1} = 0.0125$, $m_{x1} = 0.21 \times 10^{-5}$,$n_{x1} = 0.61 \times 10^{-3}$, and $m_2 = 1 \times 10^{-3}$, $n_{d2} = 0.0126$,$m_{d2} =$

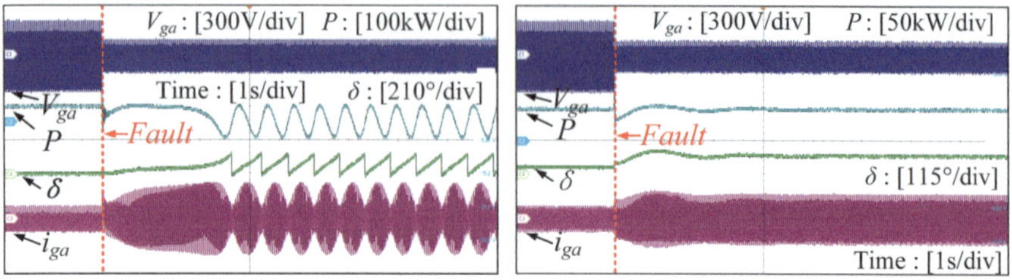

(a) Traditional GFM method

(b) Proposed method with m_x=4e-6,m_d=0,n_x=4e-4,n_d=1e-4

Fig. 5.25 Waveforms of grid voltage, active power, phase angle and current when grid voltage drops to 0.44 p.u. with highly inductive impedance: $L_g = 4$ mH and $R_g = 0.08\ \Omega$

0.15×10^{-5}, $m_{x2} = 0.22 \times 10^{-5}$, $n_{x2} = 0.64e - 3$. As Fig. 21 shows, around 0.6 s, the line resistance is $R_1 = R_2 = 0.2\ \Omega$, and line inductance changes from $L_1 = 0.7$ mH, $L_2 = 0.96$ mH to $L_1 = 4$ mH, $L_2 = 4.5$ mH, i.e. the ratio of X/R changes from about 6 to 1. Around 1.4 s, the ratio of X/R changes from about 1 to 6. It is observed that the introduced method still works and the system possesses good dynamic performance when the X/R of the line impedance changes within a certain range. The selection range of control parameters is relatively wide and has certain robustness.

5.3.2 *Control Hardware-in-Loop Experiment*

Figure 5.25 shows the comparative waveforms of grid voltage, active power, phase angle and current when grid voltage drops to 0.44 p.u. with highly inductive impedance: $L_g = 4$ mH and $R_g = 0.08\ \Omega$: (a) traditional GFM method and (b) the introduced method. Figure 5.26 shows the comparative waveforms when grid voltage drops to 0.25 p.u. with complex resistive-inductive impedance: $L_g = 2$ mH and $R_g = 0.5\ \Omega$. Figure 5.27 shows the comparative waveforms when grid voltage drops to 0.3 p.u. with highly resistive impedance: $L_g = 0.1$ mH and $R_g = 2\ \Omega$. Based on the results, it is observed that the introduced method which follows the parameter design rule as concluded in Sect. 5.2 help enhance large-signal stability under grid voltage sags. Besides, it is consistent with the results of phase portrait analysis.

5.4 Conclusion

To clarify the quantitative power coupling relationship and corresponding control design under different line impedances, power coupling index (PCI)-oriented stability analysis and interaction control design for inverter-based microgrid under complex *R*/*X* ratio are conducted. To this issue, a generalized PCI is designed based

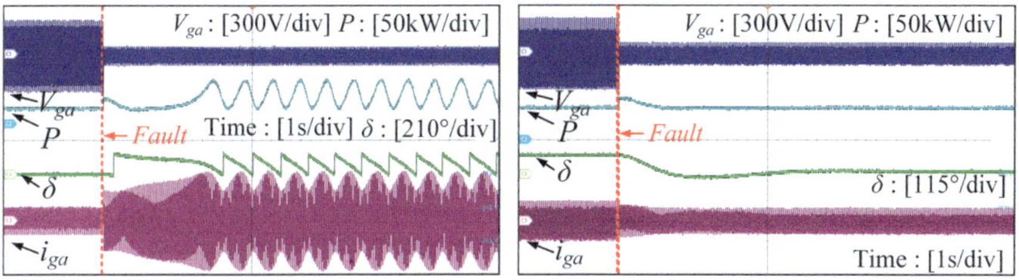

(a) Traditional GFM method

(b) Proposed method with m_x=2e-6,m_d=5e-6,n_x=2e-4,n_d=3e-4

Fig. 5.26 Waveforms of grid voltage, active power, phase angle and current when grid voltage drops to 0.25 p.u. with complex resistive-inductive impedance: $L_g = 2$ mH and $R_g = 0.5\ \Omega$

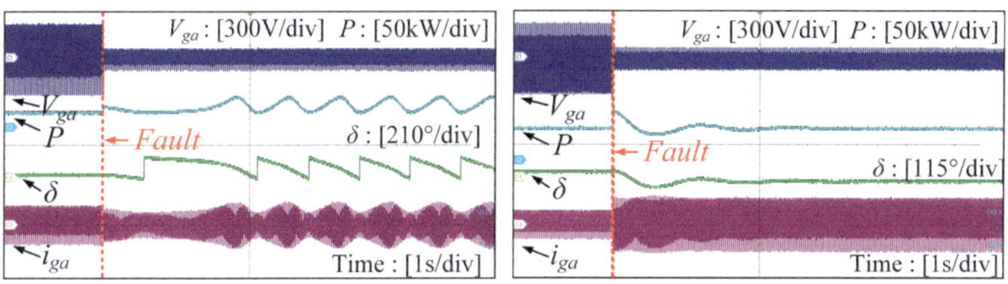

(a) Traditional GFM method

(b) Proposed method with n_x=m_x=0,m_d=1e-5,n_d=1e-3

Fig. 5.27 Waveforms of grid voltage, active power, phase angle and current when grid voltage drops to 0.25 p.u. with complex resistive-inductive impedance: $L_g = 0.1$ mH and $R_g = 2\ \Omega$

on Relative Gain Array (RGA) under a unified $[PQ] - [\omega V]$ interactive control framework. In particular, new perspectives of control parameter design under different line impedances are given that the design rules of droop control parameters between highly-inductive and highly-resistive line impedances are totally inverse. Comparative experiments verify the effectiveness of the introduced method and control design. The introduced method suppresses medium-frequency power oscillation under highly inductive line impedance, decreases the degree of power decoupling and speeds up adjusting time under highly resistive and complex resistive-inductive line impedance. Further, professional design for realistic distribution network, and the influence mechanism analysis of power coupling on the transient stability and the attractive region estimation under large disturbance fault are also the focus of our future research.

References

1. J.M. Guerrero, L.G. de Vicuna, J. Matas, M. Castilla, J. Miret, Output impedance design of parallel-connected UPS inverters with wireless load-sharing control. IEEE Trans. Ind. Electron.

52(4), 1126–1135 (2005)
2. A. Rodriguez-Cabero, J. Roldan-Perez, M. Prodanovic, Virtual impedance design considerations for virtual synchronous machines in weak grids. IEEE Trans. Emerg. Sel. Topics Power Electron. **8**(2), 1477–1489 (2020)
3. J. He, Y.W. Li, Analysis, design, and implementation of virtual impedance for power electronics interfaced distributed generation. IEEE Trans. Ind. Appl. **47**(6), 2525–2538 (2011)
4. X. Zhang et al., Analysis of dynamic power angle oscillation and its suppression strategy for the droop-controlled grid-connected inverter. IEEE J. Emerg. Selected Top. Power Electron. **9**(5), 5718–5731 (2021)
5. L. Zhang, L. Harnefors, H. Nee, Power-synchronization control of grid-connected voltage-source converters. IEEE Trans. Power Syst. **25**(2), 809–820 (2010)
6. S. Chen, Y. Sun, H. Han, S. Fu, S. Luo, G. Shi, A modified VSG control scheme with virtual resistance to enhance both small-signal stability and transient synchronization stability. IEEE Trans. Power Electron. **38**(5), 6005–6014 (2023)
7. E. Bristol, On a new measure of interaction for multivariable process control. IEEE Trans. Autom. Control, **AC-11**(1), 133–134 (1966)
8. F.G. Shinskey, *Process Control Systems: Application, Design, and Adjustment*, 3rd ed. (McGraw-Hill, New York. NY. USA, 1988), pp.181–203
9. M.C. Chandorkar, D.M. Divan, R. Adapa, Control of parallel connected inverters in standalone AC supply systems. IEEE Trans. Ind. Appl. **29**(1), 136–143 (1993)
10. N. Pogaku, M. Prodanovic, T.C. Green, Modeling, analysis and testing of autonomous operation of an inverter-based microgrid. IEEE Trans. Power Electron. **22**(2), 613–625 (2007)
11. Q.C. Zhong, G. Weiss, Synchronverters: inverters that mimic synchronous generators. IEEE Trans. Ind. Electron. **58**(4), 1259–1267 (2011)
12. H. Bevrani, T. Ise, Y. Miura, Virtual synchronous generators: a survey and new perspectives. Int. J. Electr. Power Energy Syst. **54**, 244–254 (2014)
13. H. Wu, X. Wang, Design-oriented transient stability analysis of grid-connected converters with power synchronization control. IEEE Trans. Ind Electron. **66**(8), 6473–6482 (2019). https://doi.org/10.1109/TIE.2018.2875669
14. D. Pan, X. Wang, F. Liu, R. Shi, Transient stability of voltage-source converters with grid-forming control: a design-oriented study. IEEE. J. Emerg. Sel. Topics Power Electron. **8**(2), 1019–1033 (2020). https://doi.org/10.1109/JESTPE.2019.2946310
15. T. Wen, X. Zou, D. Zhu, X. Guo, L. Peng, Y. Kang, Comprehensive perspective on virtual inductor for improved power decoupling of virtual synchronous generator control. IET Renew. Power Gener. **14**(4), 485–494 (2020)
16. L. Harnefors, M. Hinkkanen, U. Riaz, F.M.M. Rahman, L. Zhang, Robust analytic design of power-synchronization control. IEEE Trans. Ind. Electron. **66**(8), 5810–5819 (2019)

Chapter 6
Power Oscillation Suppression with Adaptive Virtual Inertia

6.1 Analysis of Power Oscillation of Multi VSG

6.1.1 Basic VSG Control Scheme

The basic electromechanical swing equation of the i-th VSG is defined as:

$$J_i \omega^* \frac{d\omega_i}{dt} = P^* - P_i - D_i \omega^* \left(\omega_i - \omega^*\right) \tag{6.1}$$

where ω_i represents the virtual angular frequency of the i-th VSG, ω_i is the nominal angular frequency. P^* and P_i represents the active power reference and output power of the i-th inverter, respectively. J_i is the inertia coefficient and D_i is the damping coefficient.

The voltage regulation in the VSG system is achieved through the QâŁ“V droop control so that the inverter has the primary voltage regulation characteristic. It is described as

$$V_i = V^* + n_i \left(Q^* - Q_i\right) \tag{6.2}$$

where V^* are the reference voltage amplitudes of the VSG, n_i is the droop coefficient, Q^* and Q_i are the reference and output reactive powers, respectively.

6.1.2 Power Oscillation Problem in Multi-VSG

To analyze the power oscillation problem in multi-VSG, transfer function models for multiple VSG-based inverters will be established to analyze the related dynamic response characteristics.

For simplicity, assume that the line impedance is mainly inductive (if not, the virtual impedance is introduced), so that the active power and the reactive power

X. Hou et al., *Self-organized Operation and Stability Control of Electronized Power Networks*, https://doi.org/10.1007/978-981-96-9067-1_6

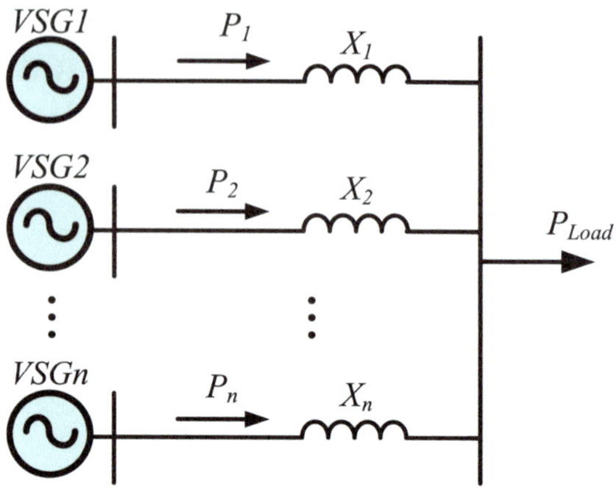

Fig. 6.1 The n-VSG electrical system

of the VSG are approximately decoupled. In Fig. 6.1, the small-signal model of the power transfer equation of i-th VSG is

$$\Delta P_i = \frac{V_i V_{pcc}}{X_i} \cos \delta_{i0} \Delta \delta_i = y_i \Delta \delta_i \tag{6.3}$$

where X_i is the total reactance, V_i is the amplitude of inverter output voltage, V_{pcc} is the amplitude of PCC voltage, δ_i is the phase difference between the inverter and PCC. Phase difference variation $\Delta\delta_i$ can be obtained as

$$\Delta \delta_i = \frac{1}{s}\left(\Delta \omega_i - \Delta \omega_g\right) \tag{6.4}$$

where $\Delta\omega_g$ represents the angular frequency variation of PCC.

Thus, the small-signal model of (6.1) can be expressed as

$$\frac{d\Delta\omega_i}{dt} = \frac{-y_i \Delta\delta_i - D_i \omega^* \Delta\omega_i}{J_i \omega^*} \tag{6.5}$$

In case of load change of ΔP_{Load}, it can be obtained that

$$\sum_{i=1}^{n} \Delta P_i = \Delta P_{Load} \tag{6.6}$$

Then the transfer function model can be obtained according to (6.3)–(6.6)

$$G(s) = \frac{\Delta P_i}{\Delta P_{Load}} = \frac{G_i \left(J_i \omega^* s + D_i \omega^*\right)}{\sum_{m=1}^{n} G_m \left(J_m \omega^* s + D_m \omega^*\right)} \tag{6.7}$$

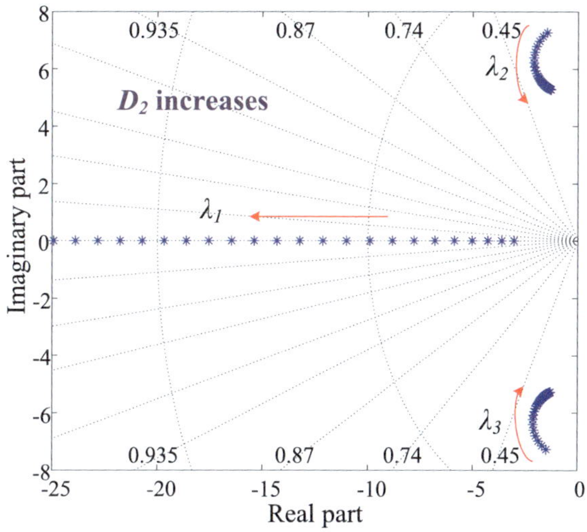

Fig. 6.2 The n-VSG electrical system

$$G_i = \frac{y_i}{J_i \omega^* s^2 + D_i \omega^* s + y_i} \tag{6.8}$$

where $G(s)$ represents the influence of load on the output power of the i-th inverter. Obviously, the output power of each paralleled VSG-based inverter is not only related to the load demand, but also affected by the parameters of other inverters as (6.7) shows. To observe conveniently, a two-VSG system is taken as an example where the line impedance and inertia coefficient are set to be equal. The loci of poles of $G_{l-l}(s)$ as D_2 increases is represented in Fig. 6.2. The power oscillation exists owing to a pair of underdamped conjugate dominant poles $\lambda_{2,3}$. When the damping coefficient D_2 gets too large, poles $\lambda_{2,3}$ move to the imaginary axis. It is unfriendly to system stability. Besides, as the increase of D_2 make the response speed decrease, the adjustment time gets longer [1]. Thus, it is necessary to study methods to suppress power oscillations. By analyzing $G(s)$, it is found that if the parameters satisfy the following relations

$$J_1 : J_2 : \cdots : J_n = D_1 : D_2 : \cdots : D_n = \frac{1}{X_1} : \frac{1}{X_2} : \cdots : \frac{1}{X_n} \tag{6.9}$$

$G(s)$ will become a proportional component [2]. When load demand changes, the inverter's output power will change proportionally without an oscillation process. The conclusion (6.9) can be regarded as a constraint for parameter design. However, the constraint is too strict to be practical. To relax the constraint, the mechanisms for power oscillation were analyzed in [3]. At the moment of load changes, the angular velocity of VSG cannot change abruptly and will maintain its original. Thus,

$$\frac{\dot{\omega}_i}{\dot{\omega}_j} \approx \frac{\Delta P_i}{J_i \omega^*} \frac{J_j \omega^*}{\Delta P_j} \approx \frac{X_i}{X_j} \frac{J_j}{J_i} \tag{6.10}$$

Further, loose constraint can be expressed as

$$J_1 : J_2 : \cdots : J_n = \frac{1}{X_1} : \frac{1}{X_2} : \cdots : \frac{1}{X_n} \tag{6.11}$$

From (6.10), the different angular accelerations are considered as the initial system oscillation excitation. Thus, if the angular acceleration of each VSG is exactly equal, the initial system oscillation excitation will be zero. Otherwise, different angular accelerations will directly cause periodical power angle oscillation and active power oscillation between different inverters.

6.2 Adaptive Virtual Inertia

6.2.1 Adaptive Virtual Inertia

From (6.5) and (6.10), it can be found that adjusting inertia coefficient dynamically is a direct way to change the angular acceleration and reduce the oscillation. Based on this idea, we present an adaptive virtual inertia control to solve the problem of the power oscillation.

To describe clearly the principle of the introduced algorithm, the frequency dynamic response of the two-VSG system is shown in Fig. 6.3. A disturbance on the load is put at t_1. As shown in Fig. 6.3, the typical oscillation process is divided into four situations, and the relationship between J and the angular velocity variation is analyzed. Since the angular velocity and power variation characteristics in each interval are different, the required rotor inertia is also different. Here, the variation trend of virtual inertia of VSG1 is mainly analyzed. The frequency ω_2 is taken as

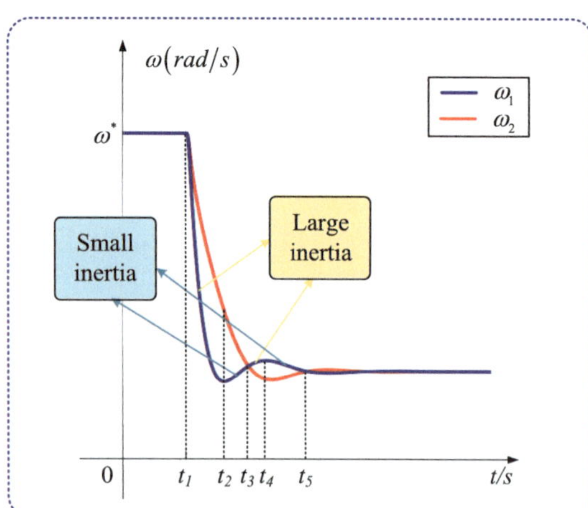

Fig. 6.3 Oscillating curve of the rotor angular frequency

Table 6.1 Design principle of virtual inertia

Frequency deviation	Change rate $d\omega_i/dt$	J change
$\omega_i < \omega_e$	< 0	Increase
$\omega_i < \omega_i$	> 0	Decrease
$\omega_i > \omega_e$	> 0	Increase
$\omega_i > \omega_i$	< 0	Decrease

a static reference of ω_1. The VSG1 system should have a slow response when ω_1 deviates from ω_2, and thus large inertia should be adopted. On the other hand, small inertia should be adopted to accelerate system dynamics when ω_1 returns to ω_2. Thus, in the dynamic process, it is beneficial to promote the consistency of response curves of ω_1 and ω_2, so as to achieve the effect of suppressing frequency oscillation.

Based on the analysis above, the design principle of virtual inertia at different operation states applicable for multi-VSG system is obtained as shown in Table 6.1, where ω_1 is local angular frequency of the i-th inverter and ω_1 is the angular frequency of neighbors of the i-th inverter.

According to the design principle shown in Table 6.1, a concise and unified mathematical equation of the adaptive virtual inertia is constructed as follows

$$J_i = J_{0i} + k_i \sum_{j \in N_i} (\omega_i - \omega_j) \frac{d\omega_i}{dt} \tag{6.12}$$

where N_i represents the set of the neighbors of the i-th inverter. From (6.12), the constructed inertia has two terms. The first term J_{0i} is the nominal constant inertia, and the second term is the adaptive compensation inertia. k_i is a positive inertia compensation coefficient, which can adjust the response speed of the frequency dynamic. Actually, the adaptive virtual inertia algorithm considers ω_j as a reference value for real-time updates. The total moment of inertia is modified based on the relative angular velocity $(\omega_i - \omega_j)$ and the change-rate of local angular frequency in real-time. Especially, in the nominal steady-state $(\omega_i - \omega_j)$, the second term of adaptive compensation inertia would be 0, and the total inertia is equal to J_{0i}. Conventionally, the virtual inertia J is used to slow down the change rate of COI of frequency and thus raise the level of the overall inertia of the system. Here, the adaptive term only helps improve frequency dynamic performance and suppress the power oscillation but does not affect the total inertia of the system.

6.2.2 *Practical Control Scheme Without Derivative Action*

In (6.12), it is worth noting that the adaptive inertia value would be inaccurate if we calculate it directly since the frequency derivative is sensitive to measurement noise [2, 4, 5]. Thus, we need to find a practical and effective method to address the issue.

Define the slip frequency ω_{si} as

$$\omega_{si}=\omega_i - \omega^* \tag{6.13}$$

where ω^* is the nominal angular frequency. Thus,

$$\omega_i - \omega_j=\omega_{si} - \omega_{sj} \tag{6.14}$$

Substituting the constructed inertia (6.12) into the typical VSG control (6.1) yields

$$\left(J_{0i} + k_i \sum_{j\in N_i} (\omega_{si} - \omega_{sj})\frac{d\omega_{si}}{dt}\right)\frac{d\omega_{si}}{dt} = P^* - P_i - D_i\omega_{si} \tag{6.15}$$

where simplification is made since the ω^* multiplied by J and D is not the key parameter.

Rewriting (6.15) yields

$$k_i \sum_{j\in N_i} (\omega_{si} - \omega_{sj})\dot{\omega}_{si}^2 + J_{0i}\dot{\omega}_{si} + D_i\omega_{si} = P_{res} \tag{6.16}$$

$$P_{res}=P^* - P_i \tag{6.17}$$

where P_{res} is the reserved power, which implies the difference between nominal power and actual output power.

Obviously, (6.16) is a quadratic equation in the variable $\dot{\omega}_s$. According to the Vieta Theorem, two roots are solved.

$$\dot{\omega}_{si}=\frac{-J_{0i} \pm \sqrt{{J_{0i}}^2 - 4k_i \sum\limits_{j\in N_i} \left(\omega_{si} - \omega_{sj}\right)\left(D_i\omega_{si} - P_{res}\right)}}{2k_i \sum\limits_{j\in N_i} \left(\omega_{si} - \omega_{sj}\right)} \tag{6.18}$$

As both $\dot{\omega}_{si} \sum\limits_{j\in N_i} \left(\omega_{si} - \omega_{sj}\right) > 0$ and $\dot{\omega}_{si} \sum\limits_{j\in N_i} \left(\omega_{si} - \omega_{sj}\right) < 0$ may exist in (6.18), only one root of (6.18) is effective, derived as follow

$$\begin{aligned}\dot{\omega}_{si}&=f\left(\omega_{si}, \omega_{sj}, P_{res}\right)\\ &=\frac{-J_{0i} + \sqrt{{J_{0i}}^2 - 4k_i \sum\limits_{j\in N_i} \left(\omega_{si} - \omega_{sj}\right)\left(D_i\omega_{si} - P_{res}\right)}}{2k_i \sum\limits_{j\in N_i} \left(\omega_{si} - \omega_{sj}\right)}\end{aligned} \tag{6.19}$$

To avoid the singular point ($\omega_{si} - \omega_{sj} = 0$), rewrite (6.19) by numerator rationalization

$$\dot{\omega}_{si} = f\left(\omega_{si}, \omega_{sj}, P_{res}\right) = \frac{-2(D_i\omega_{si} - P_{res})}{J_{0i} + \sqrt{J_{0i}{}^2 - 4k_i \sum_{j\in N_i}\left(\omega_{si} - \omega_{sj}\right)(D_i\omega_{si} - P_{res})}} \tag{6.20}$$

Then, the improved active power-frequency ($P - \omega$) control based on the adaptive virtual inertia algorithm is obtained by combining (6.13) and (6.20).

$$\omega_i = \omega^* + \omega_{si} = \omega^* + \int f\left(\omega_{si}, \omega_{sj}, P_{res}\right)dt \tag{6.21}$$

From (6.21), the angular frequency reference is a function of output active power and the angular frequency of neighbors. The detailed control scheme with adaptive virtual inertia is presented in Fig. 6.4. The control input is the real-time active power and the angular frequency of neighbors. The control output is the angular frequency reference. The control function (6.21) is derived from (6.12)–(6.15), and its design principle is shown in Table I. Compared with the power loop of a typical VSG, the control algorithm with adaptive virtual inertia is added. It is worth noting that the frequency derivative term is avoided in Fig. 6.4. Thus, the introduced control scheme is simple and practical.

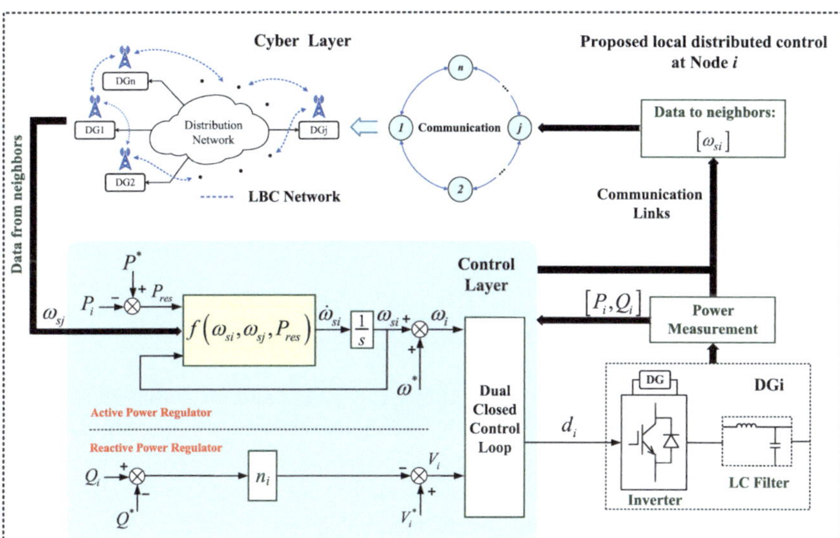

Fig. 6.4 Schematic diagram of the introduced adaptive virtual inertia control algorithm

6.2.3 Design Guideline for Compensation Coefficient

In this subsection, the design guidelines for key control parameters, such as the damping coefficient D_i, the initial inertia coefficient J_0 and the compensation coefficient k_i are discussed.

The system angular frequency should lie in the allowable range $[\omega_{min}, \omega_{max}]$. Thus, the specific design of damping coefficient D is as follows:

$$D_i \geq \frac{P_{\max} - P_{\min}}{\omega_{\max} - \omega_{\min}} \tag{6.22}$$

In the steady state, the adaptive inertia term would be 0, and the total inertia J is equal to J_0. Thus, adaptive terms are not considered when designing initial inertia coefficient J_0. The dynamic of the nominal steady-state is obtained from (6.15)

$$J_{0i}\ddot{\delta}_i + D_i\dot{\delta}_i + \frac{V_i V^*}{X_i}\sin\delta_i = P^* \tag{6.23}$$

Linearization of (6.23) at the steady-state point yields

$$J_{0i}\Delta\ddot{\delta}_i + D_i\Delta\dot{\delta}_i + \frac{V_i V^* \cos\delta_{i0}}{X_i}\Delta\delta_i = 0 \tag{6.24}$$

The natural frequency ω_n and damping ratio ζ are obtained as

$$\omega_n = \sqrt{\frac{V_i V^* \cos\delta_{i0}}{J_0 X_i}};\ \zeta = \frac{D_i}{2}\sqrt{\frac{X_i}{J_0 V_i V^* \cos\delta_{i0}}} \tag{6.25}$$

The damping factor$\zeta \in [0.1, 1.414]$ should be met to get a satisfactory transient response [6]. The initial inertia coefficient J_0 should be chosen as

$$\frac{0.125 {D_i}^2 X_i}{(V^*)^2} \leq J_{0i} = \frac{{D_i}^2 X_i}{4\zeta^2 V_i V^* \cos\delta_{i0}} \leq \frac{25 {D_i}^2 X_i}{(V^*)^2} \tag{6.26}$$

The more detailed design of the damping coefficient and the initial inertia coefficient can be seen in [7]. In this section, the inertia compensation coefficient k_i is mainly discussed. In (6.19), the angular acceleration $\dot{\omega}_s$ must be a real number rather than an imaginary number to ensure the validity of the introduced control. Hence, the following condition must hold identically.

$${J_{0i}}^2 - 4k_i \sum_{j\in N_i} \left(\omega_{si} - \omega_{sj}\right)\left(D_i\omega_{si} - P_{res}\right) \geq 0 \tag{6.27}$$

Thus,

$$\begin{cases} {J_{0i}}^2 \geq 4k_i \sum\limits_{j\in N_i} \left(\omega_{si} - \omega_{sj}\right)\left(D_i\omega_{si} - P_{res}\right); when \\ \omega_{si} = \omega^* - \omega_{\max};\ P_{res} = P^* - P_{\min};\ \sum\limits_{j\in N_i} \left(\omega_{si} - \omega_{sj}\right) = \sum\limits_{j\in N_i} \left(\omega_{\min} - \omega_{\max}\right); \\ {J_{0i}}^2 \geq 4k_i \sum\limits_{j\in N_i} \left(\omega_{si} - \omega_{sj}\right)\left(D_i\omega_{si} - P_{res}\right); when \\ \omega_{si}{=}\omega^* - \omega_{\min};\ P_{res} = P^* - P_{\max};\ \sum\limits_{j\in N_i} \left(\omega_{si} - \omega_{sj}\right) = \sum\limits_{j\in N_i} \left(\omega_{\max} - \omega_{\min}\right); \end{cases} \tag{6.28}$$

In the steady-state as shown in (6.15), there exists

$$\begin{cases} P^* - P_{\min} = -D\left(\omega^* - \omega_{\max}\right) \\ P^* - P_{\max} = -D\left(\omega^* - \omega_{\min}\right) \end{cases} \tag{6.29}$$

Combining (6.28) and (6.29), the range of the inertia compensation coefficient k_i is given by

$$0 \leq k_i \leq \frac{{J_{0i}}^2}{8E} \tag{6.30}$$

where

$$E = \max\left\{\sum_{j\in N_i} \left(\omega_{\min} - \omega_{\max}\right)\left(P_{\min} - P^*\right), \sum_{j\in N_i} \left(\omega_{\max} - \omega_{\min}\right)\left(P_{\max} - P^*\right)\right\} \tag{6.31}$$

In(6.12), a relatively large value of compensation coefficient k_i is favorable to exhibit the effectiveness of adaptive inertia control. Thus, ki should be chosen as an upper bound according to (6.30).

6.3 Stability Analysis

The stability of the Multi-VSG under the introduced control algorithm will be investigated based on the Lyapunov stability theorem in this section.

6.3.1 Stability Analysis Without Time Delay

The network model of the multi-machine system is described by the nodal admittance equation. By eliminating the load nodes, the reduced network can be described as [6]

$$P_i = V_i^2 G_{ii} + \sum_{m \neq i}^{n} V_i V_m G_{im} \cos(\delta_i - \delta_m) + \sum_{m \neq i}^{n} V_i V_m B_{im} \sin(\delta_i - \delta_m) \quad (6.32)$$

where $\mathbf{V}_i = V_i e^{j\delta_i}$ is the voltage at node i, and $Y_{im} = G_{im} + jB_{im}$ are the elements of the reduced admittance matrix.

Simplifying (6.32) as

$$P_i = P_{0i} + \sum_{m \neq i}^{n} k_{im} \sin(\delta_i - \delta_m) \quad (6.33)$$

where $k_{im} = V_i V_m B_{im}$ is the magnitude of the power-angle characteristic for the transfer equivalent branch and

$$P_{0i} = V_i^2 G_{ii} + \sum_{m \neq i}^{n} V_i V_m G_{im} \cos(\delta_i - \delta_m) \quad (6.34)$$

Generally, P_{0i} depends on power angles and is not constant during the transient period when the rotors are swinging. Stability analysis of the multi-machine system by the Lyapunov direct method is straightforward assuming that P_{0i} is a constant. That assumption in practice means that transmission losses in transfer branches are assumed to be constant and added to the equivalent loads.

According to (6.15)–(6.16), the dynamic of the introduced control algorithm can be expressed as

$$k_i \sum_{j \in N_i} \left(\omega_{si} - \omega_{sj}\right)(\dot{\omega}_{si})^2 + J_{0i}\dot{\omega}_{si} + D_i\omega_{si} = P^* - P_i \quad (6.35)$$

Let $\dot{\delta}_{si} = \omega_i$ the simplified multi-VSG grid model resulting from (6.32) and (6.33) can be summarized in the following set of state-space equations:

$$\begin{aligned} k_i \sum_{j \in N_i} \left(\dot{\delta}_{si} - \dot{\delta}_{sj}\right)\left(\ddot{\delta}_{si}\right)^2 + J_{0i}\ddot{\delta}_{si} + D_i\dot{\delta}_{si} \\ = P^* - P_{0i} - \sum_{m=1, m \neq i}^{n} k_{im} \sin(\delta_i - \delta_m) \end{aligned} \quad (6.36)$$

To introduce the Lyapunov function, in the first step, we need to define the potential and kinetic energy functions that are presented as:

$$\begin{cases} E_k = \frac{1}{2}\sum_{i=1}^{n} J_{0i}\omega_{si}^2 \\ E_p = -\sum_{i=1}^{n} (P_i^* - k_{ii})(\delta_i - \delta_i^*) - \sum_{i=1}^{n-1}\sum_{m=i+1}^{n} k_{im}\left[\cos\delta_{im} - \cos\delta_{im}^*\right] \end{cases} \quad (6.37)$$

Now, the Lyapunov function is obtained as a sum of two energy functions:

$$V_1 = E_k + E_p \tag{6.38}$$

If $\delta_{im} < \frac{\pi}{2}$ for all VSGs, energy function V_1 defined in (6.38) satisfies the condition of being a Lyapunov function that has been discussed in detail in [6, 8]. The derivative of V_1 is expressed as

$$\frac{dV_1}{dt} = -\sum_{i=1}^{n} D_i \dot{\delta}_i^2 - \sum_{i=1}^{n} \dot{\delta}_i k_i \ddot{\delta}_i^2 \sum_{j\in\Omega} \left(\dot{\delta}_i - \dot{\delta}_j\right) \frac{\pi}{2} \tag{6.39}$$

Thus,

$$\frac{dV_1}{dt} = -\dot{\delta}^T \left(D + \frac{1}{2}(KL + LK)\right) \dot{\delta} = -\dot{\delta}^T C \dot{\delta} \tag{6.40}$$

where L is a Laplace matrix that depends on the communication topology and

$$\begin{cases} \dot{\delta}^T = [\delta_1, \delta_2 \cdots \delta_n]^T \\ D = diag\{D_1, D_2 \cdots D_n\} \\ K = diag\left\{k_1 \ddot{\delta}_1^2, k_2 \ddot{\delta}_2^2 \cdots k_n \ddot{\delta}_n^2\right\} \end{cases} \tag{6.41}$$

The next proof employs the following two lemmas:

Definition 1 Let A be a $n \times n$ matrix with complex or real elements with eigenvalues $\lambda_1, \ldots, \lambda_n$. Then the spectral radius $\rho(A)$ of A is $\rho(A) = \max_{1\le i\le n} |\lambda_i|$, i.e., the largest absolute value (or complex modulus) of its eigenvalues.

Lemma 1 *Let $A \in C^{n\times n}$ with spectral radius $\rho(A)$ and for any matrix norm $\|\cdot\|$,we have $\rho(A) \le \|A\|$.*

Lemma 2 *If $\lambda_1 \le \lambda_2 \le \cdots \le \lambda_n$ are the eigenvalues of a real symmetric matrix A, and $\beta_1 \le \beta_2 \le \cdots \le \beta_n$ are the eigenvalues of a real symmetric matrix B, we obtain*

$$\lambda_i + \beta_1 \le \eta_i \le \cdots \le \lambda_i + \beta_n \tag{6.42}$$

where $\eta_1 \le \eta_2 \le \cdots \le \eta_n$ are the eigenvalues of matrix $A + B$. (The proof is presented in [9])

Lemma 3 *For a symmetric matrix $M = \begin{bmatrix} A & B \\ C & D \end{bmatrix}$, the following three conditions are equivalent.*

(1) M is a positive definite matrix.
(2) A and $D - CA^{-1}B$ are both positive definite matrices.
(3) D and $A - BD^{-1}C$ are both positive definite matrices.

Proof According to Lemma 1, we have

$$\rho\left(\frac{1}{2}(KL+LK)\right) \leq \left\|\frac{1}{2}(KL+LK)\right\| \leq \|K\|\,\|L\| \leq k_{\max}\,\|L\| \tag{6.43}$$

where k_{max} represents the maximum elements of the diagonal matrix K.

Further, designing the control parameter satisfying

$$k_{\max}\,\|L\| < d_{\min} \tag{6.44}$$

i.e.

$$\rho\left(\frac{1}{2}(KL+LK)\right) < d_{\min} \tag{6.45}$$

where d_{min} represents the minimum damping coefficient. Thus, according to lemma 2

$$\lambda\left(D + \frac{1}{2}(KL+LK)\right) > 0 \tag{6.46}$$

i.e. the eigenvalues of the symmetric matrix C are positive. Combining (6.43)–(6.46) yields

$$\frac{dV_1}{dt} = -\dot{\delta}^T C\dot{\delta} < 0 \tag{6.47}$$

Thus, the energy function of the system decreases over time.

The stability of the system is proved.

6.3.2 Stability Analysis with Time Delay

Considering positive time delay τ, the system model (6.36) can be expressed as

$$\begin{aligned} k_i \sum_{j\in N_i}\left(\dot{\delta}_{si}(t) - \dot{\delta}_{sj}(t-\tau)\right)\left(\ddot{\delta}_{si}(t)\right)^2 + J_{0i}\ddot{\delta}_{si}(t) + D_i\dot{\delta}_{si}(t) \\ = P^* - P_{0i} - \sum_{m=1,m\neq i}^{n} k_{im}\sin\left(\delta_i(t) - \delta_m(t)\right) \end{aligned} \tag{6.48}$$

A Lyapunov function is reconstructed as

$$V = V_1 + V_2 \tag{6.49}$$

where $V_i = E_i + E_p$ and $V_2 = \int_{t-\tau}^{t}\dot{\delta}^T(s)\,R\dot{\delta}(s)\,ds$ .R is a random positive definite matrix.

Taking the time derivative of V gives that

$$\begin{aligned}\frac{dV_1}{dt}&=-\dot{\delta}^T D\dot{\delta}-\left[\dot{\delta}^T,\dot{\delta}_\tau^T\right]\begin{bmatrix} nK & \frac{K(L-nI)}{2}\\ \frac{(L-nI)K}{2} & 0\end{bmatrix}\begin{bmatrix}\dot{\delta}\\ \dot{\delta}_\tau\end{bmatrix}\\ \frac{dV_2}{dt}&=\dot{\delta}^T R\dot{\delta}-\dot{\delta}_\tau^T R\dot{\delta}_\tau\end{aligned} \tag{6.50}$$

where I is an unit matrix. Thus,

$$\begin{aligned}\frac{dV}{dt}&=-\left[\dot{\delta}^T,\dot{\delta}_\tau^T\right]\begin{bmatrix} D+nK-R & \frac{K(L-nI)}{2}\\ \frac{(L-nI)K}{2} & R\end{bmatrix}\begin{bmatrix}\dot{\delta}\\ \dot{\delta}_\tau\end{bmatrix}\\ &=-\left[\dot{\delta}^T,\dot{\delta}_\tau^T\right]Q\begin{bmatrix}\dot{\delta}\\ \dot{\delta}_\tau\end{bmatrix}\end{aligned} \tag{6.51}$$

Theorem 1 *The system (6.48) is asymptotically stable and the system stability is independent of time delay τ if there exist positive definite matrix R satisfying the inequality (6.52).*

$$\frac{1}{4}k^2{}_{\max}\|L-nI\|^2\left\|R^{-1}\right\|<d_{\min}-\varepsilon_{\max} \tag{6.52}$$

where ε_{max} represents the maximum eigenvalue of matrix R.

Proof According to the compatibility of matrix norm

$$\frac{1}{4}k^2{}_{\max}\|(L-nI)\|^2\left\|R^{-1}\right\|\geq\|W\|\geq\rho(W) \tag{6.53}$$

where $W=\frac{K(L-nI)}{2}R^{-1}\frac{(L-nI)K}{2}$. According to Definition 1 and Lemma 2,

$$\begin{aligned}&\lambda\left[(D+nK-R)-W\right]\\ &\qquad\geq d_{\min}-\varepsilon_{\max}-\mu_{\max}\geq d_{\min}-\varepsilon_{\max}-\rho(W)\end{aligned} \tag{6.54}$$

where μ_{max} represents the maximum eigenvalue of matrix W.

Suppose that inequality (6.52) is satisfied. Then,

$$\lambda\left[(D+nK-R)-W\right]>0 \tag{6.55}$$

Thus, matrices R and $\left[(D+nK-R)-\frac{K(L-nI)}{2}R^{-1}\frac{(L-nI)K}{2}\right]$ are both positive definite matrices. According to Lemma 3, the symmetric matrix Q is positive. Thus, the energy function of the system decreases over time. The stability of the system is proved and is independent of time delay.

6.4 Simulation Results

In this section, to validate the effectiveness of the introduced method, simulations have been carried out in Matlab/Simulink. The circuit of the multi-VSG system is shown in Fig. 6.5 and system parameter values are listed in Tables II in Appendix A.

6.4.1 Comparisons with Constant Inertia Control

Figure 6.5 shows a power system with four VSGs. It is assumed that the load fluctuates at $t = 1$ s and $t = 4$ s. Figure 6.6 shows the output power and the frequency of the system, respectively. The constant inertia coefficient is designed as $J_{1,2,3,4} = 400$, $J_2 = 150$, $J_3 = 250$ and $J_4 = 800$. Obviously, the waveforms indicate that serious power oscillation and undesired frequency dynamic response performance occurs with conventional VSG control after load fluctuation. Unlike SGs, the inverter-based VSG with low overcurrent capability are more susceptible to damage owing to active power oscillation. Moreover, the poor frequency dynamic response performance, such as high rate of change of frequency (RoCoF) and large frequency deviation, will result in undesirable load-shedding or massive blackouts [10–12].

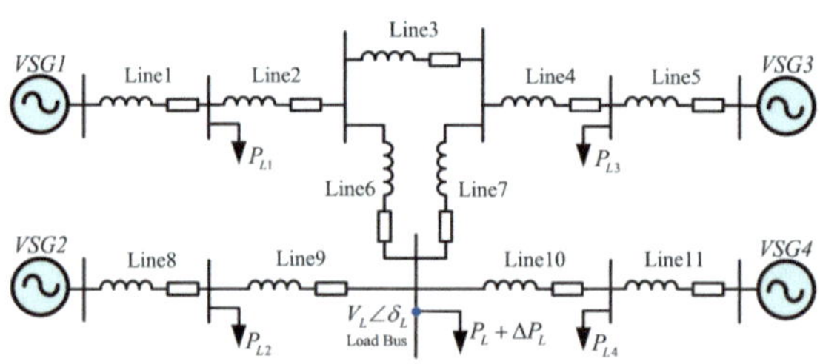

Fig. 6.5 The four-VSG simulated test system

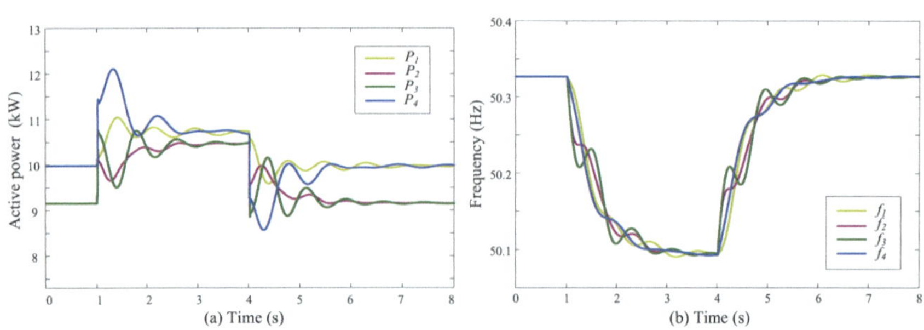

Fig. 6.6 Output active powers (**a**) and system frequency (**b**) of the system with a constant inertia coefficient

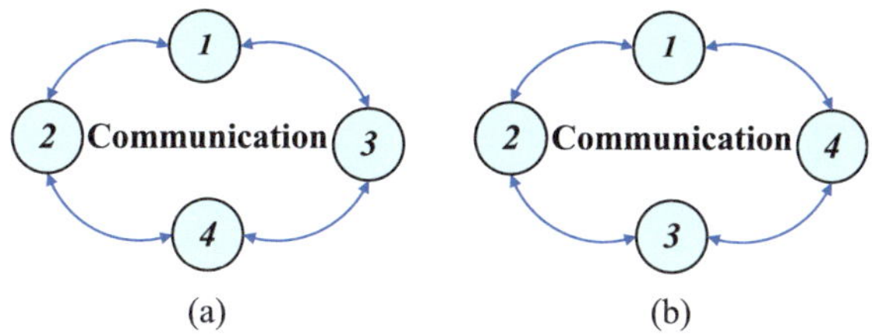

Fig. 6.7 communication topology (**a**) and communication topology (**b**)

To verify the validity of the introduced control, two different distributed communication topologies as shown in Fig. 6.7 are adopted. Figures 6.8 and 6.9 show the output active power, the system frequency and the variation trend of total inertia coefficient under different communication topology. Actually, it is obvious that choosing different neighbors has little impact on the performance on the adaptive inertia that can be ignored by comparing Fig. 6.8 with Fig. 6.9. The introduced distributed adaptive control strategy can suppress power oscillation and improve frequency performance in any distributed communication topology with at least one spanning tree [12, 13]. In addition, the topology is chosen such that in case of any link failure the remaining network still contains at least one spanning tree. This redundancy is required to ensure link-failure resiliency. Neighbors are chosen based on economic factors in this section. Two VSGs that are geographically close to each other are neighbors.

Since different communication topology has little impact on the system performance, here, the performance with the introduced method under communication topology a is mainly discussed. Figure 6.8 presents the output active power and the frequency of the system after applying the adaptive virtual inertia control where the adaptive coefficient is set as $k_1 = k_2 = k_3 = k_4 = 1000$. Figure 6.8c shows the variation waveform of the virtual inertia coefficient after load fluctuation. In Fig. 6.6a, the instantaneous impact power of VSG4 is 1400 W with conventional VSG control. It is larger compared with the instantaneous impact power 800 W in Fig. 6.8a. It is evident that the impact active power is reduced in Fig. 6.8a which is more friendly to power electronic inverters and the power oscillation is suppressed between each VSG. Besides, the frequency dynamic response performance is greatly improved as shown in Fig. 6.8b that helps enhance the frequency stability of the system. At the moment of a sudden change in frequency, we can see from Fig. 6.8c that the variation trend of the virtual inertia coefficient is favorable for each frequency to approach each other. To be specific, a VSG with a relatively high initial inertia coefficient decreases inertia, and a VSG with a relatively low initial inertia increases inertia. The inertial compensation term is obtained by comparing its own frequency with adjacent frequencies in real-time. It is worth noting that the inertia coefficient goes back to its initial value at a steady state.

Fig. 6.8 **a** Output active powers, **b** system frequency, and **c** the variation trend of total inertia coefficient of the system with k_i=1000 under communication topology a

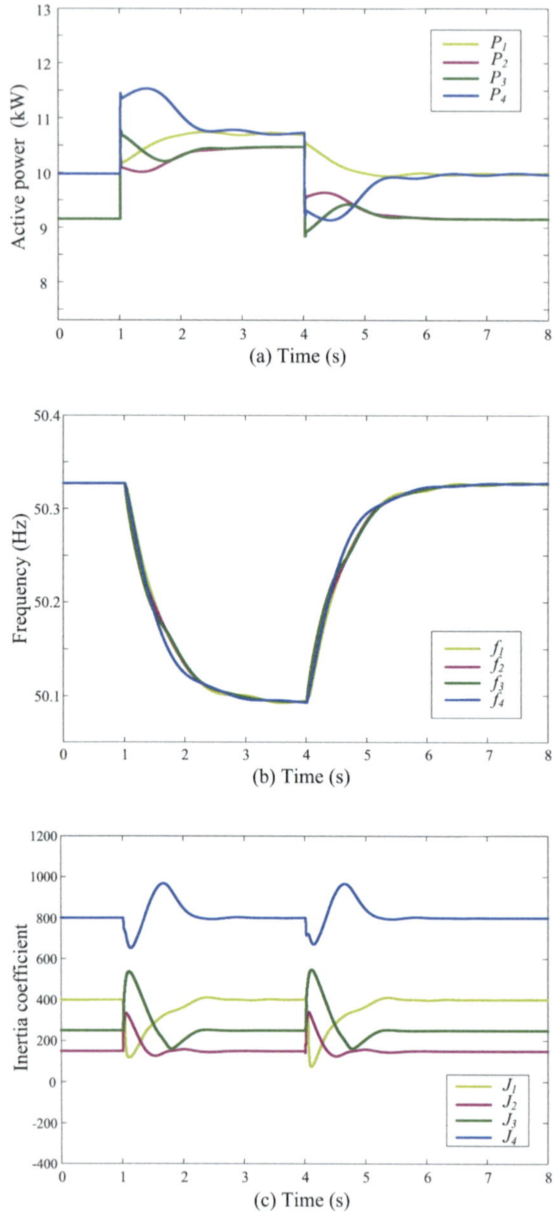

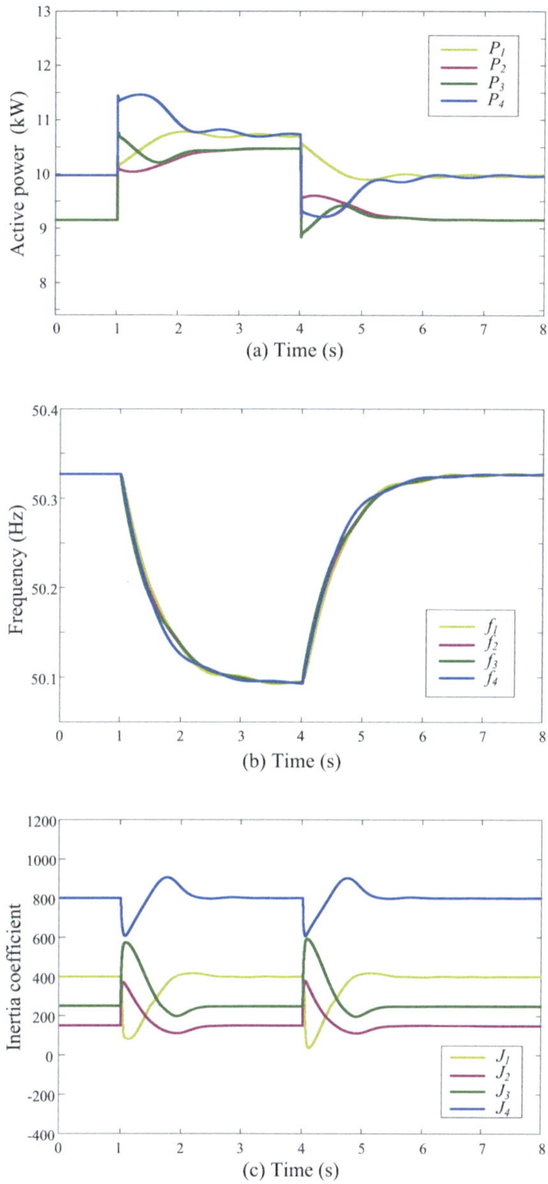

Fig. 6.9 **a** Output active powers, **b** system frequency, and **c** the variation trend of total inertia coefficient of the system with $k_i = 1000$ under communication topology b

6.4.2 Comparison with Different Inertia Coefficients

In order to investigate the influence of the key control parameter on system performance, i.e. inertia compensation coefficient k_i, a set of different k_i values (0, 200, 500, 1000, 2000) are applied in the introduced control method. In order to clearly see the variation of system output power and frequency waveform with the change of k_i, the waveform of VSG3 is listed in this part to illustrate the effectiveness of the control strategy. As shown in Fig. 6.10a, the output active powers oscillation has been effectively suppressed. Additionally, with the increase of the inertia compensation coefficient k_i, the power oscillation suppression effect is better. At the same time, the maximum frequency deviation and maximum RoCoF has been reduced as shown in Fig. 6.10b. In other words, the frequency dynamic response performance has been greatly improved. Besides, it takes less time to reach steady state. Conventionally, the inertia coefficient is designed to slow down the RoCoF. It is worth mentioning that the introduced adaptive inertia is designed to suppress the frequency oscillation and does not change the general trend of frequency change. The variation trend of the total inertia coefficient with different adaptive virtual inertia coefficient (k_i) is shown in Fig. 6.10c. With larger k_i, the range of inertia will increase in the transient process. Thus, a better effect of oscillation suppression can be obtained.

6.4.3 Damping Performance Under Short-Circuit Fault

In this case, it is assumed that a three-phase short circuit on transmission Line 2, near VSG1 occurs at $t = 1$ s and is cleared at $t = 1.1$ s. In order to better observe the damping ability of the introduced control strategy, the output active powers and system frequency of VSG1 with conventional VSG control and introduced adaptive virtual inertia control are presented in Fig. 6.11. It is obvious that the introduced method has a good damping performance on power and frequency oscillations. Besides, the maximum frequency deviation is reduced to a large extent and the system adjustment time is also reduced.

6.4.4 Robustness Performance Under Noises

To test the robustness performances of introduced control, high-frequency noise is imposed on the detection of the system operating frequency. Figures 6.12a, b show the frequency without and with adaptive virtual inertia control, respectively. Apparently, even though high-frequency noise exists in the system frequency detection, the introduced control method also helps improve frequency dynamic response, and thus suppress power oscillation. Obviously, the introduced method is not sensitive to high-frequency noises.

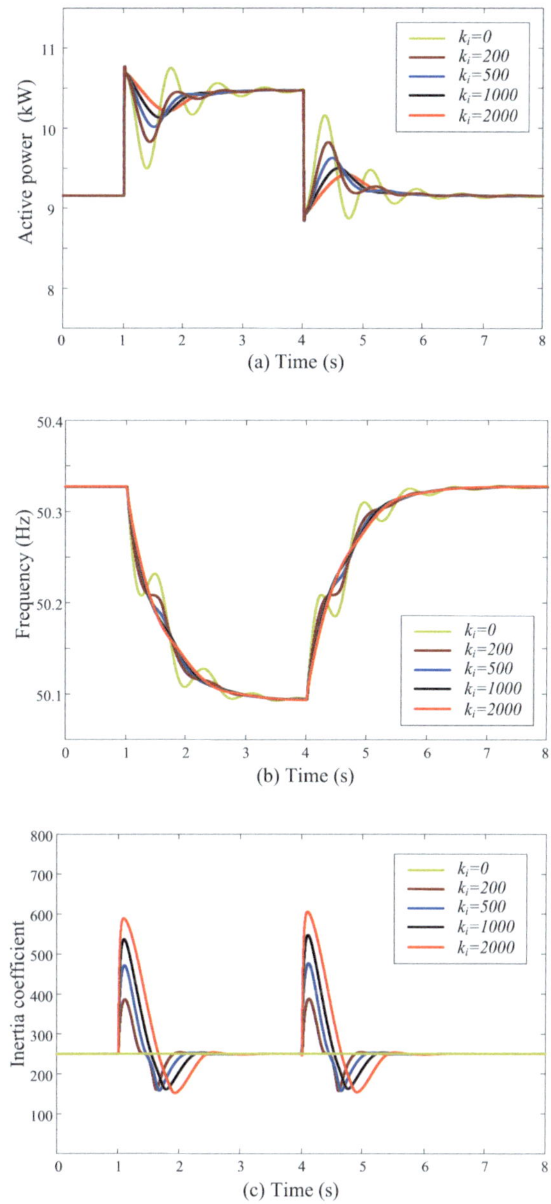

Fig. 6.10 **a** Output active power, **b** system frequency, and **c** the variation trend of total inertia coefficient of the system with different adaptive virtual inertia coefficient (k_i)

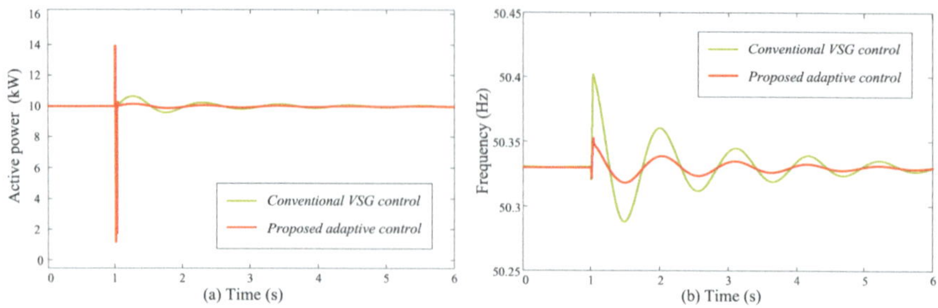

Fig. 6.11 **a** Output active powers, and **b** system frequency of VSG1 with conventional VSG control and introduced adaptive virtual inertia control under short-circuit fault

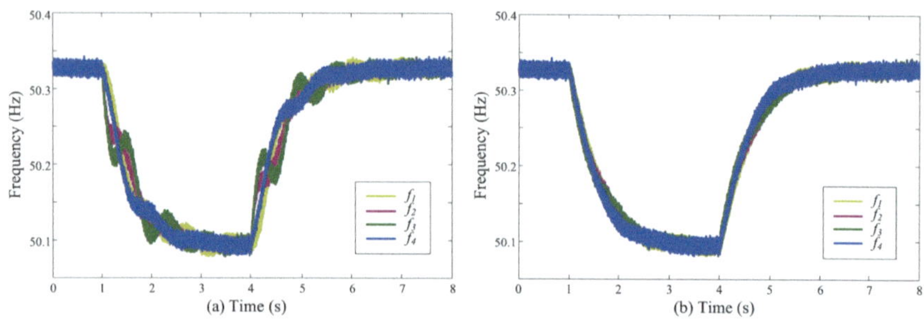

Fig. 6.12 System frequency without **a** and with **b** adaptive virtual inertia coefficient

6.4.5 Communication Delay Analysis

This section analyzes the impact of transmission latency on the system performance of the introduced method, which is critically important. Since the control strategy introduced in this chapter relies on distributed communication, i.e. the information interaction between VSG neighbors. Figure 6.13a and b show the output active power and the frequency response waveform of VSG3 in the system under different delay times with $k_i = 1000$. In the absence of communication delay, the adaptive inertia does not change the overall trend in the frequency and only helps suppress the frequency oscillation. As the communication delay increases, the overall RoCoF of the system decreases. Thus, the dynamic response performance is still improved. Moreover, it is obvious that the power oscillation is effectively suppressed even in the presence of communication delay and the range of permissible communication delays is wide as shown in Fig. 6.13. It should be noted that depending on the purpose of using the controller, the allowable latency is different. To improve stability and state estimation accuracy of the electrical networks, the latency of about 100

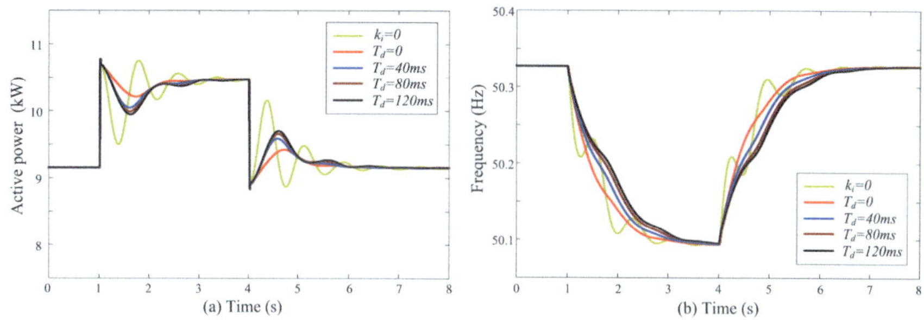

Fig. 6.13 **a** Output active powers and **b** system frequency of the system with adaptive virtual inertia method under different time delay

ms is acceptable, which is well-suited for today's technologies even for large networks [14, 15]. Therefore, in the introduced control method, with the low amount of information and especially simple calculations, the delay time can easily be implemented in less than 100 ms. Consequently, the effectiveness of the introduced control strategy is verified within the allowable delay range.

6.4.6 Communication Failure Analysis

In this case, the performance of the introduced distributed control strategy is tested in the case of a single communication link failure. The latency time due to the communication delay is set to be $T_d = 80$ ms and the adaptive inertia coefficient is set to be $k_i = 1000$. It is assumed that load fluctuates at $t = 1$ s and $t = 4$ s and a link failure occurs between VSG2 and VSG3. Figure 6.14a and b shows the output active power and the system frequency of VSGs. Compared with Fig. 6.6 where $k_i = 0$, it is obvious that the power oscillation of the system is suppressed and frequency

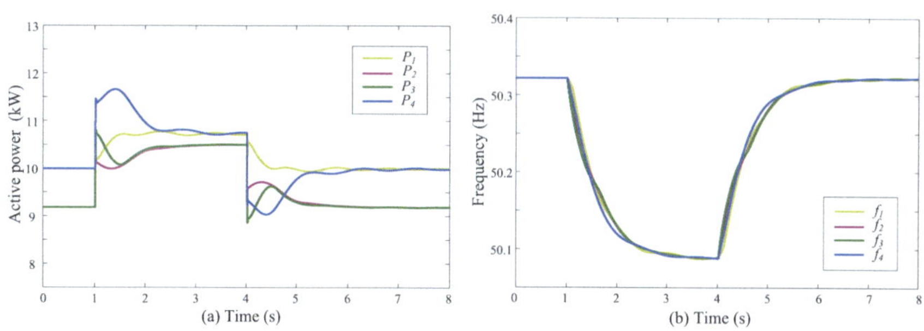

Fig. 6.14 System frequency without (**a**) and with (**b**) adaptive virtual inertia coefficient under single communication link failure

dynamic response is improved even in the case of single line communication failure. Compared with Fig. 6.8 without link failure, it is apparent that the more information obtained, the better the effect of power oscillation suppression. In addition, increasing the adaptive coefficient is also an important means to improve the suppression effect of power oscillation.

6.4.7 Performance Under Nonlinear Load

In this case, the performance of the introduced distributed control strategy is tested under nonlinear load. The PL as shown in Fig. 6.5 is replaced by an uncontrolled rectifier bridge. Figure 6.15a shows the FFT analysis of load voltage, the THD of which has reached 5.12%. Figure 6.15b, c present the output active power and the frequency of the system with the adaptive virtual inertia control. It is obvious that the power oscillation of the system is suppressed and frequency dynamic response is improved. Thus, the effectiveness of the introduced method has been verified under harmonic load.

6.5 Hardware-in-Loop Results

In this section, the introduced adaptive virtual inertia control is verified by real-time HIL test based OPAL-RT (OP4510). The structure of four-VSG grid shown in Fig. 6.5 has been studied.

6.5.1 Conventional VSG Control Method

Figure 6.16 shows the output power and the frequency of the system under load fluctuation, respectively. The constant inertia coefficient is designed as $J_{1,2,3,4} = 400$, $J_2 = 150$, $J_3 = 250$ and $J_4 = 800$. It should be noted that we set $f_i^{'} = (f_i - 50)$ Hz in order to better observe the dynamic process of frequency clearly as shown in Fig. 6.16b. With conventional VSG control, the power oscillation of the system can be observed in Fig. 6.16a when load fluctuates, and the instantaneous impact power is very large. Besides, the frequency oscillation phenomenon is also obvious which results in large frequency deviation.

6.5.2 VSG Control Method

With the introduced adaptive virtual inertia control method (k_i=1000), the improved output active power, the frequency of the system and the variation trend of total inertia coefficient are shown in Fig. 6.17. It is evident that the impact active power is reduced

Fig. 6.15 **a** Harmonic analysis of the load voltage, **b** output active powers, and **c** system frequency of the system with the introduced method

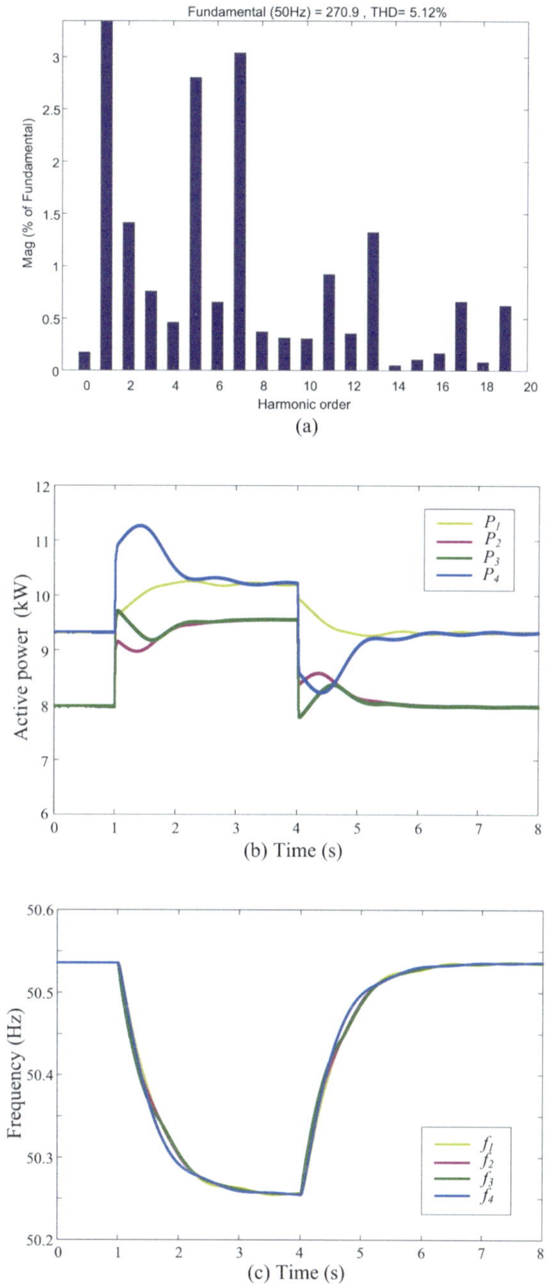

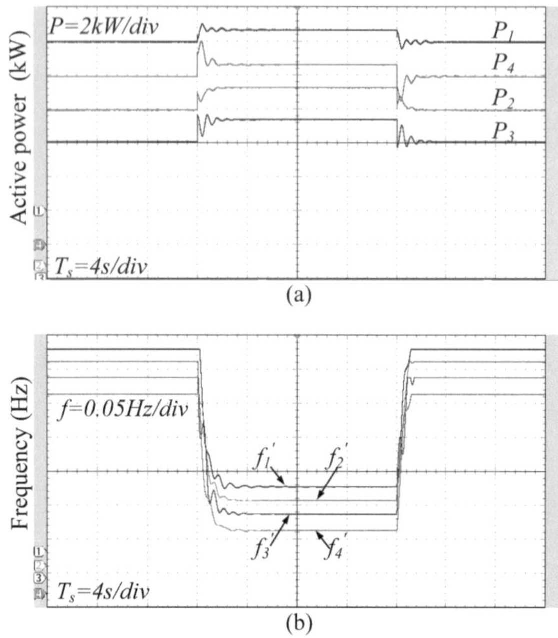

Fig. 6.16 **a** Output active powers and **b** system frequency of the system with a constant inertia coefficient

in Fig. 6.17a which is more friendly to power electronic inverters and the power oscillation is suppressed between each VSG. Obviously, the introduced method has smaller impact current during post-contingency dynamics which is desirable for a converter [11, 16]. The maximum frequency deviation and the maximum RoCoF has been reduced as shown in Fig. 6.17b. In other words, the frequency dynamic response performance has been greatly improved. The hardware-in-loop results are consistent with the simulation results compared with Figs. 6.6 and 6.8.

6.5.3 VSG Control Method Under Time Delay

This case verifies the effectiveness of the introduced adaptive virtual inertia control method under time delay. The delay time is set as 100 ms, which is acceptable and well-suited for large networks [14, 15]. Figure 6.18 shows the output active power and the frequency response waveform with k_i=1000. Obviously, even in the case of communication delays, the impact active power is reduced which is more friendly to power electronic inverters and the power oscillation is suppressed between each VSG. Besides, the frequency dynamic response performance is greatly improved that helps enhance the frequency stability of the system.

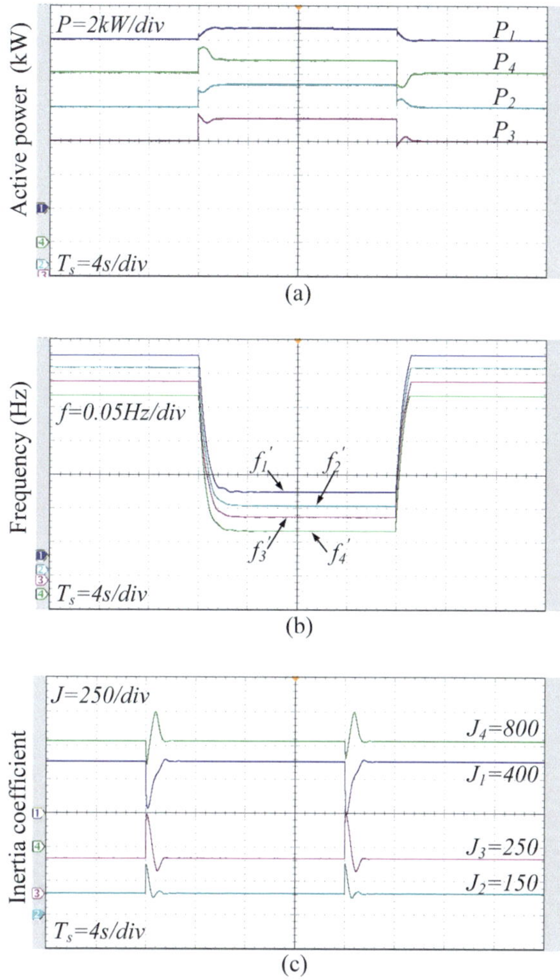

Fig. 6.17 **a** Output active power, **b** system frequency, and **c** the variation trend of total inertia coefficient of the system with the introduced method (k_i=1000)

6.6 Conclusion

This section introduces an adaptive virtual inertia control method for stability enhancement of a multi-VSG grid. The advantages of the presented control method include: (1) The presented method contributes to power oscillation suppression and dynamic frequency response improvement in the multi-VSG grid. (2) It is a practical control algorithm that avoid the direct frequency derivative action which introduces

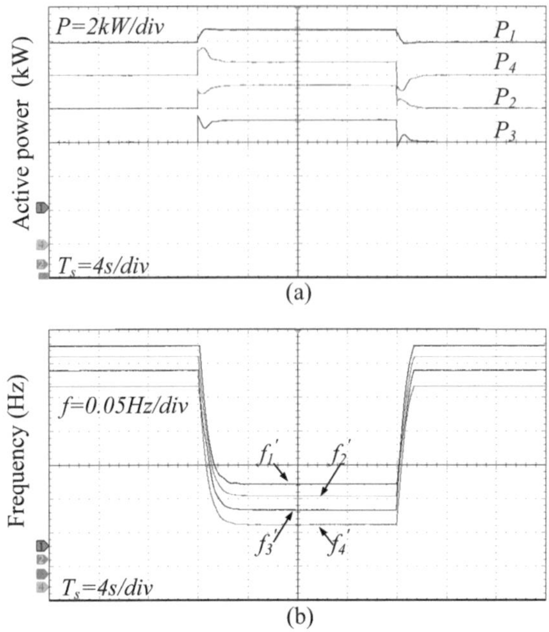

Fig. 6.18 **a** Output active powers and **b** the frequency of the system with the introduced method with time delay T_d=100 ms

high-frequency noise. Moreover, it has good robustness performance even under noise. (3) The presented method has a high communication delay margin and can still operate stably under a large communication delay. Besides, it is not susceptible to any single link failure, which leads to a more reliable control framework. The effectiveness of the presented method has been verified by simulation results and HIL results.

Appendix A

See Table 6.2.

Table 6.2 Parameters of the test system

Line parameters		
Line	Resistance (Ω)	Inductance (mH)
Line 1	0.8	12
Line 2	0.4	6
Line 3, Line 4, Line 6, Line 7, Line10	0.8	3
Line 5	0.5	1
Line 8	0.2	5
Line 9	0.2	3.5
Line 11	0.8	4
System parameters		
Parameter	Symbol	Value
Nominal frequency	f	50 Hz
Nominal voltage	V	380 V
Rated active power	P	10 kW
Rated reactive power	Q	2 kVar
Load reactive power	$P_{L1}, P_{L2}, P_{L3}, P_{L4}$	10 kW
Control parameters		
	VSG1	
Power filter time constant	τ	1/60
Droop damp coefficient	D_L	500
Initial inertia coefficient	J_L	400
$Q-V$ droop coefficient	η_L	0.001
	VSG2	
Power filter time constant	τ	1/60
Droop Damp coefficient	D_2	900
Initial inertia coefficient	J_2	150
$Q-V$ droop coefficient	η_2	0.001
	VSG3	
Power filter time constant	τ	1/60
Droop damp coefficient	D_3	900
Initial inertia coefficient	J_3	250
$Q-V$ droop coefficient	η_3	0.001
	VSG4	
Power filter time constant	τ	1/60
Droop damp coefficient	D_4	500
Initial inertia coefficient	J_4	800
$Q-V$ droop coefficient	η_4	0.001

References

1. X. Ningyi, W. Yue, L. Mingxuan, W. Wenti, W. Ning, L. Jiliang, An optimal control method of virtual angular acceleration to improve transient response based on virtual synchronous generator, in *2017 IEEE 3rd International Future Energy Electronics Conference and ECCE Asia (IFEEC 2017—ECCE Asia)*, pp. 1559–1563
2. M. Mao, C. Qian, Y. Ding, Decentralized coordination power control for islanding microgrid based on PV/BES-VSG. CPSS Trans. Power Electron. Appl. **3**(1), 14–24 (2018)
3. Z. Shuai, W. Huang, Z. J. Shen, A. Luo, and Z. Tian, Active power oscillation and suppression techniques between two parallel
4. A. Karimi et al., Inertia response improvement in AC microgrids: a fuzzy-based virtual synchronous generator control. IEEE Trans. Power Electron. **35**(4), 4321–4331 (2020)
5. M. Choopani, S.H. Hosseinain, B. Vahidi, A novel comprehensive method to enhance stability of multi-VSG grids. Int. J. Electr. Power Energy Syst. **104**, 502–514 (2019)
6. J. Machowski, Z. Lubosny, J. W. Bialek, and J. R. Bumby, Power System Dynamics: Stability and Control. (John Wiley & Sons, 2020)
7. X. Hou, Y. Sun, X. Zhang, J. Lu, P. Wang, J.M. Guerrero, Improvement of frequency regulation in VSG-based AC microgrid via adaptive virtual inertia. IEEE Trans. Power Electron. **35**(2), 1589–1602 (2020)
8. M. Choopani, S.H. Hosseinian, B. Vahidi, New transient stability and LVRT improvement of multi-VSG grids using the frequency of the center of inertia. IEEE Trans. Power Syst. **35**(1), 527–538 (2020)
9. C. D. Meyer, Matrix Analysis and Applied Linear Algebra (Siam, 2000)
10. M.-S. Debry, G. Denis, T. Prevost, F. Xavier, A. Menze, Maximizing the penetration of inverter-based generation on large transmission systems: the migrate project, in *6th Solar Integration Workshop* (2017)
11. F. Milano, F. Dörfler, G. Hug, D.J. Hill, G. Verbič, Foundations and challenges of low-inertia systems (Invited Paper). Power Systems Computation Conference (PSCC) **2018**, 1–25 (2018)
12. V. Nasirian, A. Davoudi, F.L. Lewis, J.M. Guerrero, Distributed adaptive droop control for DC distribution systems. IEEE Trans. Energy Convers. **29**(4), 944–956 (2014)
13. V. Nasirian, S. Moayedi, A. Davoudi, F.L. Lewis, Distributed cooperative control of DC microgrids. IEEE Trans. Power Electron. **30**(4), 2288–2303 (2015)
14. M.F.M. Arani, Y.A.I. Mohamed, E.F. El-Saadany, Analysis and mitigation of the impacts of asymmetrical virtual inertia. IEEE Trans. Power Syst. **29**(6), 2862–2874 (2014)
15. W.K. Chai et al., An information-centric communication infrastructure for real-time state estimation of active distribution networks. IEEE Trans. Smart Grid **6**(4), 2134–2146 (2015)
16. L. Huang, H. Xin, Z. Wang, L. Zhang, K. Wu, J. Hu, Transient stability analysis and control design of droop-controlled voltage source converters considering current limitation. IEEE Trans. Smart Grid **10**(1), 578–591 (2019)

Chapter 7
Power Oscillation Suppression with Decentralized Mutual Damping Control

7.1 Decentralized Mutual Damping Control Method

7.1.1 Decentralized Mutual Damping Control Method

This section presents a communication-free mutual damping method to work out the trouble of power oscillation on the basis of VSG control. Without loss of generality, as shown in Fig. 7.1, the research is based on the network topology with n nodes.

The control algorithm is shown in Fig. 7.2 and can be expressed as

$$J_i \frac{d\omega_i}{dt} = P^* - P_i - D_i \left(\omega_i - \omega^*\right) - m_i \frac{dP_i}{dt} \tag{7.1}$$

where m_i is the mutual damping coefficient. From (7.1), to address the power oscillation problem, a simple solution is to increase parameter D_i. However, when both inertia and power-sharing are required in applications, other methods have to be used. The fourth term on the right-hand side of (7.1) is the key to the introduced solution.

Under the assumptions of highly inductive lines, the output power can be expressed as nodal injection power [1], i.e.

$$P_i \approx \sum_{j=1, j\neq i}^{N} b_{ij} \sin\left(\delta_i - \delta_j\right) \tag{7.2}$$

where $b_{ij} = V_i V_j B_{ij}$ and B_{ij} is the mutual susceptance between nodes i and j.

X. Hou et al., *Self-organized Operation and Stability Control of Electronized Power Networks*, https://doi.org/10.1007/978-981-96-9067-1_7

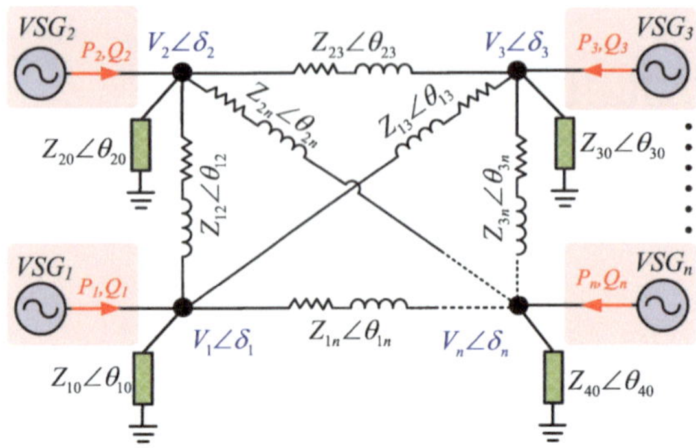

Fig. 7.1 Network topology with n nodes

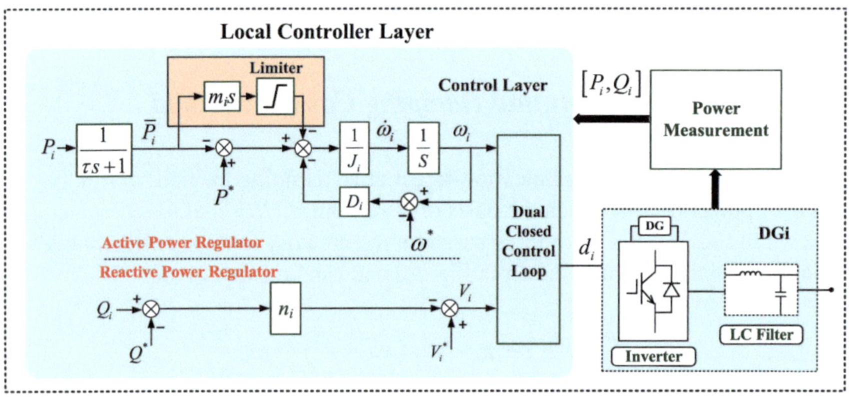

Fig. 7.2 Control block diagram of the decentralized mutual damping method

According to (7.2), the derivative of output power can be calculated as

$$\frac{dP_i}{dt} = \sum_{j=1,j\neq i}^{N} V_i V_j B_{ij}\cos\left(\delta_i - \delta_j\right)\left(\omega_i - \omega_j\right) + \sum_{j=1,j\neq i}^{N} \frac{dV_i}{dt} V_j B_{ij}\sin\left(\delta_i - \delta_j\right) + \sum_{j=1,j\neq i}^{N} \frac{dV_j}{dt} V_i B_{ij}\sin\left(\delta_i - \delta_j\right) \quad (7.3)$$

Assume that the phase difference between nodes is small and changes little, i.e. $\delta_i - \delta_j \approx 0$. Besides, the output voltage amplitude of the VSG is considered to change slowly in general [2]. Therefore, the second and the third term of (7.3) can be neglected. In this way, the derivative term of local active power can be approximately expressed as

$$\frac{dP_i}{dt} = \sum_{j=1, j\neq i}^{N} V_i V_j B_{ij} \cos\left(\delta_i - \delta_j\right)\left(\omega_i - \omega_j\right) \tag{7.4}$$

For convenience, we define

$$a_{ij} = V_i V_j B_{ij} \cos\left(\delta_i - \delta_j\right) \tag{7.5}$$

where $a_{ij} > 0$. Thus, (7.4) can be represented as

$$\frac{dP_i}{dt} = \sum_{j=1, j\neq i}^{N} a_{ij}\left(\omega_i - \omega_j\right) \tag{7.6}$$

Thus, the introduced control algorithm is equivalent to

$$J_i\dot{\omega}_i = P^* - P_i - D_i\left(\omega_i - \omega^*\right) - m_i \sum_{j=1, j\neq i}^{N} a_{ij}\left(\omega_i - \omega_j\right) \tag{7.7}$$

Cooperative control with single-integrator dynamics is known as [3]

$$\dot{x}_i = -\sum_{j=\in N_i} c_{ij}\left(x_i - x_j\right) \tag{7.8}$$

where x_i represents the state of node i, N_i represents neighbors of node i in the graph topology, and c_{ij} represents the graph edge weights. It is widely used in distributed control method [4]. Each node's control input relies upon the contrast between its state and every one of its neighbors x_j. Notice that assuming these states are no different either way. This protocol actuates all states to an identical value [3]. At the same time, $\dot{x}_i = 0$.

Comparing (7.7) with (7.8), the additional mutual damping term introduces the difference between its frequency and others' frequency. It contributes to the convergence of the angular frequency in the transient process that could suppress oscillation effectively.

The two-VSG framework is taken as an illustration to clarify the rule of the technique. As shown in Table 7.1, in the dynamic process of load fluctuation, when $\omega_1 > \omega_2$, the mutual damping term makes $d\omega_1/dt$ decreases and $d\omega_2/dt$ increases. On the contrary, when $\omega_1 < \omega_2$, the mutual damping term makes $d\omega_1/dt$ increases and $d\omega_2/dt$ decreases. In this way, whether $\omega_1 > \omega_2$ or $\omega_1 < \omega_2$, the mutual damping term helps facilitate the conformance of dynamic curves of angular frequency ω_1 and ω_2. Thus, the frequency dynamic response was effectively improved, which further suppress power oscillation. It is worth noting that $(\omega_i - \omega_j)$ is obtained indirectly by taking the derivative of the local active power.

Table 7.1 Principle of mutual damping

	State of angular frequency	$\omega_1 > \omega_2$	$\omega_1 < \omega_2$
VSG1	Mutual damping term	> 0	< 0
	$d\omega_1/dt$ change	decrease	increase
VSG2	Mutual damping term	< 0	> 0
	$d\omega_2/dt$ change	increase	decrease

At the point when the system arrives at a steady-state, the frequency of each VSG will in general be uniform, and the mutual damping term automatically decreases to zero. The principle of the method can also be extended to multi-VSG systems.

7.1.2 Stability Analysis

The stability with the decentralized mutual damping method in a multi-VSG grid is researched on the basis of the Lyapunov stability theorem in this part. The output power P_i is predigested as

$$P_i \approx \sum_{j=1, j\neq i}^{N} b_{ij} \sin\left(\delta_i - \delta_j\right) \tag{7.9}$$

For convenience, we define

$$\begin{cases} \omega_{si} = \omega_i - \omega^\circ \\ \dot{\delta}_{si} = \omega_{si} \end{cases} \tag{7.10}$$

According to (7.7) and (7.9), the improved control equation is represented as

$$\begin{aligned} & J_i\dot{\omega}_{si} + D_i\omega_{si} \\ & = P^* - \sum_{j=1, j\neq i}^{N} b_{ij} \sin\left(\delta_i - \delta_j\right) - m_i \sum_{j=1, j\neq i}^{N} a_{ij}\left(\omega_{si} - \omega_{sj}\right) \end{aligned} \tag{7.11}$$

The potential energy functions and kinetic energy functions are expressed as [5, 6]:

$$\begin{cases} E_k = \frac{1}{2}\sum\limits_{i=1}^{N} J_i {\omega_{si}}^2 \\ E_p = -\sum\limits_{i=1}^{N} (P_i{}^* - P_{i0})(\delta_i - \delta_i{}^\circ) - \sum\limits_{i=1}^{N-1}\sum\limits_{j=i+1}^{N} k_{ij}\left[\cos\delta_{ij} - \cos{\delta_{ij}}^\circ\right] \end{cases} \tag{7.12}$$

where $\delta_{ij} = \delta_i - \delta_j$ and$\delta_{ij}^{\circ} = \delta_i^{\circ} - \delta_j^{\circ}$.

Further, the sum of the potential energy functions and kinetic energy functions as (7.13) shows constructs the Lyapunov function.

$$V{=}E_k + E_p \tag{7.13}$$

For every VSG, in the case of $\delta_{ij} < \frac{\pi}{2}$, V characterized in (7.13) fulfills the positive definite condition of being a Lyapunov function [5, 6].

Combining (7.11)–(7.13) yields that

$$\frac{dV}{dt} = -\sum_{i=1}^{n} D_i {\omega_{si}}^2 - \sum_{i=1}^{n} \omega_{si} \sum_{j\in\Omega} m_i a_{ij} \left(\omega_{si} - \omega_{sj}\right) \tag{7.14}$$

Since $a_{ij} = a_{ji}$ from (7.5), (7.14) can be expressed as

$$\frac{dV}{dt} = -\sum_{i=1}^{n} D_i {\omega_{si}}^2 - m_i \sum_{i=1}^{n} \sum_{j\in\Omega} a_{ij} \left(\omega_{si} - \omega_{sj}\right)^2 < 0 \tag{7.15}$$

As seen in (7.15), as long as the mutual damping coefficient is greater than 0, the system will be sTable Moreover, under the action of mutual damping, the transient energy of the system will reduce faster with larger m_i. The proof above is correct only under ideal circumstances that the pure differentiator is applied in (7.1). It provides insight into addressing the power oscillation

7.2 Design Guideline for Key Parameters

In practical applications, the pure differentiator in (7.1) tends to be noisy. Therefore, a low pass filter before the differentiator is adopted that helps avoid the effect of high-frequency measurement noise [7]. This subsection will be centered on the impact of the low-pass filter parameter τ and the mutual damping coefficient m_i on system stability.

After introducing the low-pass filter, (7.1) can be rewritten as

$$J_i \frac{d\omega_i}{dt} = P^* - \bar{P}_i - D_i \left(\omega_i - \omega^*\right) - m_i \frac{d\bar{P}_i}{dt} \tag{7.16}$$

where $\bar{P}_i$ represents the active power after low-pass filtering and

$$\bar{P}_i = \frac{1}{\tau s + 1} P_i \tag{7.17}$$

where τ is the power filter time constant.

Based on (7.16) and (7.17), the small-signal model of the system can be expressed as

$$\Delta\dot{\omega}_i = \frac{1}{J_i}\left(\Delta P^* - \Delta\bar{P}_i - D_i\Delta\omega_i - m_i\Delta\dot{\bar{P}}_i\right) \tag{7.18}$$

$$\Delta\bar{P}_i = \frac{1}{\tau s + 1}\Delta P_i \tag{7.19}$$

From (7.2) and (7.19), it yields that

$$\Delta\dot{\bar{P}}_i = \frac{1}{\tau}\left(\Delta P_i - \Delta\bar{P}_i\right) \tag{7.20}$$

$$\Delta P_i = \sum_{j=1, j\neq i}^{N} a_{ij}{}^{\circ}\left(\Delta\delta_i - \Delta\delta_j\right) \tag{7.21}$$

where $a_{ij}{}^{\circ} = b_{ij}\cos\delta_{ij}{}^{\circ}$. Thus, the whole state-space model of the system can be expressed as

$$\Delta\dot{\mathbf{x}} = \mathbf{A}_{3n\times 3n}\Delta\mathbf{x} + \mathbf{B}_{3n\times n}\Delta\mathbf{u} \tag{7.22}$$

$$\mathbf{A} = \begin{bmatrix} -\mathbf{JD} & -\frac{m}{\tau}\mathbf{JL} & \left(\frac{m}{\tau} - 1\right)\mathbf{JE} \\ \mathbf{E} & \mathbf{0} & \mathbf{0} \\ \mathbf{0} & \frac{1}{\tau}\mathbf{L} & -\frac{1}{\tau}\mathbf{E} \end{bmatrix} \tag{7.23}$$

$$\mathbf{L}_{n\times n} = \begin{bmatrix} \sum\limits_{j=0, j\neq 1}^{n} a_{1j}{}^{0} & -a_{12}{}^{0} & \cdots & -a_{1(n-1)}{}^{0} & -a_{1n}{}^{0} \\ -a_{21}{}^{0} & \sum\limits_{j=0, j\neq 2}^{n} a_{2j}{}^{0} & \cdots & -a_{2(n-1)}{}^{0} & -a_{2n}{}^{0} \\ \vdots & \vdots & \ddots & \vdots & \vdots \\ -a_{(n-1)1}{}^{0} & -a_{(n-1)2}{}^{0} & \cdots & \sum\limits_{j=0, j\neq (n-1)}^{n} a_{(n-1)j}{}^{0} & -a_{(n-1)n}{}^{0} \\ -a_{n1}{}^{0} & -a_{n2}{}^{0} & \cdots & -a_{n(n-1)}{}^{0} & \sum\limits_{j=0, j\neq n}^{n} a_{nj}{}^{0} \end{bmatrix} \tag{7.24}$$

where $\Delta\mathbf{x} = [\Delta\omega\ \Delta\delta\ \Delta\bar{\boldsymbol{P}}_{\mathrm{i}}]$, $\Delta(\mathbf{u}) = \Delta\mathbf{P}^*$, $\mathbf{B} = [\mathbf{J}\ \mathbf{0}_{\mathbf{n}\times\mathbf{n}}\ \mathbf{0}_{\mathbf{n}\times\mathbf{n}}]$ and where $\mathbf{J}$ is an $n\times n$ diagonal matrix, whose diagonal elements are $j_{ii} = 1/J_i$, $\mathbf{D}$ is an $n\times n$ diagonal matrix, whose diagonal elements are $d_{ii} = D_i$, $\mathbf{E}$ is a $n\times n$ unit matrixes.

7.2.1 Participation Factor Analysis

In general, to investigate the association between system variables and the predominant oscillatory modes and evaluate the correlative degree of participation of system variables in a mode, Participation Factors (PFs) is employed. A four-VSG system is taken as an example. Figure 7.3 and Table 7.2 show the participation factors analysis with (a) the traditional VSG method and (b) the decentralized mutual damping method in this chapter. The damping ratios of the dominant oscillatory modes eigenvalue is small with the conventional VSG method, and they are improved a lot with the introduced method. Besides, the oscillation frequency of the three pairs of dominant modes is generally a few hertz. $\lambda_{1,2}$ are mainly related with $\Delta\omega_2$, $\Delta\delta_2$, $\Delta\omega_4$ and $\Delta\delta_4$, i.e. the angular frequency ω and phase angle δ of VSG2 and VSG4. $\lambda_{3,4}$ are mainly related with the ω and δ of VSG1 and $\lambda_{5,6}$ are mainly related with the ω andδ of VSG3. It means that these two states can be chosen as feedback information to raise the damping ratio (ζ) of the dominant modes. From (7.23), m and τ are the key coefficients of the phase angle feedback. Thus, rational design of m and τ helps increase damping and suppress power oscillation.

7.2.2 Sensitivity Analysis

Sensitivity analysis is conducted to identify appropriate ranges for key control parameters of the system. Figure 7.4a shows the track of the dominant poles of **A** as m_i increases from 0 to 0.8. Compared with traditional VSG control, i.e. $m_i = 0$, the introduced method increases the overall damping of the system, since three pairs of dominant poles ($\lambda_{1,2}$, $\lambda_{3,4}$ and $\lambda_{5,6}$) get away from the imaginary axis and move towards the real axis as m_i increases. However, when mi is too large, the other

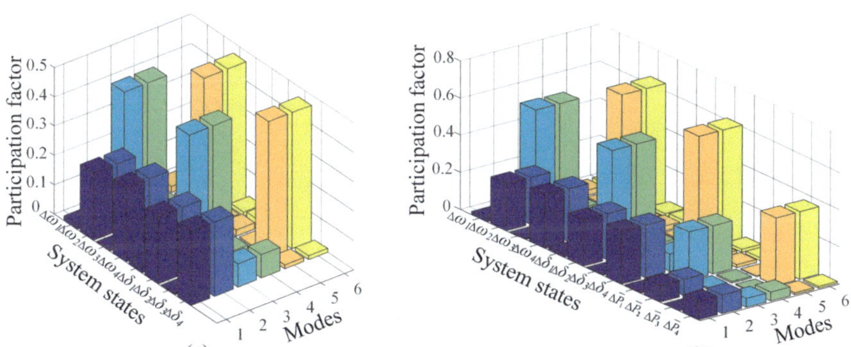

Fig. 7.3 Participation analysis of the system states to the dominant modes with **a** VSG method and **b** the decentralized mutual damping method

three pairs of non-dominant poles away from the imaginary axis begin to leave away from the real axis. In a word, parameter mi should be big, so that ($\lambda_{1,2}$, $\lambda_{3,4}$ and $\lambda_{5,6}$) can obtain desired damping ratio. Meanwhile, to avoid large imaginary parts of non-dominant poles, m_i should not be too large.

Figure 7.4b shows the track of the dominant poles of A as power filter time constant τ decreases from 1 to 1/200. When τ is too large, three pairs of dominant poles (($\lambda_{1,2}$, $\lambda_{3,4}$ and $\lambda_{5,6}$)) is exceptionally near the imaginary axis. It causes the system effortless to oscillate. Since the large time constant τ makes $(s/(\tau s + 1))$ impossible to be approximated as a differentiator, it affects the damping effect of the introduced control method. As τ decreases, $(s/(\tau s + 1))$gets closer to the differentiator and ($\lambda_{1,2}$, $\lambda_{3,4}$ and $\lambda_{5,6}$) leave away from the imaginary axis. In this way, the damping is increased. Besides, the design of τ determines the cut-off frequency of the filter. In the case of the unbalance loads and grid voltage, the instantaneous output power P_i pulses at twice the line frequency ($2f_{line}$) [8]. To reduce the impact of these double-line-frequency ripples caused by it, the filter's cut-off frequency must be much less than ($2f_{line}$) [9]. Combing the modal analysis above, the cut-off frequency should also be designed greater than the oscillation frequency. Thus,

$$2\pi f_{low} < \frac{1}{\tau} < 4\pi f_{line} \tag{7.25}$$

where f_{line} and f_{low} represent the line frequency and the oscillation frequency, respectively.

Based on the analysis of the impact of m_i and τ on the system stability, a comprehensive design of m_i and τ is required. Thus, the relationship between m_i, $1/\tau$ and the minimum damping ratio of the system eigenvalues is presented in Fig. 7.4c. In Fig. 7.4c, the negative damping ratio means the system is unsTable According to the constraint (7.25) and the desired damping ratio of the system, the ranges of mi and τ can be obtained from Fig. 7.4c.

Table 7.2 Participation analysis of the system

	Eigenvalue	Oscillation frequency/Hz	Damping ratio	Mainly related state variable
Conventional VSG method	$\lambda_{1,2}$	1.6	0.05	$\Delta\omega_2$, $\Delta\delta_2$, $\Delta\omega_4$, $\Delta\delta_4$
	$\lambda_{3,4}$	2.2	0.09	$\Delta\omega_1$, $\Delta\delta_1$
	$\lambda_{5,6}$	3.9	0.10	$\Delta\omega_3$, $\Delta\delta_3$
Introduced method	$\lambda_{1,2}$	1.6	0.22	$\Delta\omega_2$, $\Delta\delta_2$, $\Delta\omega_4$, $\Delta\delta_4$
	$\lambda_{3,4}$	2.4	0.33	$\Delta\omega_1$, $\Delta\delta_1$
	$\lambda_{5,6}$	5.1	0.37	$\Delta\omega_3$, $\Delta\delta_3$

Fig. 7.4 **a** Dominant poles of **A** as m_i increases from 0 to 0.8, **b** Dominant poles of **A** as τ decreases from 1 to 1/200, **c** the relationship between m_i, $1/\tau$ and minimum damping ratio of the system eigenvalues

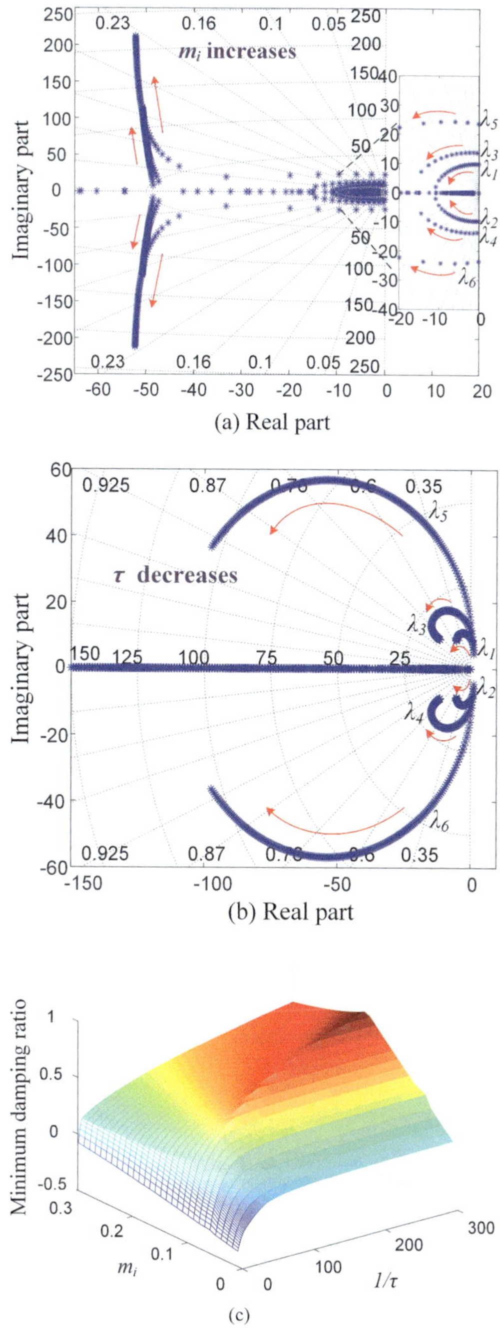

7.3 Simulation Results

To verify the availability of the method introduced in this chapter, simulation results implemented are presented in this section. Parameters used in simulations are summarized in Table 7.3. Figure 7.5 depicts the schematic of the three-phase simulated system.

7.3.1 *Comparison with Conventional VSG Control*

In this case, the simulation results of active power and frequency are presented. The system loads at t = 1 s and unloads at t=5 s. As can be seen in Fig. 7.6, the serious oscillation between all sources is fully reflected in the output active power and frequency, when the inverters employ conventional VSG control. However, when the frequency dynamic performance is poor (such as big frequency deviation), it makes the system at risk of unexpected load-shedding or extensive blackouts.

As observed in Fig. 7.7a, when the inverters employ the introduced VSG control method, the power oscillation has been deeply restrained. Figure 7.7b, c show system

Table 7.3 Parameters of simulated test system

Symbol	Value
Reference frequency f	50 Hz
Reference voltage V	311 V
Reference active power P	8 kW
Reference reactive power Q	2 kvar
Load power P_T	7.2 kW
Damp coefficient D_1, D_2, D_3, D_4	900
Inertia coefficient J_1, J_2, J_3, J_4	300/150/250/800
Voltage droop coefficient n_1, n_2, n_3, n_4	0.001
Mutual damping coefficient m_r	0.25
low-pass filter parameter τ	1/200

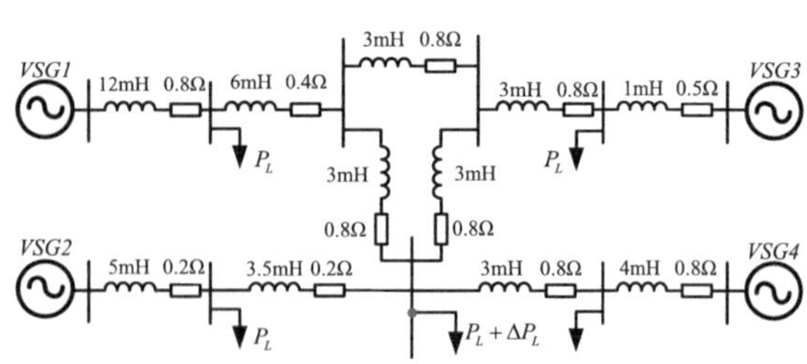

Fig. 7.5 The four-VSG three-phase simulated test system

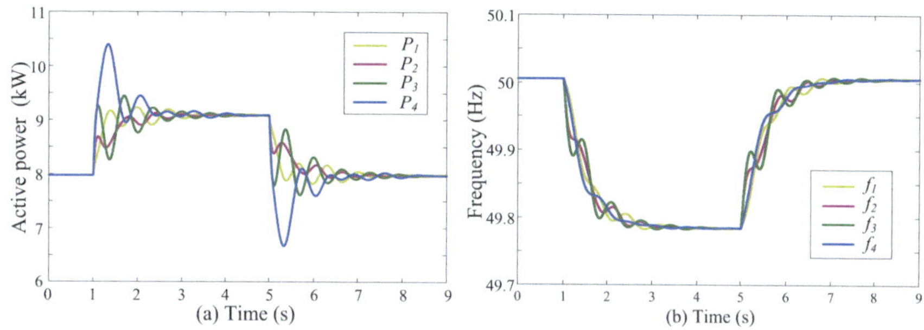

Fig. 7.6 Simulation results of active power and frequency with conventional VSG control

frequency and the additional term $m_i(d\bar{P}_i/dt)$ after a limiter, respectively. For simplicity, we define $e_i = m_i(d\bar{P}_i/dt)$. From Fig. 7.7b, a sudden load change occurs at 1 s. After about 1.04 s, $m_i(d\bar{P}_i/dt)$ reflects the frequency difference. e_1 and e_4 with relatively large angular frequency (ω_1, ω_4) is positive and thus decreases its angular frequency. At the same time, the additional term $e_2$and e_3 with relatively small angular frequency (ω_2, ω_3) is negative, and thus increase its angular frequency. As all angular frequencies get closer, the additional term e_i of all VSGs gets smaller. After about 1.3 s, when ω_3 is the minimum angular frequency, e_3 becomes negative. At the same time, e_1, e_2, and e_4 become positive. It is observed that the additional term $m_i(d\bar{P}_i/dt)$ approximately reflects the difference between its frequency and other's frequency, as (7.6) describes. In this way, the additional term helps promote frequency consistency and improve frequency dynamic performance.

For quantity analysis, we define percentage overshoot $PO = 100\% \times |\Delta Z|/Z$, where Z represents the steady-state value of output active power and ΔZ represents the maximum power deviation from the steady-state value. Table 7.4 shows the comparative quantity analysis of conventional VSG and the introduced method, taking VSG4 as an example (the blue line). The comparison indicators include transient time and the percentage overshoot (PO). From Table IV, the introduced method effectively reduces the transient time from 4s to 3s, and makes the system reach the steady-state faster. Besides, it decreases the PO. The PO value of the active power of VSG4 with the introduced method is 9.3%, and is reduced by 5.8% compared with the one that employs the conventional VSG method. Therefore, the method introduced in this chapter achieves power and frequency oscillation suppression. Moreover, the additional decentralized mutual damping term only works during the transient process and returns to zero at a steady state. Consequently, the power-sharing effect is not affected.

For further comparison, the frequency deviation $f_i - f_j$ during the transient process is displayed in Fig. 7.8. The full line represents the frequency deviation with the introduced method. From Fig. 7.8, the frequency deviation reaches up maximally to 0.06 Hz when the inverters employ the conventional VSG control method. However, when the inverters employ the introduced VSG control method, the frequency deviation drops maximally to 0.025 Hz. Thus, the introduced method helps promote frequency consistency in the dynamic process.

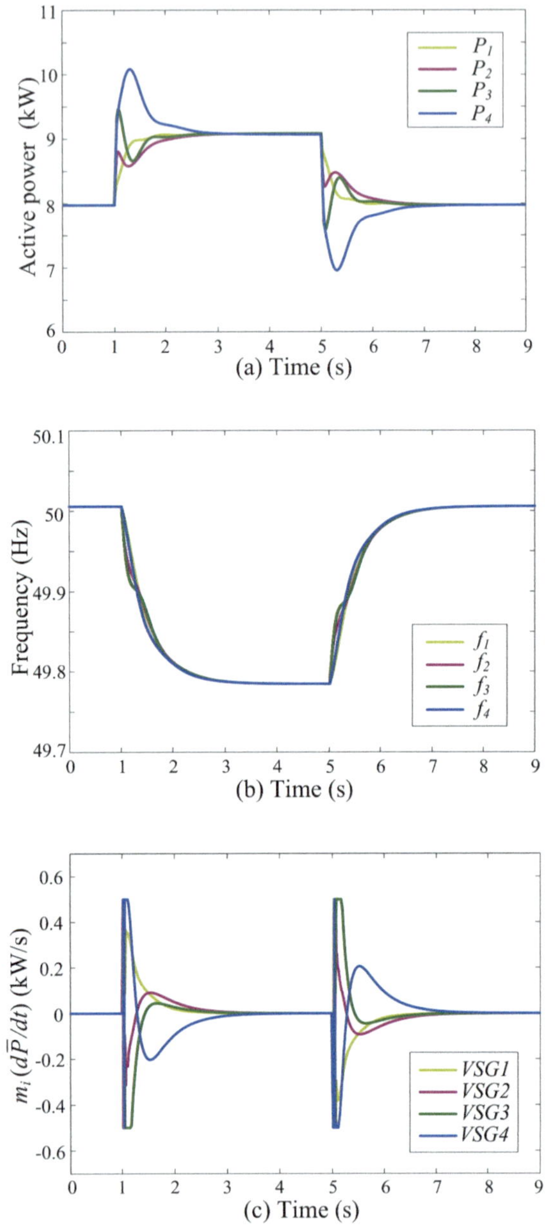

Fig. 7.7 Simulation results of active power, frequency and $m_i(dP/dt)$ with the introduced decentralized mutual damping method

Table 7.4 Quantity analysis comparisons of active power of VSG4

Comparison term	Conventional VSG	Introduced method
Transient time (s)	4	3
PO (%)	15.1	9.3

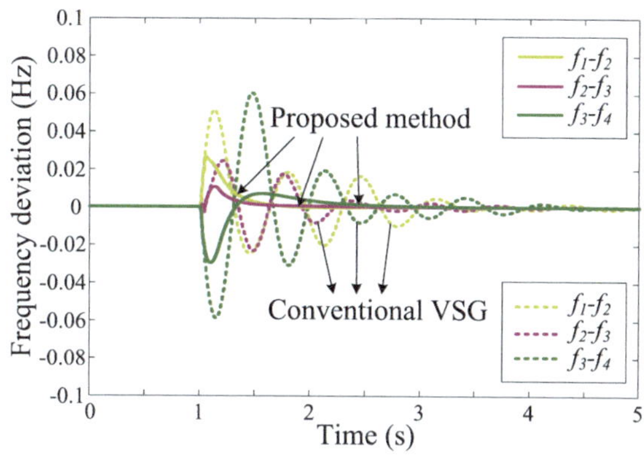

Fig. 7.8 Comparison of the frequency deviation with the conventional VSG method (dashed line) and the introduced method (full line)

7.3.2 *Performance Under Short-Circuit Fault*

This case studies the transient stability of the introduced method during a short circuit. Figure 7.9 shows the output active power of the system with the conventional VSG method (a) and the introduced method (b), respectively. A three-phase short circuit fault happens at 1 s, and is removed at 1.5 s. It is seen that the transient energy will diminish quicker than that with the ordinary VSG strategy. The simulation results are consistent with theoretical analysis.

7.3.3 *Comparison with Different Mutual Damping Coefficients*

The influence of key control parameter m_i on the dynamic performance is analyzed in this case. Three different sets of m_i are applied: without mutual damping term m_i=0, small mutual damping coefficient m_i=0.05, big mutual damping coefficient m_i=0.25. To observe the comparison effect, only the waveforms of VSG3 (power and frequency) are presented. In Fig. 7.10a, it is seen that the suppression effect of power oscillation gets better when m_i increases. n like manner, the setting time is significantly reduced. As displayed in Fig. 7.10b, frequency oscillation is suppressed preferably. Simultaneously, a smaller frequency deviation is obtained.

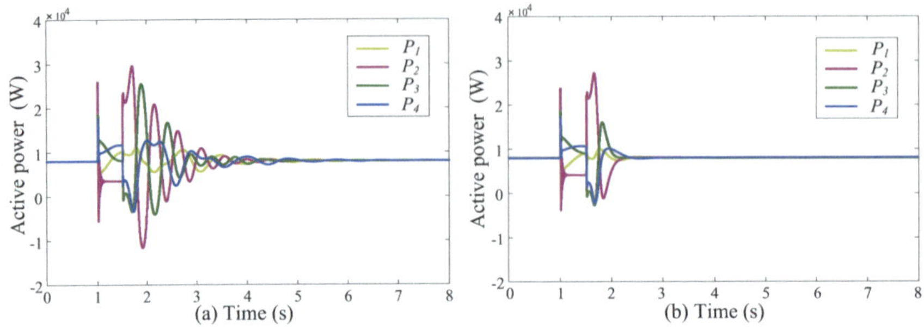

Fig. 7.9 Comparison results under short-circuit fault

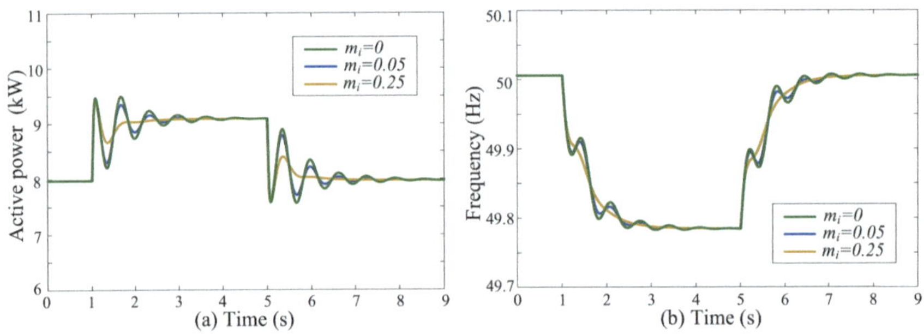

Fig. 7.10 Simulation results with different mutual damping coefficients (m_i)

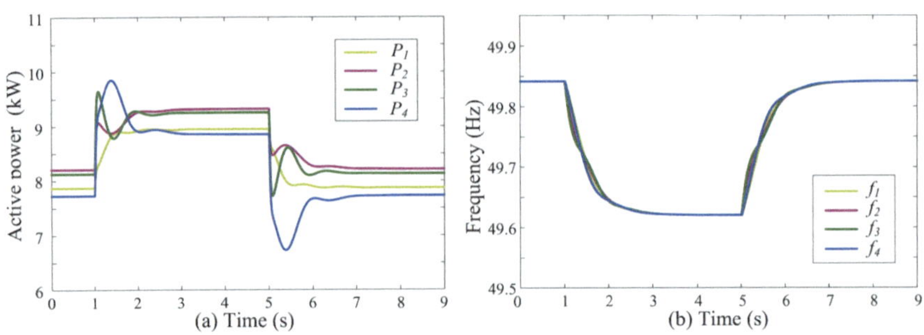

Fig. 7.11 Simulation results with the centralized method [10, 11] under time delay T_d=50 ms

7.3.4 Comparison with Centralized Control Method

The purpose of this part is to verify the advantages of the introduced method compared with the centralized method [5, 12]. In [5, 12], COI frequency is taken as a reference frequency and a PID controller is introduced to control the input power of VSGs to reduce oscillations of VSGs. Figure 7.11 shows active power (a) and system frequency (b) under time delay T_d=50 ms (latency must be considered for systems

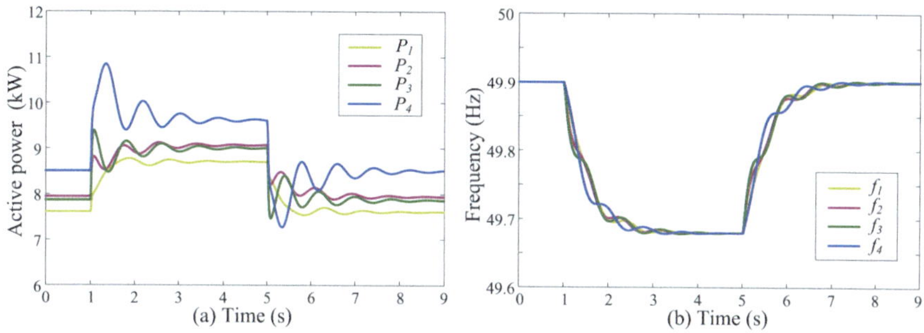

Fig. 7.12 Simulation results with the centralized method [5, 12] under single link failure

Table 7.5 Comparison with the current methods

Methods	Advantages	Disadvantages
Virtual impedance method [10, 11, 15]	Suppress power oscillations	Need exact line impedance information
	Without communication	Designed only for star-shaped topologies
Centralized [5, 12]/distributed [13, 16] methods	Suppress power oscillations	Communication is required
	Effective for any topologies	Increased costs, and performance deteriorates in the case of communication delay and faults
Decentralized method [14, 17]	Suppress power oscillations	Parameter design is complicated
	Without communication	
Introduced method	Suppress power oscillations Without communication Effective for any topologies	

with communication requirements). Compared with Fig. 7.6, it has a good effect on suppressing power oscillations. However, the integral term $(q_{is} \int (\omega_i - \omega_{COI})dt)$ makes the power no longer shared proportionally.Besides, the effect of power oscillation suppression deteriorates under single link failure as Fig. 7.12 shows. Any other method that requires communication has the problem, such as distributed method [4, 13]. Moreover, as seen in Table 7.5, the introduced method has obvious advantages.

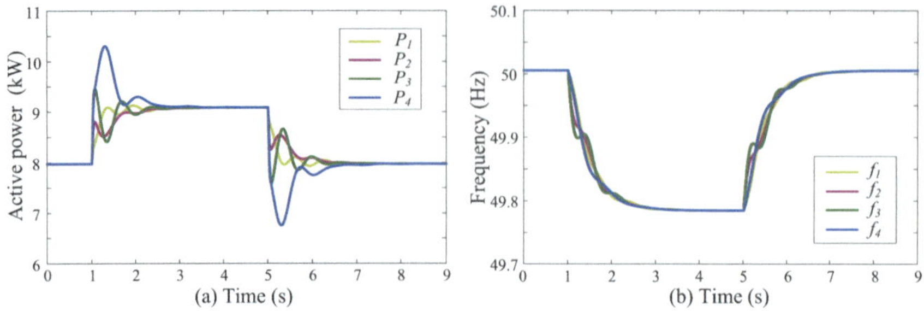

Fig. 7.13 Simulation results of active power and frequency with the decentralized method [14]

7.3.5 Comparison with Decentralized Control Method

In [14], a decentralized supplementary PSS method is introduced. The angular frequency deviation $\Delta\omega$ is introduced to the reactive power control loop through the transfer function $T_{pss}(s) = K_{pss}(\frac{1+sT_2}{1+sT_1})(\frac{1+sT_4}{1+sT_3})$ as an additional damping signal. The waveforms of power and frequency have been shown in Fig. 7.13. It also has a good effect of suppressing power oscillation. However, the supplementary feedback term $T_{pss}(s)$ contains many parameters that increases the difficulty of parameter design.

7.4 Experimental Results

The experimental setup is displayed in Fig. 7.14. It is set up to confirm the viability of the method introduced in this chapter. The test system incorporates three parallel single-stage voltage source inverters and a public load. The experimental parameters are listed in Table 7.6.

7.4.1 Conventional VSG Method

Figure 7.15 displays the waveform of the current of VSG1 when the load active power is increased from 435 W to 1020 W. It is seen that the system current oscillates after increasing the load. Figure 7.16 shows the active power (a) and system frequency (b), respectively. The load demand has an increase around 5 s. With the conventional VSG control method, power oscillations and frequency oscillations among DGs occur.

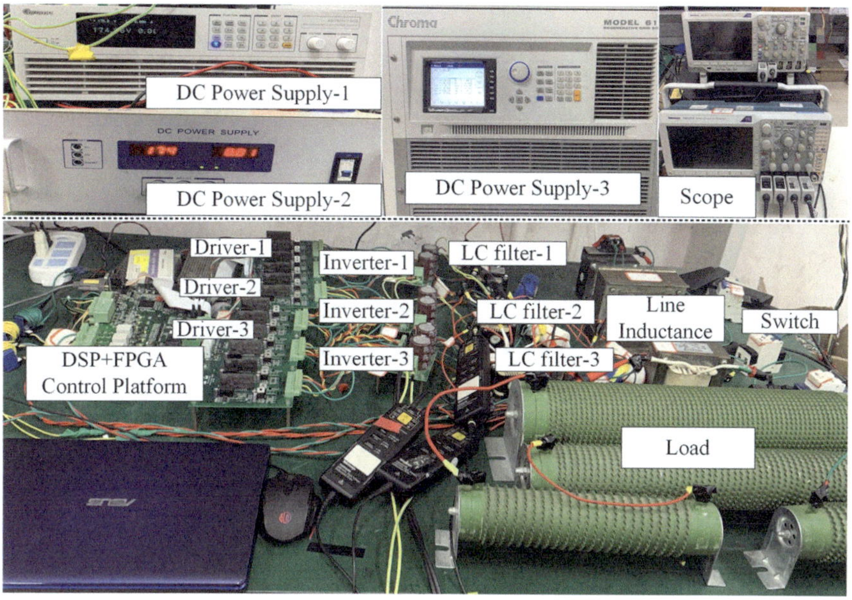

Fig. 7.14 Experimental prototype

Table 7.6 Parameters of experimental system

Parameters	Values	Parameters	Values
Nominal voltage V^*	110 V	Damping coefficient $D_1/D_2/D_3$	380
Line inductance L_{line}	3 mH	Inertia coefficient $J_1/J_2/J_3$	40/20/5
Filter inductance L_f	1.5 mH	Filter capacitor C_f	20 μF
m_r	0.05	τ	1/20

7.4.2 Decentralized Mutual Damping Control Method

To confirm the viability of the method introduced in this section, the key coefficient m_i is set as 0.05. Figure 7.17 shows the experimental waveforms of current when the system goes to a steady state. Compared with Fig. 7.15, the oscillation on current is suppressed and the transient time of the system is greatly reduced with the introduced method. Additionally, the viability of the introduced strategy can be observed by comparing Figs. 7.16 and 7.18, i.e. the waveforms of active power and frequency.

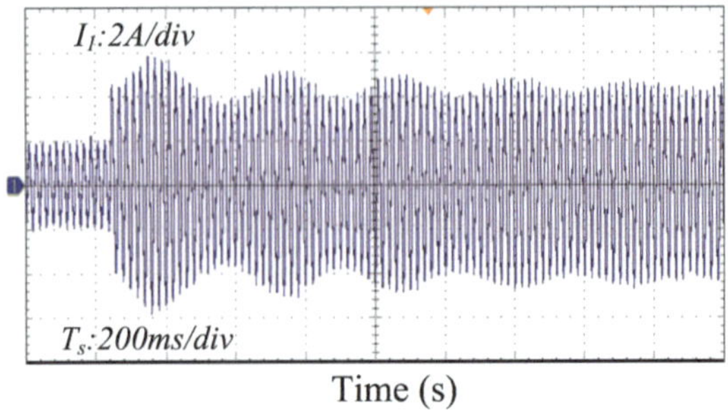

Fig. 7.15 Waveform of current of VSG1 with the conventional VSG method

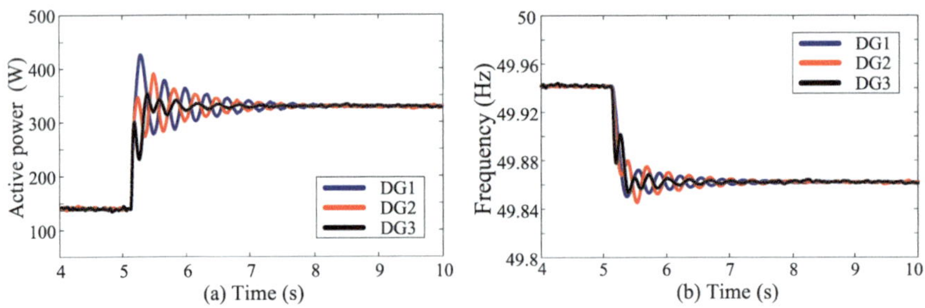

Fig. 7.16 Experimental results of active power and frequency with the conventional VSG method

Fig. 7.17 Waveform of current of VSG1 with the introduced VSG method

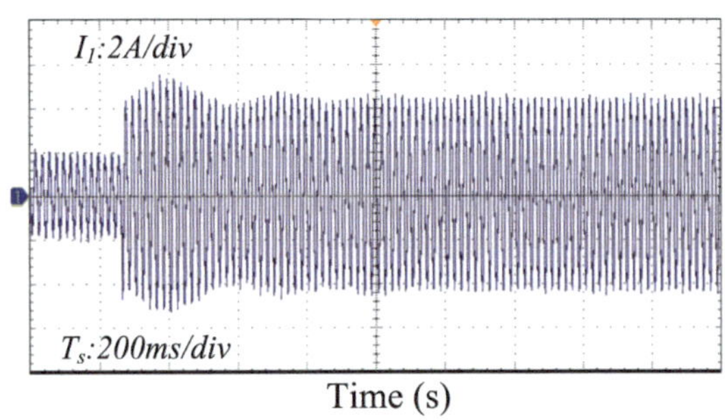

7.5 Conclusion

To deal with the power oscillation problem that occurs in multiple VSGs, a fully decentralized mutual damping method is introduced. It indirectly obtained the difference between each angular frequency by introducing the derivative of the local power. In this way, through this method, frequency dynamic performances can be

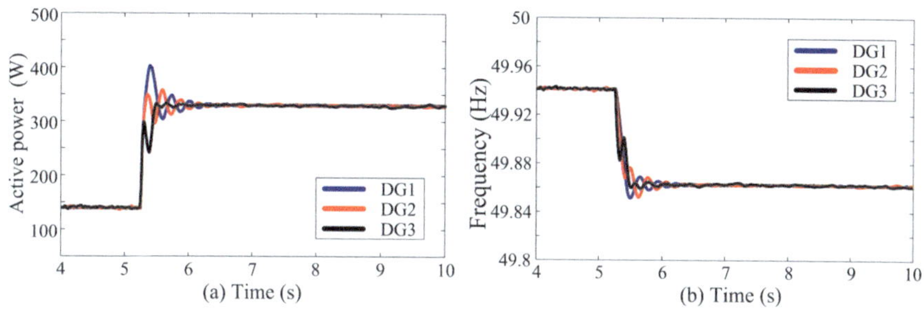

Fig. 7.18 Experimental results of active power and frequency with the introduced VSG method

improved and power oscillations can be suppressed. Besides, the power-sharing effect is not affected at all. The transient stability is verified by the Lyapunov stability theory, and the parameter design guideline is provided. Finally, the simulation results and experiment results have demonstrated the effectiveness of the introduced mutual damping control method.

References

1. J. Liu, Y. Miura, T. Ise, Comparison of dynamic characteristics between virtual synchronous generator and droop control in inverter-based distributed generators. IEEE Trans. Power Electronics **31**(5), 3600–3611 (2015)
2. B.K. Poolla, S. Bolognani, F. Dörfler, Optimal placement of virtual inertia in power grids. IEEE Trans. Autom. Control **62**(12), 6209–6220 (2017)
3. F.L. Lewis, H. Zhang, K. Hengster-Movric, A. Das. *Cooperative Control of Multi-agent Systems: Optimal and Adaptive Design Approaches*. Springer Science & Business Media (2013)
4. V. Nasirian, S. Moayedi, A. Davoudi, F.L. Lewis, Distributed cooperative control of DC microgrids. IEEE Trans. Power Electronics **30**(4), 2288–2303 (2015)
5. M. Choopani, S.H. Hosseinian, B. Vahidi, New transient stability and LVRT improvement of multi-VSG grids using the frequency of the center of inertia. IEEE Trans. Power Syst. **35**(1), 527–538 (2020)
6. J. Machowski, Z. Lubosny, J. W. Bialek, J. R. Bumby, *Power System Dynamics: Stability and Control*. John Wiley & Sons (2020)
7. J. M. Guerrero, L. G. d. Vicuna, J. Matas, M. Castilla, J. Miret, A wireless controller to enhance dynamic performance of parallel inverters in distributed generation systems. IEEE Trans. Power Electronics **19**(5), 1205–1213 (2004)
8. X. Li et al., Power management unit with its control for a three-phase fuel cell power system without large electrolytic capacitors. IEEE Trans. Power Electronics **26**(12), 3766–3777 (2011)
9. H. Wu et al., Small-signal modeling and parameters design for virtual synchronous generators. IEEE Trans. Ind. Electronics **63**(7), 4292–4303 (2016)
10. J. Liu, Y. Miura, H. Bevrani, T. Ise, Enhanced virtual synchronous generator control for parallel inverters in microgrids. IEEE Trans. Smart Grid **8**(5), 2268–2277 (2017)
11. M. Mao, C. Qian, Y. Ding, Decentralized coordination power control for islanding microgrid based on PV/BES-VSG. CPSS Trans. Power Electronics Appl. **3**(1), 14–24 (2018)
12. M. Choopani, S.H. Hosseinain, B. Vahidi, A novel comprehensive method to enhance stability of multi-VSG grids. Int. J. Electrical Power Energy Syst. **104**, 502–514 (2019)

13. M. Shi, X. Chen, J. Zhou, Y. Chen, J. Wen, H. He, Frequency restoration and oscillation damping of distributed VSGs in microgrid with low bandwidth communication. IEEE Trans. Smart Grid **12**(2), 1011–1021 (2021)
14. P. Sun, J. Yao, Y. Zhao, X. Fang, J. Cao, Stability sssessment and damping optimization control of multiple grid-connected virtual synchronous generators. IEEE Trans. Energy Conversion. https://doi.org/10.1109/TEC.2021.3104348
15. Z. Shuai, W. Huang, Z.J. Shen, A. Luo, Z. Tian, Active power oscillation and suppression techniques between two parallel synchronverters during load fluctuations. IEEE Trans. Power Electronics **35**(4), 4127–4142 (2020)
16. Z. Luo, H. Han, S. Chen, Z. Wu, M. Shi, K. Zhang, Power oscillation suppression method based on distributed cooperative control of multi-VSG grids, in 2021 IEEE 12th Energy Conversion Congress & Exposition—Asia (ECCE-Asia) (2021), pp. 1875–1878
17. M. Zhao, H. Yin, Y. Xue, X.P. Zhang, Y. Lan, Coordinated damping control design for power system with multiple virtual synchronous generators based on prony method. IEEE Open Access J. Power Energy **8**, 316–328 (2021)

Chapter 8
Power Oscillation Suppression with Adaptive Virtual Impedance

8.1 Adaptive Virtual Impedance Control Method

8.1.1 Adaptive Virtual Impedance for Power Oscillation Suppression

As illustrated in Fig. 8.1, the virtual impedance is commonly employed to modify the output impedance of VSG. It drops the output voltage reference proportionally to the output current [2]. The output voltage reference can be expressed as

$$v_i = v_{ri} - Z_{vi} i_i \tag{8.1}$$

where $v_{ri} = V_{ri}\angle\delta_{ri}$ and i_i are output voltage and current of the i-th inverter, respectively. $v_i = V_i\angle\delta_i$ is the voltage reference produced by the i-th VSG control, and

$$Z_{vi} = R_{vi} + jX_{vi} \tag{8.2}$$

is the virtual impedance.

In this chapter, the role of virtual inductance is mainly considered. The schematic diagram of the introduced adaptive virtual impedance control algorithm is shown in Fig. 8.2. As [3] analyzed, the cause of power oscillation is due to different angular frequency acceleration $d\omega/dt$ during the transient process. Based on the finding, the adaptive virtual impedance is designed to facilitate the consistency of response curves of all frequencies and expressed as

$$Z_{vi} = jX_{vi} = j\left[X_{vi}{}^{*} - m_i \sum_{h\in N_i} (\omega_i - \omega_h)\right] \tag{8.3}$$

X. Hou et al., *Self-organized Operation and Stability Control of Electronized Power Networks*, https://doi.org/10.1007/978-981-96-9067-1_8

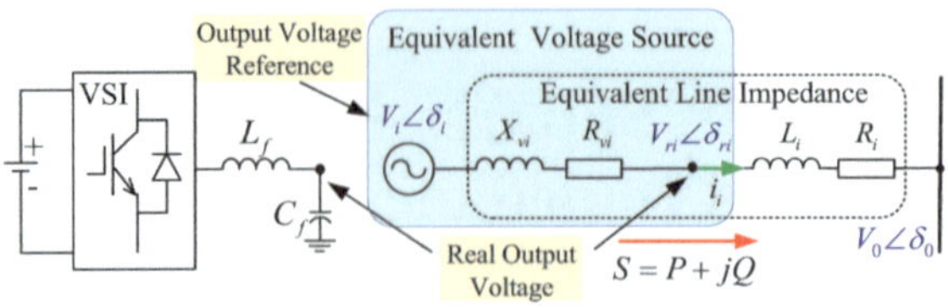

Fig. 8.1 Equivalent output voltage source considering virtual impedance

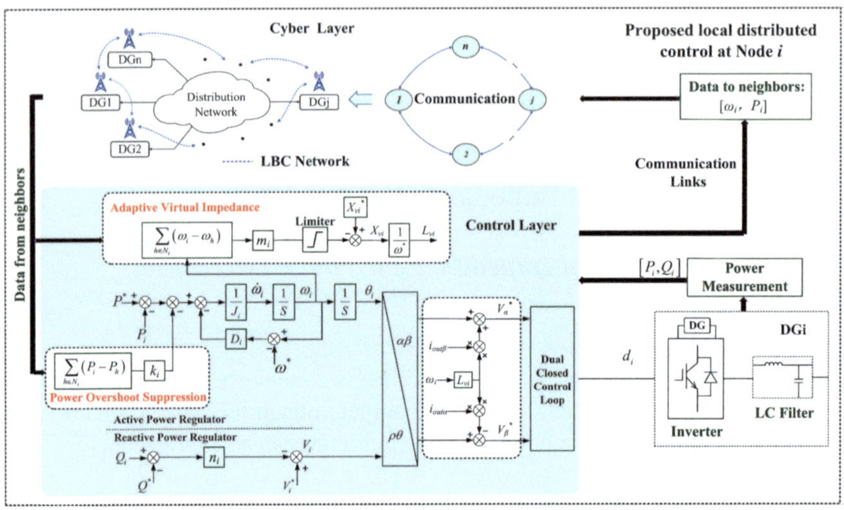

Fig. 8.2 Control block diagram

where X_{vi}^{*} is the initial virtual impedance and mi is the adaptive virtual impedance coefficient. N_i represents the set of the neighbors of the i-th inverter. The constant term of the virtual inductance is beneficial to realize power decoupling, suppress system circulating current, and reduce the influence of impedance imbalance [4, 5]. The adaptive term helps promote frequency consistency in the dynamic processes. It is observed that the reactive power sharing effect will not be affected by the adaptive term.

To illustrate the idea of adaptive virtual impedance for power oscillation suppression more distinctly, a two-VSG system is taken as an example as depicted in Table 8.1. The adaptive virtual impedance will affect the output power of each inverter. After load disturbance, the virtual impedance changes according to the frequency states. When $\omega_1 > \omega_2$, the virtual impedance X_{v1} of VSG1 decreases, and the virtual impedance X_{v2} of VSG2 increases. Then, the output power of VSG1 increases, and the output power of VSG2 decreases. The angular frequency change rate $d\omega/dt$

Table 8.1 Principle of adaptive virtual impedance

Frequency deviation X_{v1}	$\omega_i > \omega_s$	$\omega_i < \omega_s$
Virtual impedance X_{v1}	Decrease	Increase
Virtual impedance X_{v2}	Increase	Decrease
Input power P_{i1}	Increase	Decrease
Input power P_{i2}	Decrease	Increase
Change rate $d\omega_i/dt$	Decrease	Increase
Change rate $d\omega_s/dt$	Increase	Decrease

is proportional with negative sign to the output power, i.e. $d\omega_1/dt$ decreases and $d\omega_2/dt$ increases. In this way, the angular frequency of VSG1 will get close to the angular frequency of VSG2. On the contrary, when $\omega_1 < \omega_2$, the adaptive virtual impedance control law will also make the two frequencies closer together. The principle can also be extended to a multi-VSG grid. Thus, frequency dynamic response was effectively improved that further suppresses power oscillation. The total virtual impedance returns to X_{vi}^*, when the system reaches steady states.

8.1.2 Supplementary Control for Power Overshoot Suppression

Since the initial system parameters including line impedance, inertia coefficient and damping coefficient do not match, the uneven distribution of instantaneous power occurs that cause power overshoot [6]. To solve this problem, supplementary control schemes will be discussed. As shown in Fig. 8.2, a power correction term $\Delta P_i = -k_i(P_i - P_j)$ is added to maintain active power balance in the transient process. Thus, the modified VSG control can be expressed as

$$J_i\omega^*\frac{d\omega_i}{dt} = P^* - P_i - D_i\omega^*\left(\omega_i - \omega^*\right) - c_i\sum_{h\in N_i}(P_i - P_h) \tag{8.4}$$

where c_i is the power overshoot suppression coefficient. In this way, the active power of VSGs will move towards each other in the dynamic process.

8.2 Stability Analysis

A small-signal stability analysis has been carried out in this section.

8.2.1 Closed-Loop State-Space Model

Under the adaptive virtual impedance in (8.3), the electrical output power of i-th-VSG can be represented as

$$P_i = \frac{V_i V_0}{X'_{i0}} \sin(\delta_i - \delta_0) \tag{8.5}$$

where $X'_{i0} X_{i0} + X_{vi}$ and

$$X_{vi} = X_{vi}{}^* - m_i \sum_{h \in N_i} (\omega_i - \omega_h) \tag{8.6}$$

Thus, the electrical output power in the sense of small signal can be expressed as

$$\Delta P_i = k_{i0}(\Delta\delta_i - \Delta\delta_0) + a_{i0} \sum_{h \in N_i} m_i (\Delta\omega_i - \Delta\omega_h) \tag{8.7}$$

where $a_{i0} = \left(\frac{1}{X_{i0}}\right)^2 V_i V_0 \sin(\delta_i{}^\circ - \delta_0{}^\circ)$

The following equation can be obtained

$$\Delta\delta_0 = \frac{\sum_{i=1}^n a_{i0} \sum_{h \in N_i} (\Delta\omega_i - \Delta\omega_h) - \Delta P_{Load} + \sum_{i=1}^n k_{i0} \Delta\delta_i}{\sum_{i=1}^n k_{i0}} \tag{8.8}$$

Thus, substituting (8.8) into (8.7) yields

$$\begin{aligned} \Delta P_i &= k_{i0}\left(\Delta\delta_i - \frac{\sum_{i=1}^n (k_{i0}\Delta\delta_i)}{\sum_{i=1}^n k_{i0}}\right) + \frac{k_{i0}}{\sum_{i=1}^n k_{i0}} \Delta P_{Load} \\ &+ a_{i0} \sum_{h \in N_i} m_i (\Delta\omega_i - \Delta\omega_h) - \frac{k_{i0}}{\sum_{i=1}^n k_{i0}} \sum_{i=1}^n a_{i0} \sum_{h \in N_i} m_i (\Delta\omega_i - \Delta\omega_h) \end{aligned} \tag{8.9}$$

From the fourth term on the right hand of (8.9), ΔP_i obtains all angular frequency differences in the sense of small signal.

The small-signal model of (8.4) can be written as:

$$J_i \Delta\dot{\omega}_i = \Delta P_i{}^* - \Delta P_i - D_i \Delta\omega_i - k_i \sum_{h \in N_i} (P_i - P_h) \tag{8.10}$$

Table 8.2 Parameters applied in simulation

Line information		
	Resistance (Ω)	Inductance (mH)
Line 1 impedance	0.8	7
Line 2 impedance	0.2	4
Line 3 impedance	0.5	1
Line 4 impedance	0.8	4
Parameter value		
Reference value for frequency	f^*	50 Hz
Reference value for voltage	V^*	311 V
Reference value for active power	P^*	10 kW
Reference value for reactive power	Q^*	2 kVar
Virtual inductance	X_v	0.7Ω
Damp coefficient	D_1, D_2, D_3, D_4	800
Inertia coefficient of VSG1	J_1	300
Inertia coefficient of VSG2	J_2	150
Inertia coefficient of VSG3	J_3	250
Inertia coefficient of VSG4	J_4	800
voltage droop coefficient	n_1, n_2, n_3, n_4	0.001
Mutual damping coefficient	m_1, m_2, m_3, m_4	0.2

where we make a simplification that removes the non-critical constant ω^*.

The ordinary state-space equations take the form of $\dot{\mathbf{X}} = \mathbf{A}_{2n\times 2n}\mathbf{X} + \mathbf{B}_{2n\times 2n}\mathbf{U}$. By combining (8.9) and (8.10), the state-space equations of the system can be simplified as:

$$\begin{bmatrix} \Delta\dot{\omega} \\ \Delta\dot{\delta} \end{bmatrix} = \begin{bmatrix} \mathbf{JA}_{11} & \mathbf{JA}_{12} \\ \mathbf{A}_{21} & \mathbf{A}_{22} \end{bmatrix} \begin{bmatrix} \Delta\omega \\ \Delta\delta \end{bmatrix} + \begin{bmatrix} \mathbf{JB}_{11} & \mathbf{JB}_{12} \\ \mathbf{B}_{21} & \mathbf{B}_{22} \end{bmatrix} \begin{bmatrix} \Delta\mathbf{P}^* \\ \Delta\mathbf{P}_{\mathbf{L}} \end{bmatrix} \tag{8.11}$$

The selection of adaptive virtual impedance coefficient m_i plays an important role in obtaining the excellent dynamic response of the system. A system constitutive of two VSGs is examined to study the influence of parameters on eigenvalues of matrix $\mathbf{A}_{\mathbf{2n}\times\mathbf{2n}}$. In Table 8.2, the parameters of VSG1 and VSG2 are listed. Fig. 8.3a shows that the loci of eigenvalues of $\mathbf{A}_{\mathbf{2n}\times\mathbf{2n}}$ when the adaptive virtual impedance coefficient increases. When the adaptive virtual impedance coefficient is increased, the imaginary part of the eigenvalues reduces, and the real part increases. Obviously, the damping ratio of the system can be improved effectively by increasing the adaptive virtual impedance coefficient. To better illustrate the benefits of the introduced control approach over the traditional VSG method, Fig. 8.3b presents the loci of two different control methods when the inertia coefficient increases. The adaptive virtual impedance coefficient is set as 0.3 and 0.7 with the introduced method. Clearly,

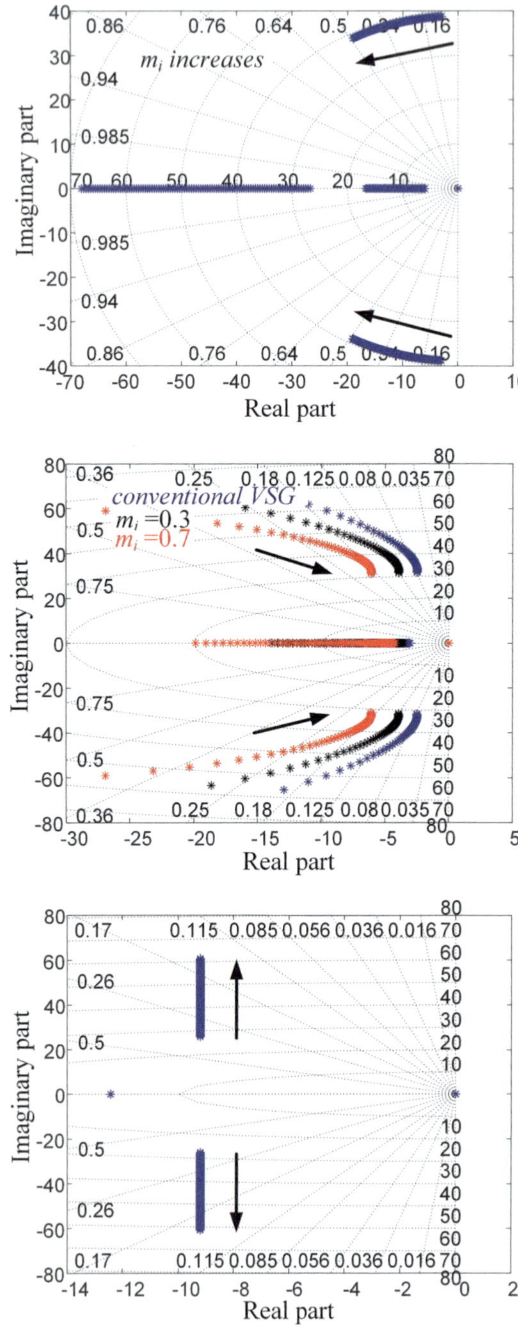

Fig. 8.3 Poles distributions with **a** adaptive virtual impedance coefficient m_i in [0, 1.6], **b** inertia coefficient J_1 of VSG1 in [20, 300], **c** power overshoot suppression coefficient c_i in [1,2]

the method effectively increases the damping of the system and suppresses power oscillation. As shown in Fig. 8.3c, the power overshoot suppression coefficient c_i has a significant effect on the eigenvalues. With the increase of the coefficient c_i, the eigenvalue moves away from the real axis, which increases the oscillation frequency and decreases the damping ratio of the system.

8.2.2 Parameters Design

In a general way, the damping coefficient D_i is designed as [7]

$$D_i \geq \frac{P_{\max} - P_{\min}}{\omega_{\max} - \omega_{\min}} \tag{8.12}$$

where $\omega_{\min}$ and $\omega_{\max}$ represent the minimum and maximum angular frequency. $P_{\min}$ and $P_{\max}$ represent the minimum and maximum active power. Besides, the damping coefficient should be designed to ensure power-sharing, i.e.

$$D_1 : D_2 : \cdots : D_n = P_1{}^* : P_2{}^* : \cdots : P_n{}^* \tag{8.13}$$

It should be noted that D_i should not be too large that will lead to a long settling time [2].

Conventionally, the design of the inertia coefficient J_i usually needs to meet the requirements of the system damping ratio [7–10]. In recent years, its design ideas go above the normal damping-to-inertia ratio hypothesis, and have taken into account more factors, such as the inertial requirements of the system, capacity constraints of the sources, and economic budget constraints [10, 11]. The system will oscillate when the inertia and damping coefficients do not match. Thus, the damping effect provided by the virtual impedance is explored in this study.

Since the virtual impedance causes voltage drop as shown in [4], the adaptive virtual impedance should not exceed its maximum setting value $X_{v\max}$. Thus, the following equation can be obtained

$$X_{vi}{}^* - m_i \sum_{h \in N_i} (\omega_i - \omega_h) \leq X_{v\max} \tag{8.14}$$

Further,

$$m_i \leq \frac{X_{v\max} - X_{vi}{}^*}{z\,(\omega_{\max} - \omega_{\min})} \tag{8.15}$$

where z denotes the amount of the neighbors of the local VSG. Based on the designed m_i, given the desired damping ratio of the system, the corresponding power overshoot suppression coefficient can be designed in terms of Fig. 8.3c.

8.3 Simulation Results

The introduced method was evaluated with simulations based on Matlab/Simulink to validate its effectiveness. Supplementary control for power overshoot suppression is discussed solely in Sect. 8.3.5. Figure 8.12 presents the schematic of the test system. Table 8.2 presents the parameters applied in the simulation.

8.3.1 Comparisons with Conventional VSG Control

A system consisting of four parallel VSGs has been tested as shown in Fig. 8.12. As loads change, the simulation results with the typical VSG strategy are shown in Fig. 8.4. After adding the introduced virtual impedance, the power and frequency waveforms can be seen in Fig. 8.5. Once the load increases rapidly at 1 s, as illustrated in Fig. 8.4a-b, longstanding oscillations in power and frequency of the system appeared. With the decreasing load at 6 s, analogous events occur. By contrast, the fluctuation of load does not induce power oscillations between VSGs when the adaptive virtual impedance is introduced, as illustrated in Fig. 8.5a. The maximum impact power of VSGs decreases to a certain extent. To quantify the ability to suppress overshoot, we define the percentage overshoot $PO = \frac{P_{\max} - P_s}{P_s} \times 100\%$, where $P_{\max}$ represents the maximum power after load disturbance and P_s represents the steady-state value of power. Through the comparative analysis of Figs. 8.4a and 8.5a, the PO of the introduced method is decreased by 7.9%. The effect of power overshoot suppression is limited, due to the initial mismatch parameters of the system including line impedance, inertia coefficient and damping coefficient that have been explained in detail in [6]. Besides, a good improvement in the frequency dynamic response is observed in Fig. 8.5b. In this way, the maximum frequency deviation and ROCOF are reduced which is friendly to the system. Simultaneously, as shown in Fig. 8.4c, because of the couplings of active and reactive power, active power oscillations will induce reactive power oscillations. Thus, the introduced method also helps suppress the reactive power oscillations as shown in Fig. 8.5c. By the way, the reactive power sharing performance is beyond the scope of this chapter.

8.3.2 Influence of Adaptive Impedance Coefficients

Four groups of parameters: m_i=0, m_i=0.03, m_i=0.16, and m_i=0.3 are applied to test the impact of the adaptive impedance coefficient. As an illustration of the effectiveness of the approach, the waveform of VSG3 is shown. In this manner, the diversification in active power and frequency can be observed obviously, when mi changes. An

Fig. 8.4 Simulation results with conventional VSG control. **a** Active power, **b** Frequency, **c** Reactive power

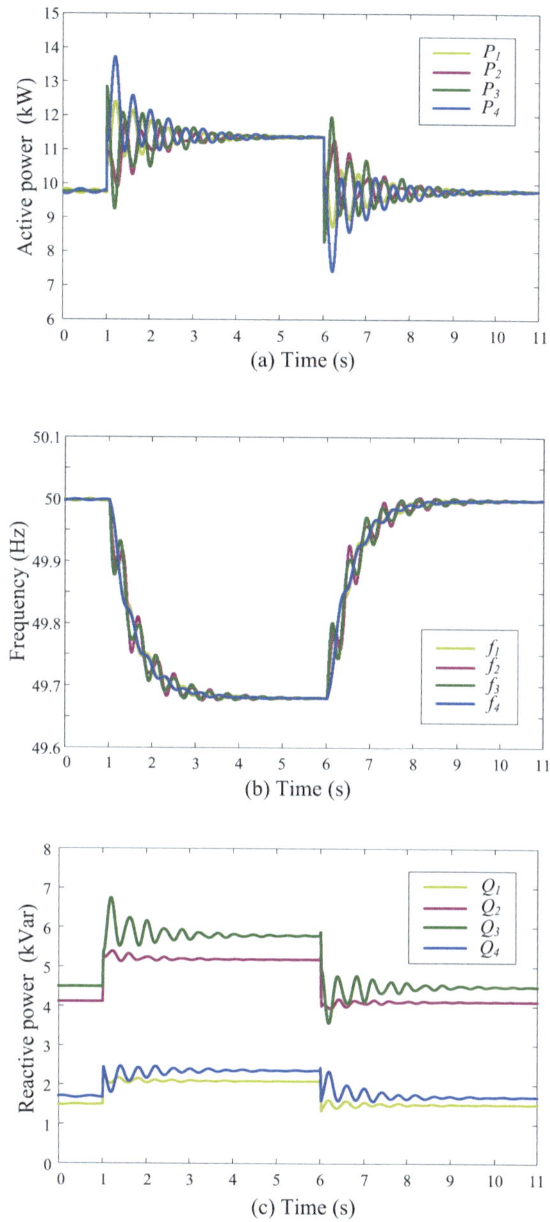

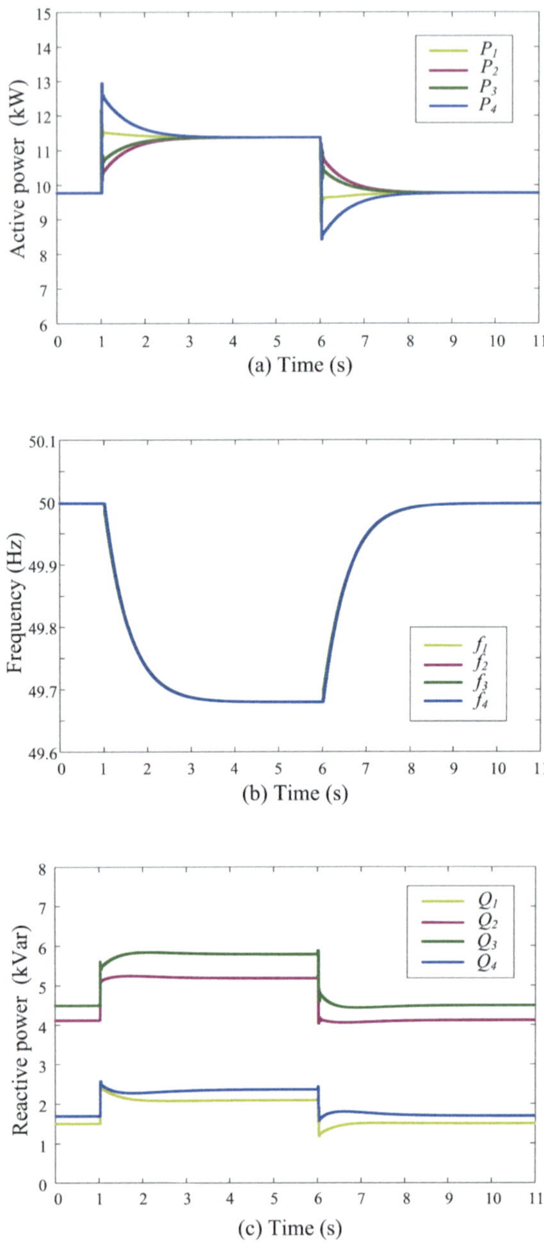

Fig. 8.5 Simulation results with the introduced method. **a** Active power, **b** Frequency, **c** Reactive power

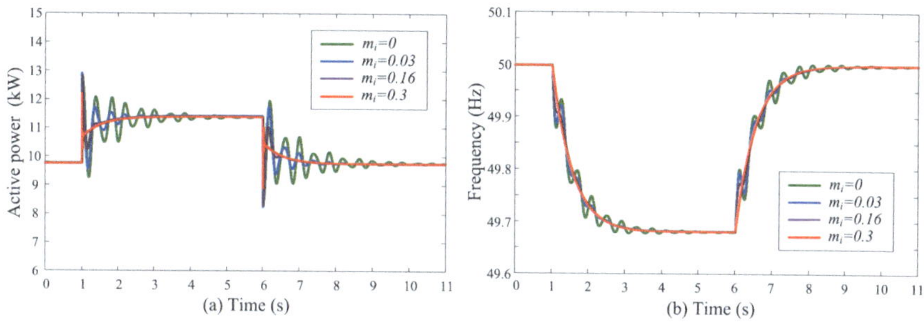

Fig. 8.6 Simulation results with m_i=0, 0.03, 0.16 and 0.3. **a** Active power, **b** Frequency

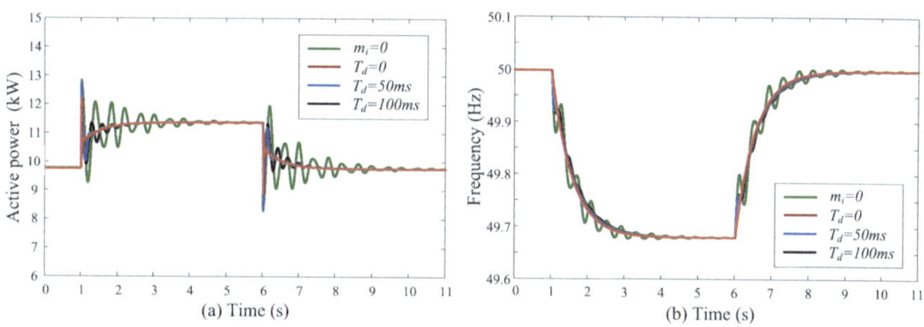

Fig. 8.7 Simulation results with T_d=0, 50 ms and 100 ms. **a** Active power, **b** Frequency

increase in m_i makes an improvement in power oscillation suppression, as Fig. 8.6a shows, and a reduction in frequency deviation is achieved, as Fig. 8.7b shows. As a consequence, the dynamic response for frequency is improved and the attenuation speed to reach a steady state is increased.

8.3.3 Influence of Time Delay

The influence of communication delay of the adaptive virtual impedance strategy is extraordinarily crucial that is analyzed as shown in Fig. 8.7. In this case, the adaptive virtual impedance coefficient m_i is set as 0.2. When communication is not delayed, the power oscillation effect of the adaptive virtual impedance is best. With the increase of the communication delay, the power oscillation suppression effect decreases to some extent. Nevertheless, the power oscillation can be effectively suppressed. It shows that the communication delay is allowed over an extensive range. Facing different purposes of making use of the controller, the allowable time delay is different. Since the introduced method only needs a simple computation, $T_d < 100$ ms is acceptable.

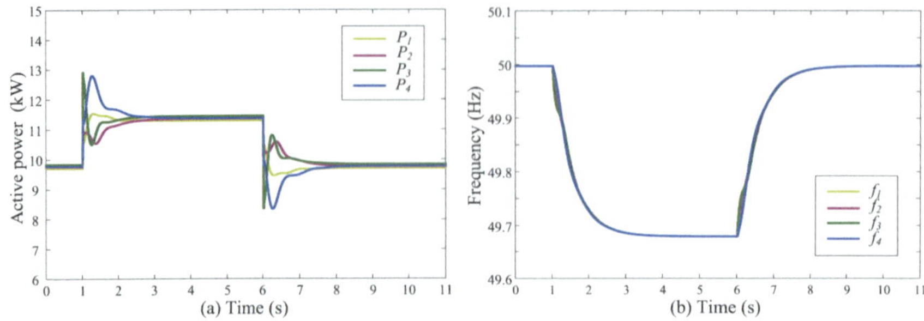

Fig. 8.8 Simulation results with the centralized method [12] under normal circumstance. **a** Active power, **b** Frequency

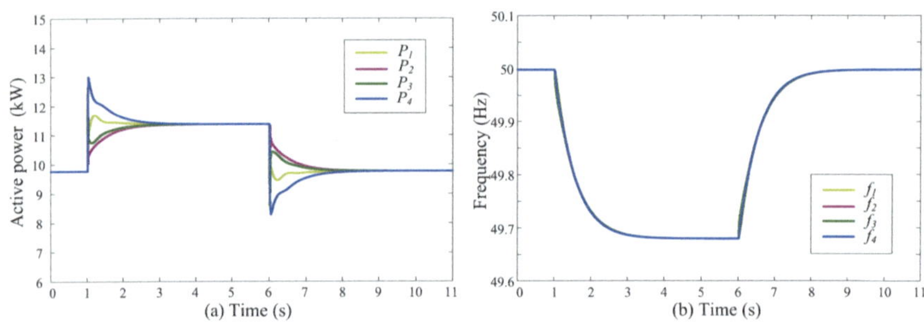

Fig. 8.9 Simulation results under single link failure with the introduced distributed method. **a** Active power, **b** Frequency

8.3.4 Comparisons with the Centralized Method

As shown in Fig. 8.8, under normal circumstances, the centralized method [12] effectively suppresses power oscillations. However, the integral parameter affects power-sharing between multiple VSGs. This deficiency also occurs in [13]. In comparison, the introduced approach has no effect on the system's steady-state performance, as Fig. 8.5 shows. In addition, the reliability of the two methods is compared. Simulation results of the introduced distributed method and the centralized method are shown in Figs. 8.9 and 8.10, respectively, when the test system suffers from single communication link failure. Distinctly, the time delay T_d has little effect on the impression of the introduced approach that shows it possesses higher reliability.

8.3.5 Supplementary Control for Overshoot Suppression

As can be seen from Figs. 8.4a and 8.5a, there exists a large power overshoot during load fluctuations. To solve this problem, supplementary control schemes will be

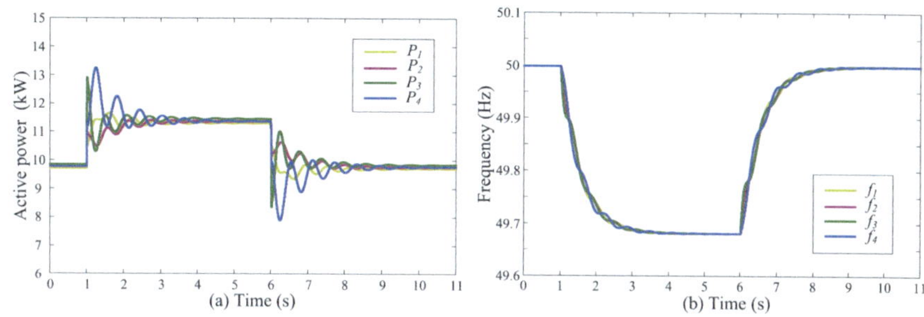

Fig. 8.10 Simulation results under single link failure with the centralized method [12]. **a** Active power, **b** Frequency

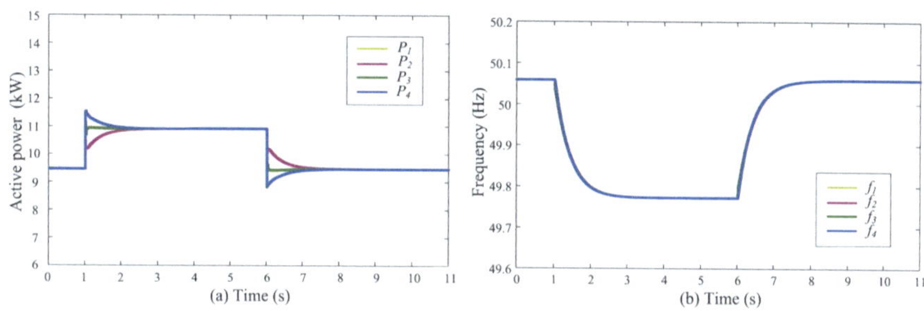

Fig. 8.11 Simulation results with the supplement control. **a** Active power, **b** Frequency

discussed. Parameter mismatch is the main reason for instantaneous active power imbalance [6]. Since increasing the initial virtual impedance X_{vi}^* can reduce the impedance mismatch, a larger X_{vi}^* will help reduce the power overshoot. Besides, a power correction term $\Delta P_i = -c_i(P_i - P_j)$ is also an effective solution to maintain active power balance. Simulation results with the supplementary method are shown in Fig. 8.11a, b, respectively. As seen, the PO is reduced to 3.6%. Compared with the result in Fig. 8.4a, it can be concluded that the supplementary method is very effective to reduce the power overshoot.

8.4 Hardware-In-Loop Results

In this part, HIL tests are presented. The physical circuits are realized by the simulator OP4510. The real configuration of equipment and schematic of the test system are shown in Fig. 8.12. The parameters applied are the same as simulation parameters.

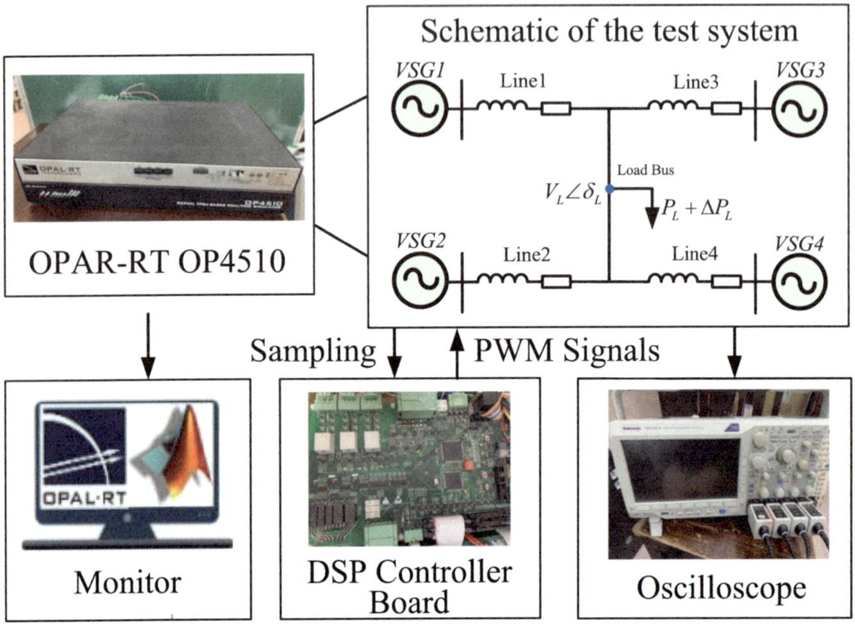

Fig. 8.12 The real configuration of equipment and schematic of the test system

8.4.1 Conventional VSG Control Method

Figure 8.13 shows the HIL results with conventional VSG control when loads fluctuate. In Fig. 8.13b, the frequency is set as $f_i^{'} = (f_i - 50)$, so that the dynamic response can be observed distinctly. When the load increases at t=12 s or drops at t=28 s abruptly, severe power oscillation and frequency oscillation exists inside the VSGs. It would make the system prone to instability. Simultaneously, because of the active and reactive power couplings, reactive power oscillations occur.

8.4.2 Adaptive Virtual Impedance Method

In this case, m_i is set as 0.2. HIL results of active power and frequency with the adaptive virtual impedance method have been displayed in Fig. 8.14 that consist of four VSGs. After introducing the virtual impedance, the oscillation attenuation speed is increased significantly. The analysis based on the introduced approach is successfully validated by these findings.

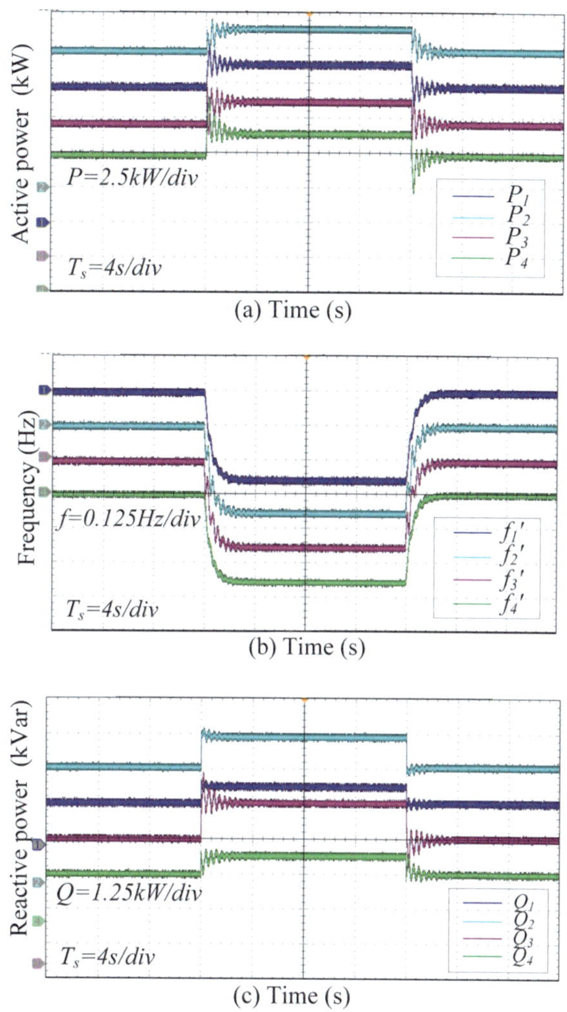

Fig. 8.13 HIL results with conventional VSG method. **a** Active power, **b** Frequency, and **c** Reactive power

8.5 Conclusion

This section presents an adaptive virtual impedance method applied for multiple parallel VSG. The following are the superiorities of the section: (1) For the power oscillation suppression, a distributed adaptive virtual impedance technique is presented. In addition to achieving power decoupling in the case of any transmission impedance, it gives aid to reducing frequency deviation that contributes to the improvement of frequency dynamic performance. (2) The simulation results verify that a wide communication delay range is allowed. A final proof of the effectiveness of the introduced control method is shown by simulation and HIL results.

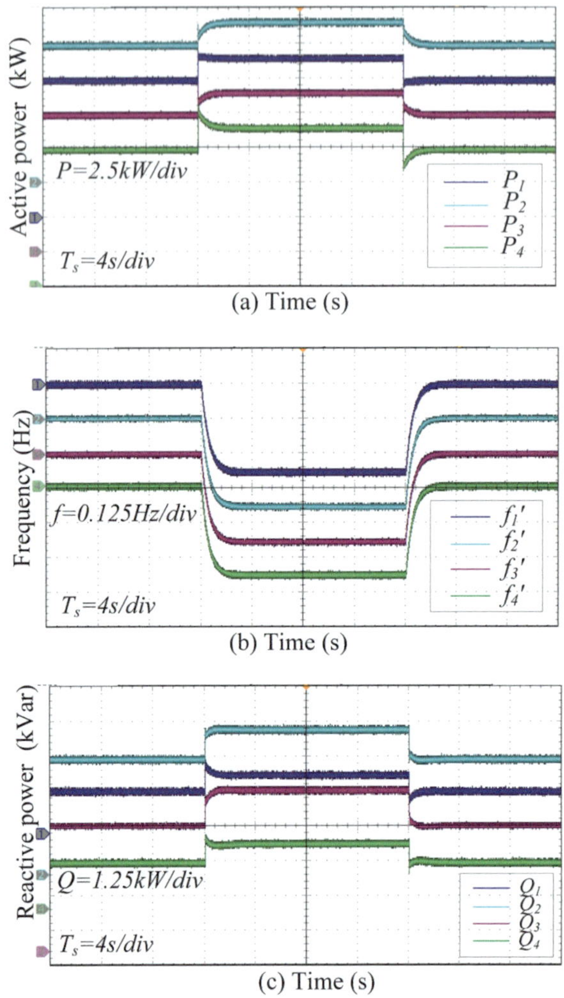

Fig. 8.14 HIL results with the introduced method. **a** Active power, **b** Frequency, and **c** Reactive power

Appendix

The matrix equation of the multi-VSG grid is

$$\begin{bmatrix} \Delta\dot{\omega} \\ \Delta\dot{\delta} \end{bmatrix} = \begin{bmatrix} \mathbf{JA}_{11} & \mathbf{JA}_{12} \\ \mathbf{A}_{21} & \mathbf{A}_{22} \end{bmatrix} \begin{bmatrix} \Delta\omega \\ \Delta\delta \end{bmatrix} + \begin{bmatrix} \mathbf{JB}_{11} & \mathbf{JB}_{12} \\ \mathbf{B}_{21} & \mathbf{B}_{22} \end{bmatrix} \begin{bmatrix} \Delta\mathbf{P}^{*} \\ \Delta\mathbf{P}_{\mathbf{Load}} \end{bmatrix} \tag{8.16}$$

where the diagonal elements of the n × n diagonal matrix J are $j_{ii} = 1/J_i$, and the diagonal elements of the n×n diagonal matrix $\mathbf{D}$ are $d_{ii} = D_i$.

In (8.16)

$$\begin{cases} \mathbf{A_{11}} = (\mathbf{I} - \mathbf{CL}) \cdot \mathbf{S} \cdot \mathbf{R} \cdot \mathbf{Q} - (\mathbf{I} - \mathbf{CL}) \mathbf{R} \cdot \mathbf{F} - \mathbf{D} \\ \mathbf{A_{12}} = (\mathbf{I} - \mathbf{CL}) \cdot \mathbf{W} \cdot \mathbf{T} \end{cases} \tag{8.17}$$

where $\mathbf{I_{n \times n}}$ and $\mathbf{L_{n \times n}}$ are a unit matrix and a Laplace matrix, respectively. $\mathbf{S_{n \times n}}$, $\mathbf{R_{n \times n}}$, $\mathbf{C_{n \times n}}$ and $\mathbf{W_{n \times n}}$ are diagonal matrices, whose diagonal elements are $s_{ii} = k_{i0} / \sum_{i=1}^{n} k_{i0}$, $r_{ii} = m_i$, $c_{ii} = c_i$ and $w_{ii} = k_{i0}$, respectively. $\mathbf{Q_{n \times n}} = [q \; q \cdots q]^T$, where $\mathbf{q_{n \times 1}}$ is a column vector,

$$\mathbf{q} = \left[(2a_{10} - a_{20} - a_{n0}) \; (2a_{20} - a_{10} - a_{30}) \cdots \left(2a_{n0} - a_{10} - a_{(n-1)0} \right) \right]^T \tag{8.18}$$

Besides, $\mathbf{T_{n \times n}}$ and $\mathbf{F_{n \times n}}$ are written in detail.

$$\mathbf{T}_{n \times n} = \frac{1}{\sum\limits_{i=1}^{n} k_{i0}} \begin{bmatrix} \sum\limits_{i \neq 1}^{n} k_{i0} & -k_{20} & \cdots\cdots & -k_{(n-1)0} & -k_{n0} \\ -k_{10} & \sum\limits_{i \neq 2}^{n} k_{i0} & \cdots\cdots & -k_{(n-1)0} & -k_{n0} \\ \vdots & & \ddots & & \vdots \\ -k_{10} & -k_{20} & \ddots & \sum\limits_{i \neq n-1}^{n} k_{i0} & -k_{n0} \\ -k_{10} & -k_{20} & \cdots\cdots & -k_{(n-1)0} & \sum\limits_{i \neq n}^{n} k_{i0} \\ -k_{n1} & -k_{n2} & \cdots\cdots & -k_{n(n-1)} & \sum\limits_{j=0, j \neq n}^{n} k_{nj} \end{bmatrix} \tag{8.19}$$

$$\mathbf{F}_{n \times n} = \begin{bmatrix} 2a_{10} & -a_{10} & 0 & \cdots & 0 & -a_{10} \\ -a_{20} & 2a_{20} & a_{20} & 0 & \cdots & 0 \\ 0 & & \ddots & & & 0 \\ \vdots & & & \ddots & & \vdots \\ 0 & & & \ldots & 2a_{(n-1)0} & -a_{(n-1)0} \\ -a_{n0} & 0 & \cdots & 0 & -a_{n0} & 2a_{n0} \end{bmatrix} \tag{8.20}$$

References

1. J. Rocabert, A. Luna, F. Blaabjerg, P. Rodríguez, Control of power converters in AC microgrids. IEEE Trans. Power Electron. **27**(11), 4734–4749 (2012)
2. C. Cheng, H. Yang, Z. Zeng, S. Tang, R. Zhao, Rotor inertia adaptive control method of VSG. Automat. Electric Power Syst. **39**(19), 82–89 (2015)
3. Z. Shuai, W. Huang, Z.J. Shen, A. Luo, Z. Tian, Active power oscillation and suppression techniques between two parallel synchronverters during load fluctuations. IEEE Trans. Power Electron. **35**(4), 4127–4142 (2020)

4. C. Dou, Z. Zhang, D. Yue, M. Song, Improved droop control based on virtual impedance and virtual power source in low-voltage microgrid. IET Generation Trans. Distrib. **11**(4), 1046–1054 (2017)
5. Y. Sun, X. Hou, J. Yang, H. Han, M. Su, J.M. Guerrero, New perspectives on droop control in AC microgrid. IEEE Trans. Ind. Electron. **64**(7), 5741–5745 (2017)
6. H. Hong, W. Gu, Q. Huang, L. Chen, X. Yuan and J. Wang, The research of transient load sharing in Islanded microgrid with heterogeneous DGs, in *Proceedings of the CSEE, Early Access*, https://doi.org/10.13334/j.0258-8013.pcsee.201044
7. X. Hou, Y. Sun, X. Zhang, J. Lu, P. Wang, J.M. Guerrero, Improvement of frequency regulation in VSG-based AC microgrid via adaptive virtual inertia. IEEE Trans. Power Electron. **35**(2), 1589–1602 (2020)
8. J. Liu, Y. Miura, T. Ise, Comparison of dynamic characteristics between virtual synchronous generator and droop control in inverter-based distributed Generators. IEEE Trans. Power Electron. **31**(5), 3600–3611 (2016)
9. J. Liu, Y. Miura, H. Bevrani, T. Ise, Enhanced virtual synchronous generator control for parallel inverters in microgrids. IEEE Trans. Smart Grid **8**(5), 2268–2277 (2017)
10. B.K. Poolla, S. Bolognani, F. Dörfler, Optimal placement of virtual inertia in power grids. IEEE Trans. Automatic Cont. **62**(12), 6209–6220 (2017)
11. L. Pagnier, P. Jacquod, Optimal placement of inertia and primary control: a matrix perturbation theory approach. IEEE Access **7**, 145889–145900 (2019)
12. M. Choopani, S.H. Hosseinain, B. Vahidi, A novel comprehensive method to enhance stability of multi-VSG grids. Int. J. Electr. Power Energy Syst. **104**, 502–514 (2019)
13. Z. Luo, H. Han, S. Chen, Z. Wu, M. Shi, and K. Zhang, Power oscillation suppression method based on distributed cooperative control of multi-VSG grids, in *2021 IEEE 12th Energy Conversion Congress and Exposition—Asia (ECCE-Asia)* (2021), pp. 1875–1878

Part III
Transient Synchronization Stability

Chapter 9
Dynamic Frequency Support by Virtual Inertia and Damping

Due to the urgent demands for large-scale integration of renewable energy sources (RESs), synchronous generators (SGs) have been replaced by power electronics interfaces [1–3]. However, the power converters lack sufficient inertia support, leaving the system more vulnerable to stability issues and operation-protection consequences.

Frequency nadir and rate of change of frequency (RoCoF) are the two important performance indicators to evaluate frequency stability. Overlarge frequency nadir and RoCoF will cause a series of adverse consequences and trigger the system operation protection like low-frequency-load-shedding (UFLS) and anti-islanding protection based on the grid codes [4]. The requirements for RoCoF relay have been loosened to avoid being triggered easily in the updated grid codes recently [4–6].

To alleviate the frequency instability, virtual inertia control methods based on power converters emerge as the most effective solutions. Virtual Synchronous Generator (VSG) control is one of the most promising virtual inertia control methods, which is achieved by mimicking the characteristics of SGs. Cooperating with additional energy storage devices alongside RESs, VSG possesses the grid-forming capacity, achieves necessary frequency regulation, and provides enough virtual inertia to reduce maximum frequency excursion and RoCoF [3, 7–17]. Recently, VSGs have been used extensively in many demonstration projects [18], for example, the 5.5MW photovoltaic microgrid demonstration project in "Tianjin China-Singapore Eco-City" and the national "Zhangbei" wind/photovoltaic/energy storage and transmission demonstration project in China.

Unlike SGs, VSGs have the most potential advantages to adjust the virtual inertia and damping coefficient flexibly [12]. Literature [8, 13, 14] studied the influence of control parameters and line parameters on dynamic and steady-state characteristics through small-signal modeling analysis. A quantitative assessment based on the 2-norm and the infinity norm to evaluate frequency stability was introduced in [15]. To explore more flexibility and benefit, adaptive virtual inertia methods were designed according to the combination of real-time frequency deviation and RoCoF in [16, 17].

X. Hou et al., *Self-organized Operation and Stability Control of Electronized Power Networks*, https://doi.org/10.1007/978-981-96-9067-1_9

However, these studies focused on a single machine connected to an infinite bus (SMIB).

Parallel VSG systems have advantages in redundancy and reliability, especially in distributed generation systems and microgrids. When a single VSG model expands to parallel VSG networks, the interaction among VSGs can hardly be ignored, especially in a non-stiff grid. The dynamics are much more complicated due to the mismatched line impedances and control parameters [19, 20]. The dynamic frequency performance was analyzed on the virtual inertia interaction among VSGs in [19], but the influence of the damping coefficient was not considered. An equivalent model based on torque analysis for the two-parallel VSG system was derived in [20], and the authors revealed how the equivalent virtual inertia and damping coefficient of the system influenced the overall dynamic frequency characteristic. However, the dynamic frequency response differences among VSGs were not considered. There are dynamic distribution differences in the frequency response of a real power system. How to quantitatively assess the differences of the dynamic frequency response characteristic influenced by the system parameters among VSGs also lacks investigation.

Besides, it is also urgent to solve the severe frequency/power oscillation and inappropriate transient power-sharing problems. In [21], Shuai summarized the excitations for frequency/power oscillation in the two-parallel VSG system. Based on the above analysis, the virtual impedance method was introduced in [22] for power oscillation and inappropriate transient power-sharing suppression. However, the parameter design for virtual impedance relies on accurate line impedance estimation/data, which is not practical in real applications. A particle swarm optimization method based on the concept of center of Inertia (COI) was introduced for transient stability enhancement in [23]. COI is a stability assessment tool that is widely used in power system analysis. The COI frequency of the parallel VSG system illustrated the dynamic behavior of the entire system. But literature [23] lacked solid theoretical analysis for voltage angle deviation constraints. Moreover, the optimization algorithm is prone to fall into local optimum. A PID compensator term was added to the basic VSG control by comparing the local frequency with COI frequency in [24, 25]. However, the existence of the integral item $\int_{t_1}^{t_2} (\omega_i - \omega_{coi})\,dt$ will result in different active power-sharing. Moreover, frequency stability was not taken into consideration.

In this chapter, the dynamic frequency characteristic of the parallel VSG systems was analyzed. The influence of virtual inertia and damping coefficient was considered to reveal the interaction of parameters on the dynamic frequency differences among VSGs. A coordinated control method was introduced to enhance the dynamic frequency performance and frequency stability of parallel VSG systems. In summary, this chapter mainly has the following contributions.

(1) The influence of virtual inertia and damping coefficient on the dynamic frequency characteristic of parallel VSG systems was systematically analyzed. (1) It revealed that the initial RoCoF for each VSG was determined by its own virtual inertia, while the damping coefficient hardly affected the initial RoCoF. The increase of the damping coefficient is conducive to frequency/power oscillation, but it influences the equilibrium point of frequency and steady-state power-sharing.

(2) A quantitative assessment for dynamic frequency performance based on the 2-norm H_2 and COI frequency was introduced. Different from [15], the introduced quantitative assessment was extended for parallel VSG systems, which can reflect the interactions on dynamic performance among VSGs. The value of H_2 is inverse with the ability to suppress the frequency dynamic deviation. The parameters principle to realize optimal dynamic frequency performance was concluded through this quantitative assessment.

(3) A coordinated control method for dynamic performance improvement and frequency stability enhancement was introduced. It can be applied in parallel VSG systems, having benefits on frequency/power oscillation suppression and frequency stability improvement, even though the parameters of VSGs do not meet the optimal design requirement.

The rest of this chapter is organized as follows. In Section II, the frequency dynamic characteristic in the parallel VSG system is analyzed. In Section III, the introduced coordinated method is introduced and the small-signal stability analysis is performed. Simulation and experimental results are provided in Sections IV and V. Conclusions are presented in Section VI.

9.1 Frequency Dynamic Characteristic Analysis In Parallel VSG

9.1.1 Basic VSG Control and the Mathematical Model

Figure 9.1 shows the block diagram of basic VSG control. Equation (9.1) is the core of the VSG model.

$$2H\omega^*\frac{\mathrm{d}\omega}{\mathrm{d}t} = P_{ref} - P_e - D\omega^*(\omega - \omega^*) \tag{9.1}$$

where P_{ref} and P_e are the references and the output active power. H and D are the virtual inertia time constant and damping coefficient of the VSG. $\omega^*$and ω are the reference and voltage angular speeds. Figure 9.2 presents the structure of the parallel VSG electrical system. P_{ei} represents the output active power of the i-th

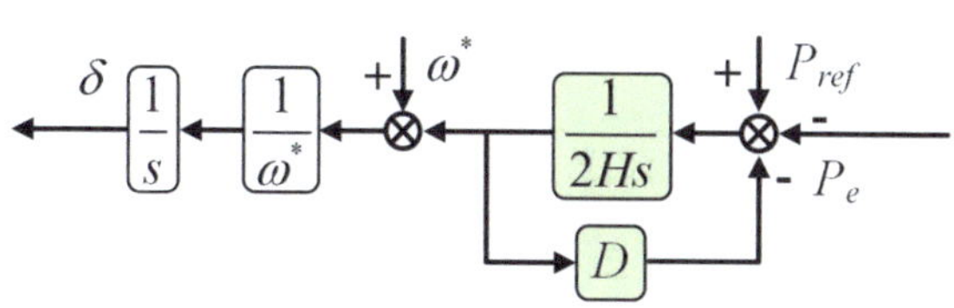

Fig. 9.1 Basic VSG control loop

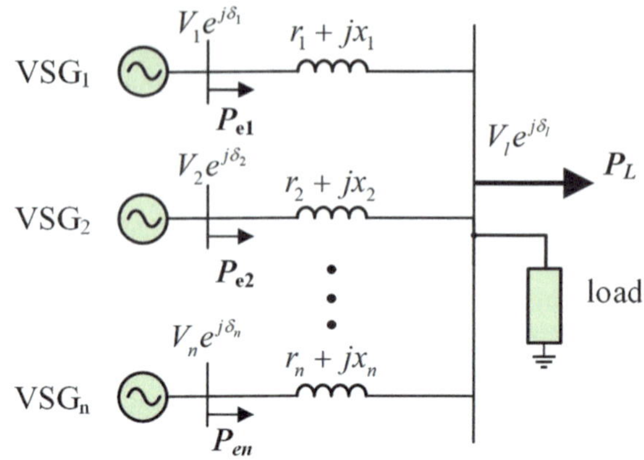

Fig. 9.2 The simplified parallel VSG system

VSG. $z_i = r_i + jx_i$ is the line impedance. P_L is the power demand of all the load. $V_l e^{j\delta_l}$ is the common bus voltage.

The power transferred from the i-th VSG to the common bus voltage can be expressed as

$$P_i = \frac{V_i V_l[r_i \cos(\delta_i - \delta_l) + x_i \sin(\delta_i - \delta_l)] - r_i V_l^2}{{r_i}^2 + {x_i}^2} \tag{9.2}$$

Linearizing (9.2)

$$\Delta P_i = k_{il}(\Delta\delta_i - \Delta\delta_l) \approx \frac{k_{il}}{s}(\Delta\omega_i - \Delta\omega_l) \tag{9.3}$$

where $k_{il} = \frac{V_i V_l[\cos(\delta_i^\circ - \delta_l^\circ) - (r_i x_i)\sin(\delta_i^\circ - \delta_l^\circ)]}{x_i[(r_i x_i)^2 + 1]}$, δ_i° and $\delta_l^\circ$are the voltage phase angle of i-th VSG and common bus in the steady-state equilibrium.

Ignoring the line losses, then

$$\sum_{i=1}^{n} \Delta P_i = \Delta P_L \tag{9.4}$$

Combining (9.1)–(9.4)

$$\begin{aligned} \Delta\omega_i(s) &= \frac{k_{il}}{2H_i\omega^* s + D_i\omega^* + k_{il}} \Delta\omega_l(s) \\ \Delta P_L(s) &= -\left[\sum_{i=1}^{n} \frac{k_{il}(2H_i\omega^* s + D_i\omega^*)}{2H_i\omega^* s + D_i\omega^* + k_{il}}\right] \Delta\omega_l(s) \end{aligned} \tag{9.5}$$

Therefore, the closed-loop transfer function of $\Delta\omega_i(s)/\Delta P_L(s)$ in the system can be obtained as (9.2).

$$\frac{\Delta\omega_i(s)}{\Delta P_L(s)} = -\frac{G_i(s)}{\sum_{m=1}^{n} G_m(s)(2H_m\omega^* s + D_m\omega^*)} \tag{9.6}$$

where, $G_i(s) = \frac{k_{il}}{2H_i\omega^* s^2 + D_i\omega^* s + k_{il}}$,$G_m(s) = \frac{k_{ml}}{2H_m\omega^* s^2 + D_m\omega^* s + k_{ml}}$,$\delta_i^\circ$ and $\delta_l^\circ$are the voltage phase angle of i-th VSG and common bus in the steady-state equilibrium.

9.1.2 Analysis of the Influence of Line Resistance

The line impedance can be viewed as highly inductive when the value of r_i/x_i is small, usually, less than 0.33:1. Thus, the line resistance on k_{il} can be neglected as

$$k_{il} \approx \frac{V_i V_l \cos(\delta_i^\circ - \delta_j^\circ)}{x_i} \tag{9.7}$$

If the line resistance is non-negligible ($r_i/x_i > 0.33 : 1$), virtual impedance is usually introduced to prevent the coupling between active power and reactive power due to resistive line impedance so that the highly inductive line impedance can be ensured. In this case, $r_i/(x_i + x_{vir}) < 0.33 : 1$, and (9.7) can still be ensured. x_{vir} represents the value of virtual impedance.

The bode diagram and step response of $\Delta\omega_1(s)/\Delta P_L(s)$ in a two-VSG system with whether considering line resistance or not is shown in Fig. 9.3. There is a highly inductive line impedance in the studied two-VSG system, r/x is usually less than

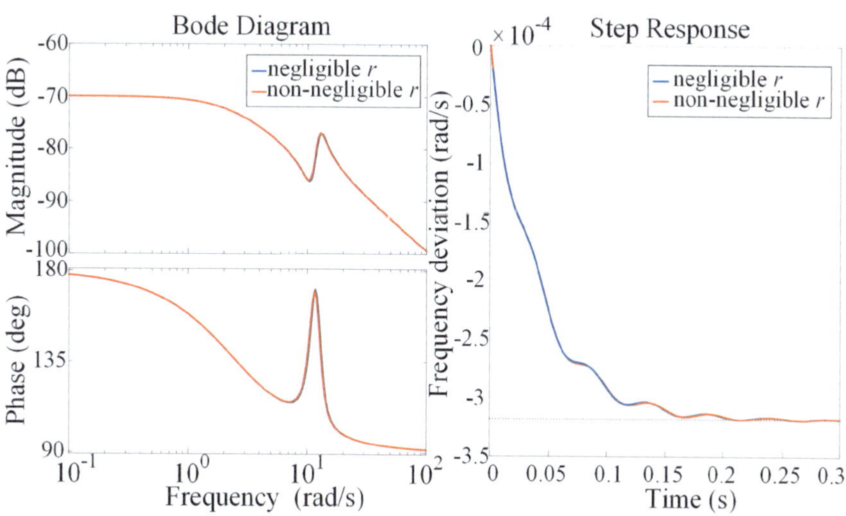

Fig. 9.3 Bode diagram and step response of $\Delta\omega_1(s)/\Delta P_L(s)$ with whether considering line resistance or not

0.33:1. Figure 9.3 shows that the bode diagrams and step response curves are highly consistent between the system with negligible line resistance and non-negligible line resistance. Thus, in a system with a highly inductive line impedance, the influence of line resistance on system frequency dynamics can be neglected.

9.1.3 Influence Analysis of Virtual Inertia and Damping

When n=1 in the system, (9.6) can be expressed as

$$\frac{\Delta\omega_1(s)}{\Delta P_L(s)} = -\frac{1}{2H_1\omega^* s + D_1\omega^*} \tag{9.8}$$

Then the step response of frequency can be expressed as

$$\Delta f_1(t) = -\frac{\Delta P_L}{2\pi\omega^* D_1}\left(1 - e^{-\frac{D_1}{2H_1}t}\right) \tag{9.9}$$

The maximum frequency nadir is $-\frac{\Delta P_L}{2\pi\omega^* D_1}$.

Through (9.9), the maximum RoCoF can be expressed as

$$\text{RoCoF}_{\max} = \frac{\mathrm{d}\Delta f_1(t)}{\mathrm{d}t}|_{t=0^+} = -\frac{\Delta P_L}{4\pi\omega^* H_1} \tag{9.10}$$

The maximum RoCoF occurs when the load changes instantaneously. From (9.8) and (9.9), there is no frequency oscillation in the single VSG system with load. The maximum frequency nadir and RoCoF are inversely proportional to the damping coefficient D and virtual inertia time constant H, respectively. It means that the increase of virtual inertia and damping coefficient are conducive to frequency stability.

In the parallel VSG system ($n \geq 2$), the dynamic frequency performance is more complicated due to the interactions caused by the parameter variations among VSGs. The time-domain analytical expression of frequency and RoCoF is difficult to be calculated. A 2-parallel VSG system was taken as an example to analyze the influence of the parameter variations on the dynamic frequency characteristic. The two VSGs are in parallel operation with the same capacity and 1:2 line impedance $r_1 : r_2 = x_1 : x_2 = 1 : 2$. According to (9.6), the bode diagram of $\Delta\omega_1(s)/\Delta P_L(s)$ can be obtained to analyze the influence of the control parameter variations on the dynamic frequency response characteristics. The magnitude-frequency curve of the bode diagram at the low-frequency region gives insight into the steady-state frequency deviation. And the magnitude-frequency curve at the high-frequency region describes the initial time domain response, which indicates the value of RoCoF at the very beginning of the transient state.

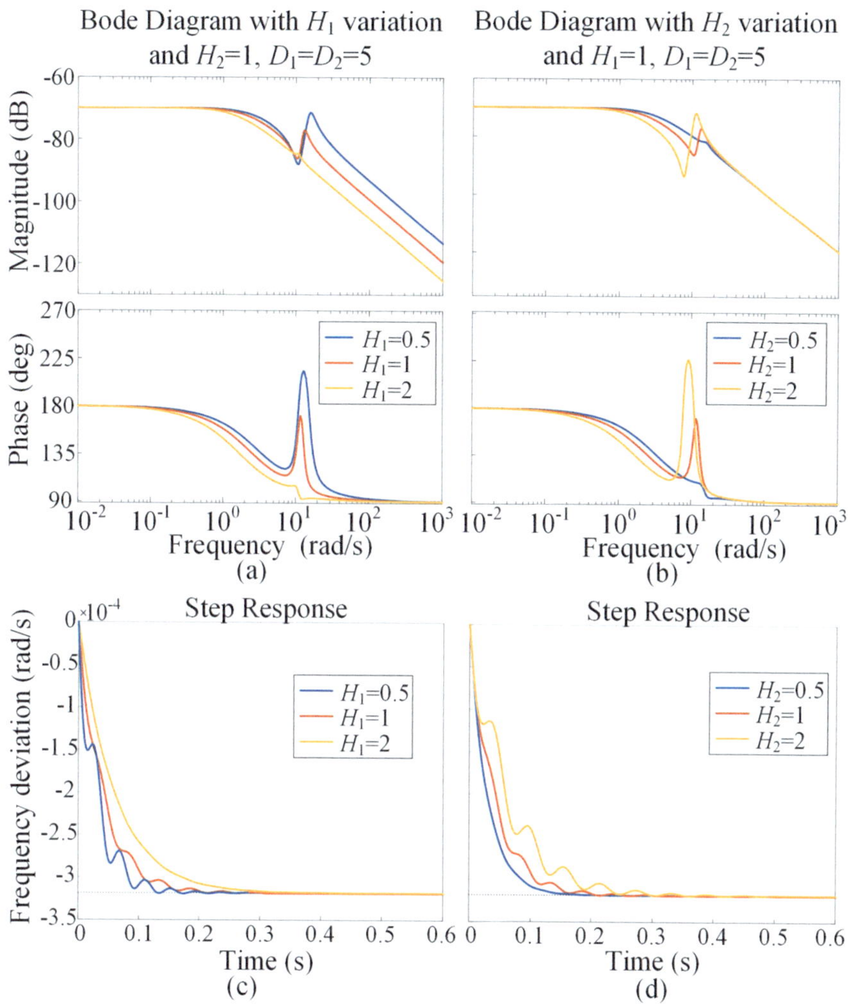

Fig. 9.4 Bode diagram and step response of $\Delta\omega_1(s)/\Delta P_L(s)$ with H variations

Figure 9.4a, c show the bode diagram and step response with variable H_1 and fixed H_2. Figure 9.4b, d show the bode diagram and step response with variable H_2 and fixed H_1. Different virtual inertia H hardly influences steady-state frequency deviation while the influence of different damping coefficients D on this is the same as long as the sum of D_1 and D_2 is equal. H_1 has an obvious influence on the high-frequency region in Fig. 9.4a, while the variation of H_2 does not influence the curves in the high-frequency region in Fig. 9.4b. It shows that larger H_1 is conducive to reducing RoCoF of VSG1, while H_2 has little influence. It indicates that the maximum RoCoF at the very beginning of the disturbance that occurs for each VSG is mainly determined by its own virtual inertia. Figure 9.4c, d verify this conclusion

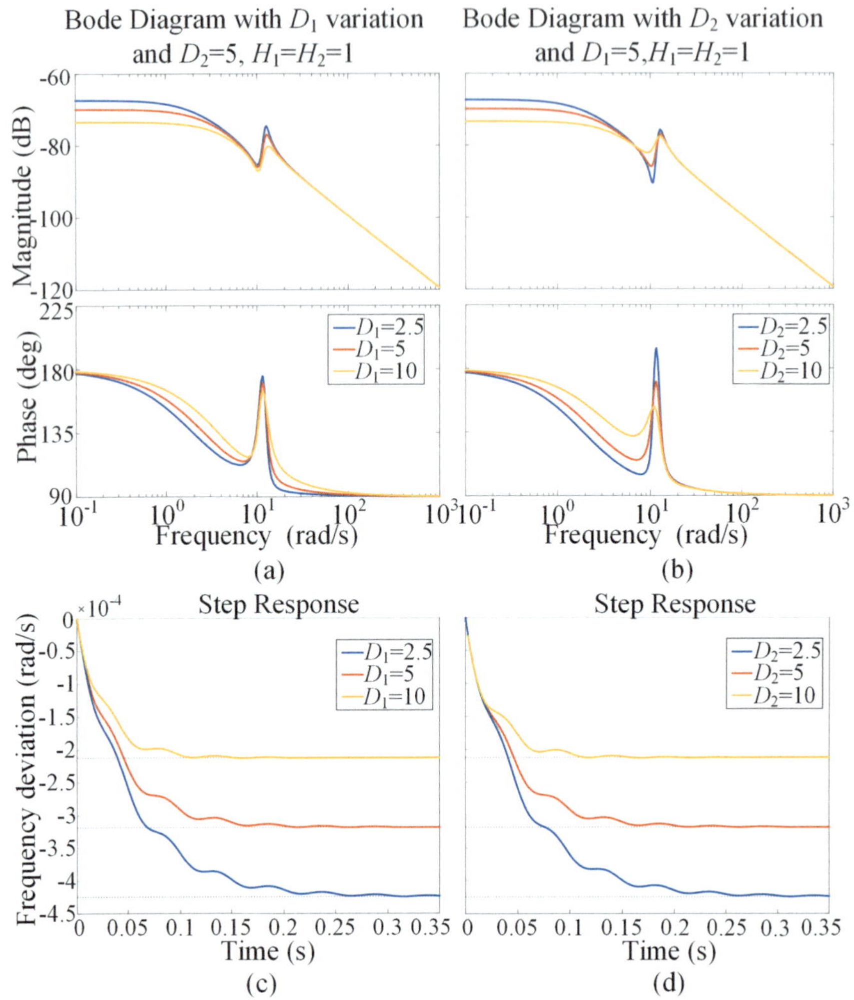

Fig. 9.5 Bode diagram and step response of $\Delta\omega_1(s)/\Delta P_L(s)$ with D variations

through step response. Moreover, the oscillation frequency and amplitude in the dynamic response of VSG1 can be reflected by the resonance peak frequency and the resonance peak value of the magnitude-frequency curve. When H_1:H_2=2:1, the resonance can be almost suppressed (seen in Fig. 9.4a yellow curve and Fig. 9.4b blue curve), and the dynamic frequency changes smoothly in step response (seen in Fig. 9.4c yellow curve and Fig. 9.4d blue curve).

Figure 9.5a, c show the bode diagram and step response with variable D_1 and fixed D_2. Figure 9.5b, d show the bode diagram and step response with variable D_2 and fixed D_1. As is shown in Fig. 9.5a, b, the magnitude-frequency curves of different D_1 and D_2 in the region (>30 rad/s) are similar. Therefore, the magnitude-frequency

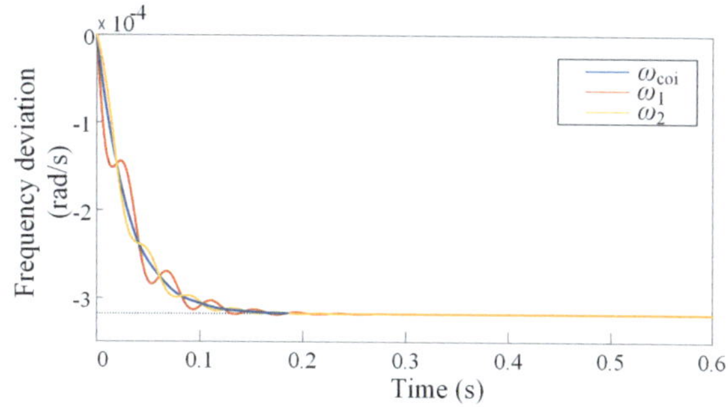

Fig. 9.6 Step load change response of the two-parallel VSGs

curves are highly coincident in the high-frequency region. Thus, the damping coefficients D_1 and D_2 have little effect on the initial RoCoF. Figure 9.5c, d verify the above conclusions through step response, as the slopes of all the step response curves are the same at the very beginning. Besides, Fig. 9.5a, b also show that the resonance peak value is reduced with the increase of damping coefficients. The curves in the low-frequency region ($<$ 1 rad/s) are variable with the variation of damping coefficients. The change of damping coefficients will influence the equilibrium point of frequency in a steady state.

9.1.4 *Quantitative Assessment for Dynamic Performance Based on the 2-Norm H_2 and COI Frequency*

To further explore the influence of control parameters on the dynamic frequency response in parallel systems, we decompose the frequency into two parts: the common-mode component and the differential-mode component. The common-mode component reflects the dynamic frequency consistency of all VSGs while the differential-mode component reflects the difference in dynamic frequency characteristics. Therefore, the dynamic frequency response can be expressed as

$$\Delta\omega_i = \Delta\omega_{coi} + \Delta\omega_{ie} \tag{9.11}$$

$$\Delta\omega_{coi} = \frac{H_1\Delta\omega_1 + \cdots + H_n\Delta\omega_n}{H_1 + \cdots + H_n} \tag{9.12}$$

where $\Delta\omega_{coi}$ represents the common-mode component based on the center of inertia (COI) and $\Delta\omega_{ie}$ represents the differential-mode component. All the frequencies will converge to the COI frequency finally. $\omega_{ie} = \omega_i - \omega_{coi}$ represents the dynamic frequency deviation towards ω_{coi}. Thus, we have

$$\Delta\omega_{ie} = \frac{H_1(\omega_i - \omega_1) + \cdots + H_n(\omega_i - \omega_n)}{H_1 + \cdots + H_n} \tag{9.13}$$

Figure 9.6 shows the response curves of ω_{coi}, ω_1, ω_1 of the two-parallel VSGs. ω_{coi} is the common-mode component of ω_1 and ω_2, which is without no oscillation under the load change. It can reflect the frequency dynamic characteristic of the overall system. The overlarge deviation is undesirable in the dynamic response.

To further describe the ability to suppress the dynamic frequency deviation related to ω_{coi}, the 2-norm H_2 of the transfer function $\Delta\omega_{ie}/\Delta P_L(s)$ is introduced as the assessment tool for VSGi. The value of H_2 is inverse with the ability to suppress the frequency dynamic deviation. The formula for calculating the 2-norm H_2 of the transfer function $G(s)$ can be expressed as

$$\| G(s)\|_2 = \sqrt{\frac{1}{2\pi}\int_{-\infty}^{\infty} \mathrm{tr}(G(j\omega)^H G(j\omega))\mathrm{d}\omega} \tag{9.14}$$

where $G(s)$ can be the transfer function matrix, $||G(s)||$ represents the value of the 2-norm H_2 of the transfer function matrix $G(s)$. tr (A) represents the trace of matrix A, and A^H represents the conjugate transpose of matrix A.

In the system with any number n of VSGs, $\Delta\omega_{ie}(s)/\Delta P_L(s)$ can be expressed as

$$G_{ie}(s) = \frac{\Delta\omega_{ie}(s)}{\Delta P_L(s)} = \frac{\sum_{j=1}^{n} H_j\left(\frac{\Delta\omega_i(s)}{\Delta P_L(s)} - \frac{\Delta\omega_j(s)}{\Delta P_L(s)}\right)}{\sum\limits_{j=1}^{n} H_j} \tag{9.15}$$

Taking a two-parallel VSG system as an example, the parameters of each VSG are assigned to k_{1l}: k_{2l} =2:1, and the transfer function of $\Delta\omega_{1e}(s)/\Delta P_L(s)$ can be expressed as

$$\frac{\Delta\omega_{ie}(s)}{\Delta P_L(s)} = \frac{H_2 s[(2H_1k_{2l} - 2H_2k_{1l})s + D_1k_{2l} - D_2k_{1l}]}{(H_1 + H_2)F(s)} \tag{9.16}$$

where $F(s) = k_{1l}(2H_1s + D_1)\left(2H_2\omega^* s^2 + D_2\omega^* s + k_{21}\right) + k_{21}(2H_2s + D_2)\left(2H_1\omega^* s^2 + D_1\omega^* s + k_{1l}\right)$

In the transfer function (9.16), the locations of zeros and poles are related to k_{1l}, k_{2l}, H_1, H_2, D_1, and D_2. The influence of H_1, H_2, D_1, D_2 on H_2 of $\Delta\omega_{1e}(s)/\Delta P_L(s)$ can be drawn from Figs. 9.7 and 9.8. When D_1:D_2=H_1:H_2 =2:1, the value of H_2 reaches the minimum. And this conclusion can be extended to the n-VSG system. The parameters should satisfy the principle H_1: H_2: ...: H_n=D_1: D_2: ...: D_n=k_{1l}: k_{2l}: ...: k_{nl} to achieve the optimal dynamic response.

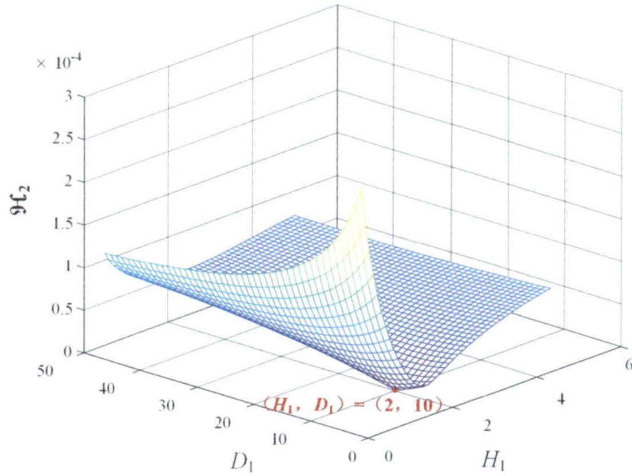

Fig. 9.7 The relationship between H_1, D_1, and H_2 of $\Delta\omega_{1e}(s)/\Delta P_L(s)$, H_2=1, D_2=5

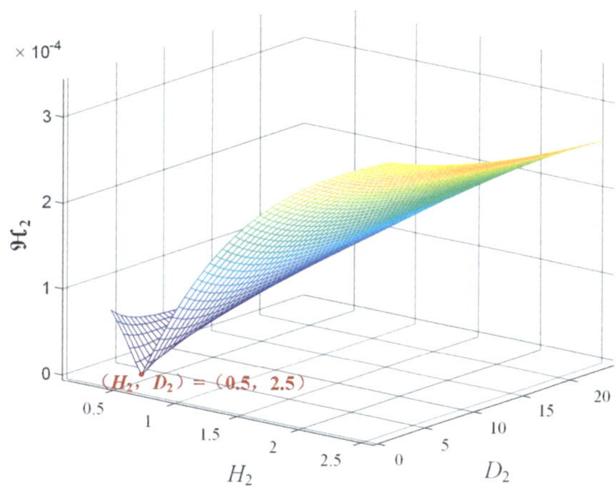

Fig. 9.8 The relationship between H_2, D_2, and H_2 of $\Delta\omega_{1e}(s)/\Delta P_L(s)$, H_1=1, D_1=5

9.1.5 Discussion

Frequency nadir, RoCoF, and ability to suppress frequency oscillations as well as dynamic deviation related to COI are major concerns in the dynamic frequency performance of the parallel VSGs.

Due to the control parameters' interactions among VSGs, the time-domain analysis of frequency and RoCoF is difficult to be obtained for the paralleled VSGs. Besides, the frequency oscillations caused by the mismatch and interaction of system parameters are prone to lead to unstable, especially in the weakly damped system. In conclusion, the dynamic frequency characteristics are summarized as follows.

(1) The maximum RoCoF for each VSG is determined by its own virtual inertia, which is independent of the interaction among other VSG's dynamics. The damping coefficient hardly affects RoCoF.

(2) ω_{coi} can represent the frequency dynamic characteristic of the overall system. And $\omega_{ie} = \omega_i - \omega_{coi}$ represents the frequency dynamic deviation towards ω_{coi}. The overlarge deviation is undesirable in the dynamic response.
(3) When parameters satisfy the principle H_1: H_2: ...: H_n=D_1: D_2: ...: D_n=k_{1l}: k_{2l}: ...: k_{nl}, the value of H_2 of $\Delta\omega_{ie}(s)/\Delta P_L(s)$ reaches the minimum and the dynamic frequency performance is optimal.

As a result, to optimize the frequency dynamic response, x_i, H_i, and D_i need to be comprehensively considered and designed. However, the accurate line impedance is hardly obtained in practice and the damping coefficient Di is usually designed based on the steady-state power-sharing. H_1: H_2: ...: H_n=D_1: D_2: ...: D_n=k_{1l}: k_{2l}: ... : k_{nl} cannot be maintained in this case all the time. Control methods to coordinately modify virtual inertia and dynamic damping are expected.

9.2 Coordinated Control

9.2.1 *Coordinated Control Based on RoCoF and COI Frequency*

According to the above analysis, the maximum RoCoF of i-th VSG in the parallel system is only determined by its own virtual inertia, while the dynamic performance is influenced by the interaction among other VSG's dynamics. Therefore, the introduced control method of i-th VSG can be designed as

$$2H_i\omega^*\frac{\mathrm{d}\omega_i}{\mathrm{d}t} = P_{ref\_i} - P_i - D_i\omega^*(\omega_i - \omega^*) - \Delta P_{coi\_i} \tag{9.17}$$

where H_i is the introduced adaptive virtual inertia constant, and ΔPcoi_iis the additional damping item related to COI frequency, respectively, which can be expressed in (9.18).

$$\begin{cases} H_i = H_{oi} + k_{adp\_i}\arctan^2(a_i\frac{\mathrm{d}\omega_i}{\mathrm{d}t}), k_{adp\_i} > 0 \\ \Delta P_{coi\_i} = k_{coi\_i}\arctan(\omega_i - \omega_{coi}), k_{coi\_i} > 0 \end{cases} \tag{9.18}$$

where H_{0i} is the initial and the introduced adaptive virtual inertia constant, respectively. k_{adp_i} and ai represent the adaptive virtual inertia coefficient and the coefficient of the non-linear adjustment, respectively. k_{adp_i} is the proportional coefficient of the additional damping item.

The introduced method contains two items that yield a 2-dimension of freedom tunable control:

(1) The adaptive virtual inertia based on local RoCoF The grid codes have specific requirements on RoCoF to avoid triggering anti-islanding protection [4]. Virtual inertia is required to increase during frequency changes. Therefore, the designed virtual inertia should increase when RoCoF tends to increase. The designed

adaptive virtual inertia H_i is altered immediately according to the output $d\omega/dt$ of the VSG active controller loop and it does not depend on the measured RoCoF and avoids extra noise. Besides, as the maximum virtual inertia H_{max} is relevant to the instantaneous power capacity limitation of the inverter's DC-link capacitor [26], the mathematical function $f\,(d\omega/dt) = arctan^2\,(ad\omega/dt)$ guarantees the upper limit for the adaptive virtual inertia, which is easy for engineers to design.

(2) The additional damping item based on COI frequency The additional damping item is designed to reduce the ω_{ie} during the dynamic process. The additional damping item ΔP_{coi_i} can be understood as transient damping. It is altered immediately according to $(\omega - \omega_{coi})$. In the steady-state, it equals 0 as $\omega_i = \omega_{coi}$, and it works during the dynamic process ($\omega_i \neq \omega_{coi}$). When $\omega_i - \omega_{coi} > 0$, the transient damping shall increase to make ω_i slowdown and approach ω_{coi}, when $\omega_i - \omega_{coi} < 0$, the transient damping shall decrease to make ω_i faster and approach ω_{coi}. The additional damping can coordinate all the VSG frequencies to follow COI frequency and improve the dynamic frequency performance of the parallel VSGs.

The mathematical function $f\,(\omega_i - \omega_{coi}) = arctan\,(\omega_i - \omega_{coi})$ is an odd function with good symmetry and monotonic increasing convergence, which is suitable for designing the principle of the additional damping item. The lower and upper limit of $arctan\,(\omega_i - \omega_{coi})$ avoids the risks that wrong calculation for $(\omega_i - \omega_{coi})$ brings.

9.2.2 Design Guidelines

The coefficient k_{adp_i} can change the upper limit for adaptive virtual inertia H_{max}, which is relevant to the instantaneous power capacity limitation of the inverter's DC-link capacitor [26].

$$4\pi\omega^* H_{\max\_i}\mathrm{RoCoF}_{\max\_i} = \Delta P_{L\_i} \propto Capacity \tag{9.19}$$

The maximum allowed grid operational $\mathrm{RoCoF}_{\max\_i}$ is required by grid codes (c.g. 2.5 Hz/s in Denmark [5, 6]).

Therefore,

$$H_{oi} + \frac{\pi^2}{4} k_{adp\_i} \le H_{\max\_i} \le \frac{\Delta P_{L\max\_i}}{4\pi\omega^*\mathrm{RoCoF}_{\max\_i}} \tag{9.20}$$

$$0 < k_{adp\_i} \le \frac{4}{\pi^2}\left(\frac{\Delta P_{L\max\_i}}{4\pi\omega^*\mathrm{RoCoF}_{\max\_i}} - H_{oi}\right) \tag{9.21}$$

Adjusting the coefficient ai can change the variation gradient of the virtual inertia according to the maximum allowed grid operational RoCoF. Usually, when RoCoF reaches the $\mathrm{RoCoF}_{\max}$, H_i equals 0.9 p.u. of the maximum $H_{\max}$.

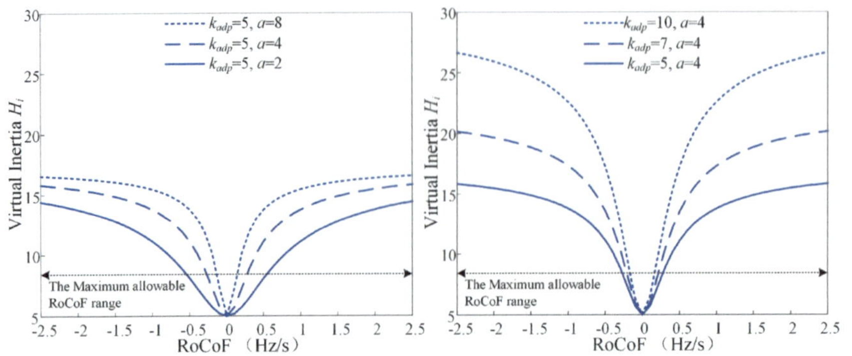

Fig. 9.9 The introduced adaptive virtual inertia curve trajectories **a** when a varies; **b** when k_{adp} varies

$$H_i = 0.9 H_{\max\_i}, \mathrm{RoCoF}_i = \mathrm{RoCoF}_{\max\_i} \tag{9.22}$$

According to (9.20)–(9.22), a_i can be derived as

$$a_i = \frac{1}{2\pi \mathrm{RoCoF}_{\max\_i}} \tan\left(\frac{0.9 H_{\max\_i} - H_{oi}}{k_{adp\_i}}\right) \tag{9.23}$$

Figure 9.9 shows the trajectories of adaptive virtual inertia variations with different k_{adp_i} and a_i. It indicates design guidelines for engineers. The selection for the proportional coefficient k_{coi_i} of the additional damping item can take the root locus as the reference, which will be discussed in the next subsection. Reasonable k_{coi_i} can guarantee the system damping ratio $\zeta \in [0.8, 1.0]$ to get a satisfactory dynamic response [27].

9.2.3 Stability Analysis

To investigate the effects of the introduced method on system stability, the small-signal model analysis is presented in this section. Owing to the control bandwidth of the inner control loop being much larger than the outer power loop, the analysis of the frequency dynamic process can neglect the fast dynamics of inner voltage/current control loops to simplify the stability analysis [28].

By eliminating the load nodes, the reduced network can be described as

$$\Delta P_i = \sum_{j \neq i}^{n} k_{ij}(\Delta\delta_i - \Delta\delta_j) \tag{9.24}$$

where i, j represents the node number of i-th and j-th VSG, $k_{il} \approx V_i V_j \cos(\delta_i^{\circ} - \delta_j^{\circ})$ x_{ij}. δ_i° and δ_j° are the voltage phase angles of i-th and j-th VSGs in the steady-state equilibrium.

According to the above analysis, the small-signal model of i-th VSG after linearization can be obtained as follows.

$$\begin{cases} \Delta\dot{\delta}_i = \Delta\omega_i \\ \Delta\dot{\omega}_i = \dfrac{-\sum\limits_{j\neq i}^{n} k_{ij}(\Delta\delta_i - \Delta\delta_j) - D_i\omega^*\Delta\omega_i - k_{coi\_i}(\Delta\omega_i - \Delta\omega_{coi})}{2H_i\omega^*} \end{cases} \tag{9.25}$$

To express (9.25) in the matrix form

$$\begin{bmatrix} \Delta\dot{\delta} \\ \Delta\dot{\omega} \end{bmatrix} = \begin{bmatrix} \mathbf{0} & \mathbf{I} \\ -\mathbf{H}^{-1}\mathbf{K_L} & -\mathbf{H}^{-1}(\mathbf{D}+\mathbf{K_{coi}}) \end{bmatrix} \begin{bmatrix} \Delta\delta \\ \Delta\omega \end{bmatrix} \tag{9.26}$$

where $\mathbf{I}$ is the unit diagonal matrix. $\mathbf{H}$, $\mathbf{K_L}$, $\mathbf{D}$, and $\mathbf{K_{coi}}$ represent the inertia matrix, power synchronization matrix, damping coefficient matrix, and additional damping matrix, respectively.

Taking the parallel VSG system in Fig. 9.11 as the study case, the root locus method is used to analyze the matrix eigenvalues of (9.26). The root locus as functions of the variations of adaptive virtual inertia H_i and proportional coefficient k_{coi_i} are illustrated in Fig. 9.10. The control parameters are listed in Table 9.2.

It is worth noting that there is an eigenvalue of 0 (λ_6) due to the rotational symmetry of the system, but it does not affect the dynamic stability of the system [29]. Conclusions can be drawn from Fig. 9.10 that the system with the introduced method is stable since all the eigenvalues are negative when the parameters are in the allowable range.

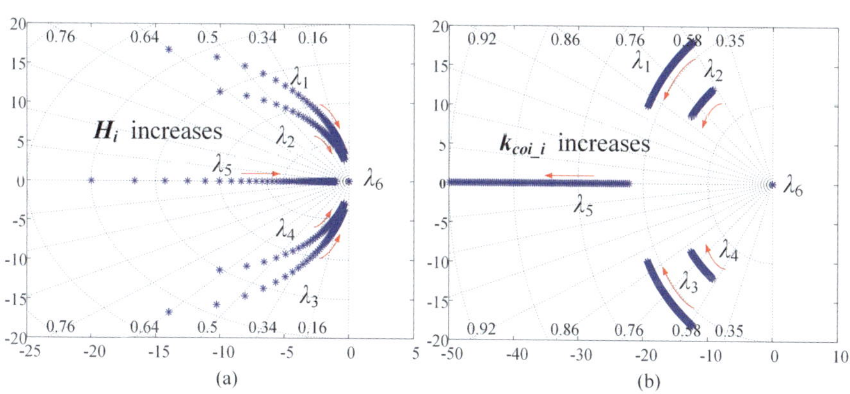

Fig. 9.10 Root locus **a** when H_i varies from 10 to 300; **a** when k_{coi_i} varies from 0 to 3000

Table 9.1 Stability region of control parameters

Methods	Design principles	Comparison with existing methods
Adaptive virtual inertia **I** [16, 17]	Virtual inertia reduces when $(\omega - \omega *)(d\omega/dt) > 0$;	✓ Dynamic performance improvement
	Virtual inertia increases when $(\omega - \omega *)(d\omega/dt) > 0$;	✓ Only suitable for a grid-connected single-machine infinite-bus power system
Adaptive virtual inertia **II** [20]	Virtual inertia increases when disturbances occur. Otherwise, virtual inertia is small	✓ Frequency stability enhancement
		× Neglect the dynamic interaction with other VSGs
Adaptive virtual inertia **III** [30]	$H = \frac{H_h k_a^2 (\frac{\omega - \omega_{ref}}{\omega_{ref}})^2 + H_o}{k_a^2 (\frac{\omega - \omega_{ref}}{\omega_{ref}})^2 + 1}$	✓ Power/frequency oscillation suppression
		× Complicated parameter design
		× Neglect the dynamic interaction with other VSGs
Dual adaptive control for virtual inertia and damping [31]	Virtual inertia increases related to $\lvert d\omega/dt \rvert$	✓ Increase inertia and damping during the dynamic process
	Damping coefficient increases related to $\lvert \omega - \omega * \rvert$	× Only suitable for a grid-connected single-machine infinite-bus power system
Intelligent virtual inertia and damping control [32, 33]	Fuzzy logic membership functions of $\Delta\delta$, $(\omega - \omega *)$, and $(d\omega/dt)$ to optimize virtual inertia and damping	✓ Inertia response improvement
		× Complicated fuzzy rule design and online computational burden
		×Neglect the dynamic interaction with other VSGs
COI-based VSG control with PID controller [24, 25]	An additional term is added to the VSG control loop: $\Delta P_{\mathrm{COI}} = q_P(\omega - \omega_{\mathrm{COI}}) + q_D(\mathrm{d}\omega/\mathrm{d}t - \mathrm{d}\omega_{\mathrm{COI}}/\mathrm{d}t) + q \int [(\omega - \omega_{\mathrm{COI}})\mathrm{d}t)$	✓ Suitable for a parallel VSG system
		× Steady-state power-sharing errors due to the integral item
Introduced coordinated method	The adaptive virtual inertia based on local RoCoF and the additional damping item based on COI frequency	✓ Suitable for multi-parallel VSG grids
		✓ Dynamic performance Improvement
		✓ Frequency stability enhancement
		✓ Low implementation complexity

Table 9.2 Parameters of the simulation test system

Parameters	Value	Parameters	Value
P_L	30 kW	H_1	14
ΔP_L	42 kW	H_2	24
Virtual impedance	$0.05 + j1.256\Omega$	H_3	10
E	311 V	$k_{coi\_1,2,3}$	1500
$\omega*$	$100 * pi\mathrm{rad/s}$	k_{P_p}	10
$N_{1,2,3}$	10000	a_i	4
$D_{1,2,3}$	6000	k in [17]	800

However, overlarge virtual inertia will make the poles move close to the imaginary axis, slow down the response speed and result in the reduction of the damping ratio. The increase of k_{coi_i} will increase the equivalent damping and enhance the system stability.

9.2.4 Compared with the Existing Methods

To further highlight the contribution of the introduced coordinated control method, a comparison with existing similar methods is presented in Table 9.1. Comparatively, the introduced method has some merits: (1) Improvement for frequency stability and dynamic performance jointly; (2) Being suitable for any multi-parallel VSG systems, like multi-VSG grids, microgrids, and so on.

9.3 Simulation Verification

To verify the feasibility and effectiveness of the introduced method, simulations based on Matlab/Simulink were carried out. A 3-parallel VSG test simulation system is shown in Fig. 9.11.

9.3.1 Comparative Simulations Verification

Comparative simulations were carried out with the basic VSG control in Fig. 9.11, the adaptive virtual inertia control method [17], and the introduced method respectively under the same condition. The system parameters are shown in Table 9.2. The load increases at 1 s and decreases at 4 s, suddenly.

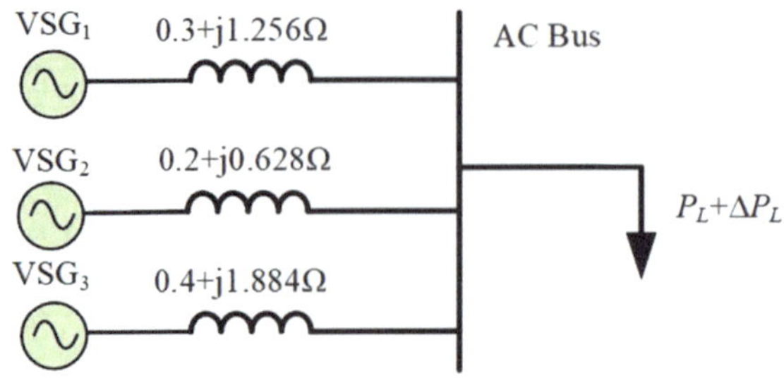

Fig. 9.11 The 3-parallel VSG simulation test system

Table 9.3 Quantitative indicators

Methods	Basic VSG method	Adaptive virtual inertia method [17]	introduced method
RFD_{max}	0.048 Hz	0.091 Hz	0.026 Hz
$RoCoF_{max}$	1.21 Hz/s	1.94 Hz/s	0.71 Hz/s
Power overshoot	30.1%	28.3%	12.8%
Settling time (5%)	1.91 s	1.74 s	1.06 s

The simulation results are shown in Fig. 9.12. With the introduced method, RoCoF is reduced during the dynamic process. Oscillations of active power and frequency are suppressed effectively. The active power overshoot is reduced and the system response speed is improved. To further verify the effectiveness of the introduced method quantitatively, the maximum frequency deviation related to other VSG (RFD), maximum RoCoF, the active power overshoot, and settling time of the dynamic performance with different control methods are listed in Tables 9.2 and 9.3.

It is worthy to notice that the maximum RoCoF at 5 s with the adaptive control method [17] is larger than the others owing to the small inertia at this moment when the frequency returns 50 Hz. Compared with the existing methods, the introduced method has significant improvements in every performance indicator.

9.3.2 *Under Different Additional Damping Items*

The selection for the proportional coefficient k_{coi_i} is vital to dynamic frequency performance improvement. As discussed previously, the root locus can be taken as the reference. Simulations are carried out with different k_{coi_i} (300, 2000, 5000) in this subsection, and the results are shown in Fig. 9.13. Figure 9.13a1 and a2 shows the results with k_{coi_i} and in this case, the system damping ratio $\zeta < 0.8$. Figure 9.13b1 and b2 shows the results with $k_{coi\_i} = 2000$ and in this case, the system damping ratio $\zeta \in [0.8, 1.0]$. And Fig. 9.13c1 and Fig. 9.13c2 shows the results with k_{coi_i} and

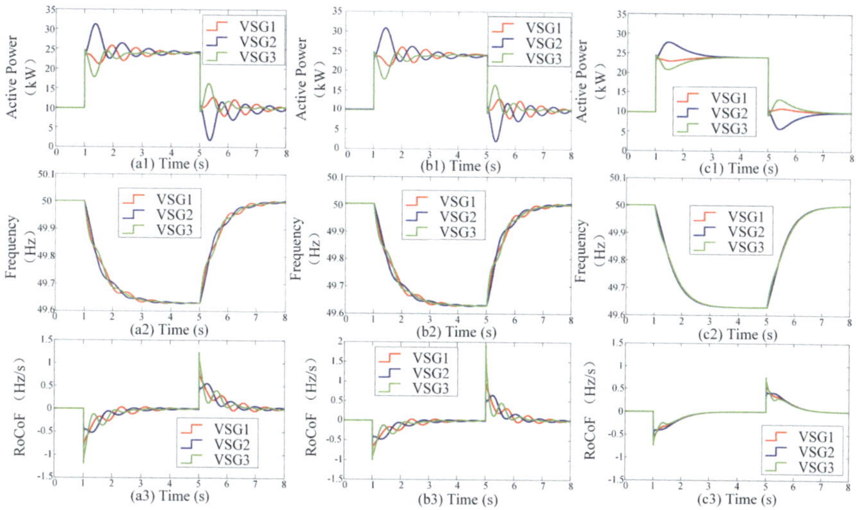

Fig. 9.12 Simulation results among the basic VSG control (**a1**, **a2**, **a3**), adaptive virtual inertia control [17] (**b1**, **b2**, **b3**), and the introduced control method (**c1**, **c2**, **c3**)

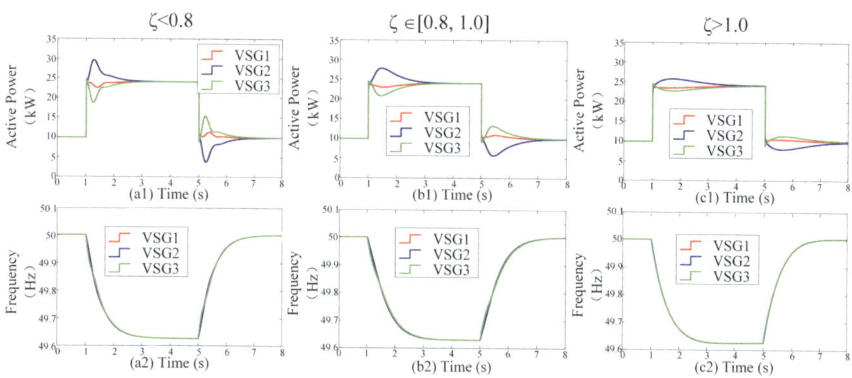

Fig. 9.13 Simulation results among the introduced control method with $k_{coi\_1,2,3}$=300 (**a1**,**a2**); with $k_{coi\_1,2,3}$=2000 (**b1**,**b2**); with $k_{coi\_1,2,3}$=5000 (**c1**,**c2**)

in this case, the system damping ratio $\zeta > 1.0$. With the increase of k_{coi_i}, the system damping ratio increases, and the power overshoot and power oscillation are further suppressed. However, the dynamic response speed is slower, and settling time also increases. Reasonable k_{coi_i} should be taken to guarantee the system damping ratio $\zeta \in [0.8, 1.0]$ to get a satisfactory dynamic response [27].

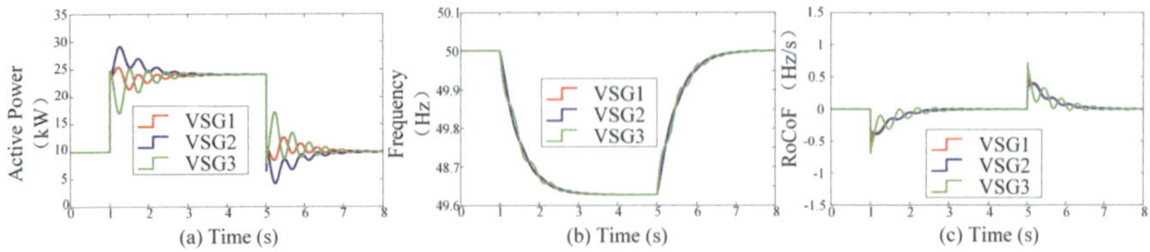

Fig. 9.14 Simulation results of the introduced method under a single communication link (VSG3) failure

9.3.3 *Under Communication Failure*

If all the communication links are disconnected, the additional damping item for power oscillation suppression is unable. This is a low-probability event. So, the situation when a single communication link failure occurs was discussed in Fig. 9.14. The communication link of VSG3 is disconnected. The suppression of power oscillation is weakened due to the failure of communication. Although communication failure will cause the failure of dynamic response improvement, the synchronization and frequency stability enhancement of the introduced method do not require communication. The system remains stable and the maximum RoCoF is not affected. Frequency stability can still be guaranteed through the introduced method.

9.3.4 *Under Short-Circuit Fault*

To further verify the frequency stability improvements of the introduced control method under extremely severe frequency accidents, a three-phase short-circuit fault occurs at t=2.0 s on the transmission line1, which is near VSG1, and the fault is cleared at t=3.0 s. The frequency and RoCoF curves of VSG1 with the basic VSG control and the introduced method are presented in Fig. 9.15.

When the short-circuit fault occurs, the maximum RoCoF of VSG1 with the basic VSG control is 5 Hz/s, which will trigger the RoCoF relay and cause a series of adverse consequences. The maximum RoCoF is reduced to 1.8 Hz/s, which is acceptable according to the grid code of Denmark. Besides, the maximum frequency deviation is reduced from 0.71 Hz to 0.39 Hz.

9.4 Experimental Verification

Parameters used in the experiment are listed in Table 9.4. Figure 9.16 shows the experimental results with the basic VSG control and the introduced method. The load demand increased at 5.3 s. With the basic VSG control method, the $RoCoF_{max}$

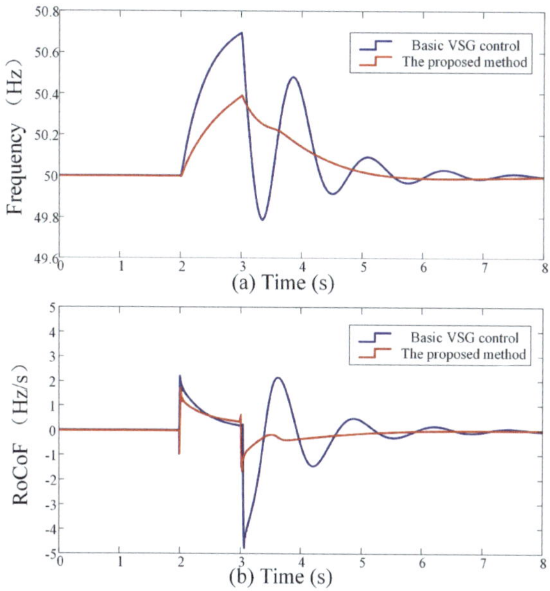

Fig. 9.15 Simulation results of VSG1 with the basic VSG control method and the introduced method under short-circuit fault. **a** Frequency. **b** RoCoF

Table 9.4 Parameters of the experimental test system

Parameters	Value	Parameters	Value
Load before switching	130 W	f	50 Hz
Load after switching	520 W	Nominal voltage E^*	175 V
Line impedance 1	1.8 mH	$D_{1,2,3}$	2
Line impedance 2	3.6 mH	$H_1/H_2/H_3$	1500
Line impedance 3	5.4 mH	$k_{adp\_1,2,3}$	10
Filter inductance L_f	1.8 mH	a_i	4
Filter capacitor C_f	30 μF	$k_{coi\,1,2,3}$	600

is about 1.58 Hz/s and the settling time is about 1.1 s. The frequency and power oscillations among VSGs can be also observed. On the contrary, as displayed in Fig. 9.16a2, b2, c2, the frequency and power oscillations are suppressed significantly with the introduced method. The $\text{RoCoF}_{\text{max}}$ is nearly 1.2 Hz/s and the settling time is about 0.4 s. The dynamic frequency performance and frequency stability are greatly improved.

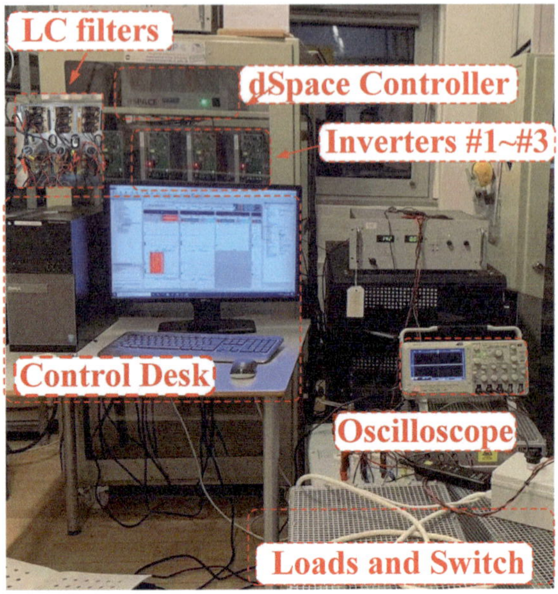

Fig. 9.16 Experimental results with the basic VSG control method (**a1**,**a2**,**a3**) and the introduced method(**b1**,**b2**,**b3**)

9.5 Conclusion

This chapter investigated the influences and interactions of different virtual inertia and damping coefficients among parallel VSGs on dynamic frequency performance. A quantitative assessment was introduced to reveal the system parameters' interaction on the dynamic frequency characteristics. The control parameters design principle was deduced to realize optimal dynamic performance. Furthermore, a coordinated control method was introduced for the parallel VSG systems. Simulation and experiment results demonstrate that the introduced method can enhance frequency stability and achieve better dynamic performances. There are still further studies required for practical applications, like communication vulnerability risks, multi-renewable-sources coordination, economic optimization, and so on.

Appendix

$$\mathbf{H} = \begin{bmatrix} 2H_1\omega^* & 0 & \cdots & 0 \\ 0 & 2H_2\omega^* & \cdots & 0 \\ \vdots & \vdots & \ddots & \vdots \\ 0 & 0 & \cdots & 2H_n\omega^* \end{bmatrix}_{n\times n} \tag{9.27}$$

$$\mathbf{D} = \begin{bmatrix} D_1\omega^* & 0 & \cdots & 0 \\ 0 & D_2\omega^* & \cdots & 0 \\ \vdots & \vdots & \ddots & 0 \\ 0 & 0 & \cdots & D_n\omega^* \end{bmatrix}_{n\times n} \tag{9.28}$$

$$\mathbf{K_L} = \begin{bmatrix} \sum\limits_{i=2,\ldots,n} k_{1i} & -k_{12} & \cdots & -k_{1n} \\ -k_{21} & \sum\limits_{i=2,\ldots,n} k_{2i} & \cdots & -k_{2n} \\ \vdots & \vdots & \ddots & \vdots \\ -k_{n1} & -k_{n2} & \cdots & \sum\limits_{i=2,\ldots,n} k_{ni} \end{bmatrix}_{n\times n} \tag{9.29}$$

$$\mathbf{K_{coi}} = \begin{bmatrix} \frac{k_{coi\_1}(\sum_{m=1}^{n} H_m - H_1)}{\sum_{m=1}^{n} H_m} & -\frac{k_{coi\_1} H_2}{\sum_{m=1}^{n} H_m} & \cdots & -\frac{k_{coi\_1} H_n}{\sum_{m=1}^{n} H_m} \\ -\frac{k_{coi\_2} H_1}{\sum_{m=1}^{n} H_m} & \frac{k_{coi\_2}(\sum_{m=1}^{n} H_m - H_2)}{\sum_{m=1}^{n} H_m} & \cdots & -\frac{k_{coi\_2} H_n}{\sum_{m=1}^{n} H_m} \\ \cdots & \cdots & \ddots & \vdots \\ -\frac{k_{coi\_n} H_1}{\sum_{m=1}^{n} H_m} & -\frac{k_{coi\_n} H_2}{\sum_{m=1}^{n} H_m} & \cdots & \frac{k_{coi\_n}(\sum_{m=1}^{n} H_m - H_n)}{\sum_{m=1}^{n} H_m} \end{bmatrix}_{n\times n} \tag{9.30}$$

References

1. R. Majumder, Some aspects of stability in microgrids. IEEE Trans. Power Syst. **3**(28), 3243–3252 (2013)
2. J. Rocabert, A. Luna, F. Blaabjerg, P. Rodriguez, Control of power converters in AC microgrids. IEEE Trans. Power Electron. **27**(11), 4734–4749 (2012)
3. K. Sakimoto, Y. Miura, T. Ise, Stabilization of a power system with a distributed generator by a Virtual Synchronous Generator function, in *8th International Conference on Power Electronics—ECCE Asia*, Jeju, South Korea (2011), pp. 1498–1505
4. L. Xiong, X. Liu, D. Zhang, Y. Liu, Rapid power compensation-based frequency response strategy for low-inertia power systems. IEEE. J. Emerg. Sel. Top. Power Electron. **9**(4), 4500–4513 (2021)
5. P. Energy, Rate of change of frequency (ROCOF)-review of TSO and generator submissions final report, in *Commission for Energy Regulation (CRU)* (Dublin, Ireland, Tech. Rep. N.20353, 2013)
6. R. O'Rourke, *Rate of Change of Frequency (RoCoF) Modification to the Grid Code*. (Commission Energy Regulation (CER), London, U.K., Tech. Rep. CER/14/081, Apr. 2014) [Online]
7. Q.C. Zhong, G. Weiss, Synchronverters: inverters that mimic synchronous generators. IEEE Trans. Ind. Electron. **58**(4), 1259–1267 (2011)
8. Q.C. Zhong, Virtual synchronous machines: a unified interface for grid integration. IEEE Power Electron. Mag. **3**(4), 18–27 (2016)
9. Z.K. Shuai, Y. Hu, Y.L. Peng, C.M. Tu, Z.J. Shen, Dynamic stability analysis of synchronverter-dominated microgrid based on bifurcation theory. IEEE Trans. Ind. Electron. **64**(9), 7467–7477 (2017)

10. H. Bevrani, T. Ise, Y. Miura, Virtual synchronous generators: a survey and new perspectives. Int. J. Electr. Power Energy Syst. **54**, 244–254 (2014)
11. F. Milano, F. Dorfler, G. Hug, D.J. Hill, G. Verbic, *Foundations and challenges of low-inertia systems (Invited paper), in 2018 Power Systems Computation Conference (PSCC)* (Ireland, Dublin, 2018), pp. 1–25
12. S. Chen, H. Han, X. Chen, Y. Sun, X. Hou, *Reviews on inertia emulation technology with power electronics, in 2020 IEEE Energy Conversion Congress and Exposition (ECCE)* (MI, USA, Detroit, 2020), pp.2101–2107
13. J. Roldan-Perez, A. Rodriguez-Cabero, M. Prodanovic, Design and analysis of virtual synchronous machines in inductive and resistive weak grids. IEEE Trans. Energy Convers. **34**(4), 1818–1828 (2019)
14. H. Wu, X. Ruan et al., Small-signal modeling and parameters design for virtual synchronous generators. IEEE Trans. Ind. Electron. **63**(7), 4292–4303 (2016)
15. L. Huang, H, Xin, et al., Quantified analysis method of frequency response characteristics for power systems with virtual inertia Automation of Electric Power Syst. **42**(8), 31–38 (2018)
16. J. Alipoor, Y. Miura, T. Ise, Power system stabilization using virtual synchronous generator with alternating moment of inertia. IEEE. J. Emerg. Sel. Top. Power Electron. **3**(2), 451–458 (2015)
17. X. Hou, Y. Sun, X. Zhang, J. Lu, P. Wang, J.M. Guerrero, Improvement of frequency regulation in VSG-based AC microgrid via adaptive virtual inertia. IEEE Trans. Power Electron. **35**(2), 1589–1602 (2020)
18. G. Liu, S. Pu, M. Wu, L. Sun, T. Zhao, B. Li, Summary of virtual synchronous machine technology demonstration project. Distribution Utilization **36**(4), 37–42 (2019). https://doi.org/10.19421/j.cnki.1006-6357.2019.04.006
19. Z. Wang, F. Zhuo, H. Yi, J. Wu, F. Wang, Z. Zeng, Analysis of dynamic frequency performance among voltage-controlled inverters considering virtual inertia interaction in microgrid. IEEE Trans. Ind. Appl. **55**(4), 4135–4144 (2019)
20. M. Chen, D. Zhou, C. Wu, F. Blaabjerg, Characteristics of parallel inverters applying virtual synchronous generator control. IEEE Trans. Smart Grid. **12**(6), 4690–4701 (2021)
21. Z. Shuai, W. Huang, Z.J. Shen, A. Luo, Z. Tian, Active power oscillation and suppression techniques between two parallel synchronverters during load fluctuations. IEEE Trans. Power Electron. **35**(4), 4127–4142 (2020)
22. Y. Peng, W. Huang, Z. Shuai, The research of transient load sharing in islanded microgrid with Heterogeneous DGs. Proceedings of the CSEE **41**(15), 5167–5179 (2021)
23. J. Alipoor, Y. Miura, T. Ise, Stability Assessment and Optimization Methods for Microgrid With Multiple VSG Units. IEEE Trans. Smart Grid. **9**(2), 1462–1471 (2018)
24. M. Choopani, S.H. Hosseinain, B. Vahidi, A novel comprehensive method to enhance stability of multi-VSG grids. Int. J. Electr. Power Energy Syst. **104**, 502–514 (2019)
25. M. Choopani, S.H. Hosseinian, B. Vahidi, New transient stability and LVRT improvement of multi-VSG grids using the frequency of the center of Inertia. IEEE Trans. Power Syst. **35**(1), 527–538 (2020)
26. L. Xiong et al., Static synchronous generator model: a new perspective to investigate dynamic characteristics and stability issues of grid-tied PWM inverter. IEEE Trans. Power Electron. **31**(9), 6264–6280 (2016)
27. K. Shi et al., Rotor inertia adaptive control and inertia matching strategy based on parallel virtual synchronous generators system. IET Gener Transm Dis. **14**(10), 1854–1861 (2020)
28. L. Xiong, X. Liu, Y. Liu, F. Zhuo, Modeling and stability issues of voltage-source converter dominated power systems: a review. CSEE J. Power Energy Syst. https://doi.org/10.17775/CSEEJPES.2020.03590
29. X. Ge, H. Han, W. Xiong, S. Mei, Z. Liu, Y. Sun, Locally-distributed and globally-decentralized control for hybrid series-parallel microgrids. Int. J. Electr. Power Energy Syst. **116**(99), 1 (2020)
30. M. Li, W. Huang, N. Tai, L. Yang, D. Duan, Z. Ma, A dual-adaptivity inertia control strategy for virtual synchronous generator. IEEE Trans. Power Syst. **35**(1), 594–604 (2020)

31. D. Li, Q. Zhu, S. Lin, X. Bian, A self-adaptive inertia and damping combination control of vsg to support frequency stability. IEEE Trans. Energy Convers. **32**(1), 397–398 (2017)
32. A. Karimi, Y. Khayat, M. Naderi et al., Inertia response improvement in AC microgrids: a fuzzy-based virtual synchronous generator Control. IEEE Trans. Power Electron. **35**(4), 4321–4331 (2020)
33. V. Thomas, K.S., S. Ashok, Fuzzy controller-based self-adaptive virtual synchronous machine for microgrid application. IEEE Trans. Energy Convers. **36**(3), 2427–2437 (2021)

Chapter 10
Quantitative Control Parameters Design Oriented to Synchronization Stability

10.1 Parameter Design Issues Oriented to Synchronization Stability

Virtual synchronous generator (VSG) control has been an attractive method for converters to provide the grid-forming ability and virtual inertia support [1]. On the one hand, the VSG mimics the behavior of the synchronous generator (SG). On the other hand, the VSG parameters design is more flexible, which makes it possible to retain and even surpass the static/dynamic characteristics of the SG.

As is widely discussed, the active power reference P^*, the virtual inertia constant H, and the damping coefficient D_p are the core parameters for the grid-connected VSG. Substantial efforts have been devoted to studying the VSG parameters settings, considering small-signal stability [2], static/dynamic performance optimization [3, 4], and physical constraints [5]. Although some insights into parameter design are given, these works hardly consider transient synchronization stability requirements. Meanwhile, the control parameter range is more restrictive and only suitable for a small H and a large D_p.

Transient synchronization stability refers to the ability that the VSG maintains synchronization with the grid under large disturbances. The influcnce of parameters on transient stability has been discussed recently. The increase of active power reference P^* narrows the transient stability margin and a VSG with a large capacity is prone to lose transient stability [6]. Virtual inertia brings the trade-off between transient stability and frequency response that large H is profitable for inertia support but small H is conducive to transient stability [7]. The damping coefficient D_p has a positive damping effect on transient stability, but it is often designed fixed due to grid code requirements [8], which restricts the increase of D_p to enhance transient stability. These above works belong to qualitative analysis. The design for these parameters to ensure transient synchronization stability remains to be elucidated, which entails quantitative studies.

X. Hou et al., *Self-organized Operation and Stability Control of Electronized Power Networks*, https://doi.org/10.1007/978-981-96-9067-1_10

10.2 Problems Description

Several assumptions are made: (i) The inner control loops are overlooked. (ii) The influence of the VSG voltage dynamic and the reactive power control are ignored. (iii) The line impedance is highly inductive. (iv) Only the transient stability issues under a light grid voltage sag, which won't trigger the overcurrent protection, are focused on.

10.2.1 System Description

Figure 10.1 shows the grid-connected VSG system. The active power control loop of VSG can be expressed as

$$2H\omega^* \frac{\mathrm{d}\omega}{\mathrm{d}t} = P^* - P - D_p\omega^*(\omega - \omega^*) \tag{10.1}$$

where H and D_p represent the virtual inertia constant, and the damping coefficient, respectively. P^* and P are the active power reference and output active power at the PCC, respectively. ω^* and ω are the frequency reference and output voltage angular frequency.

10.2.2 Grid Code Requirements on Transient Stability

Taking IEEE Standards 1547-2018 as an example, it laid down the requirements on the transient synchronization stability issues that the VSG should keep working

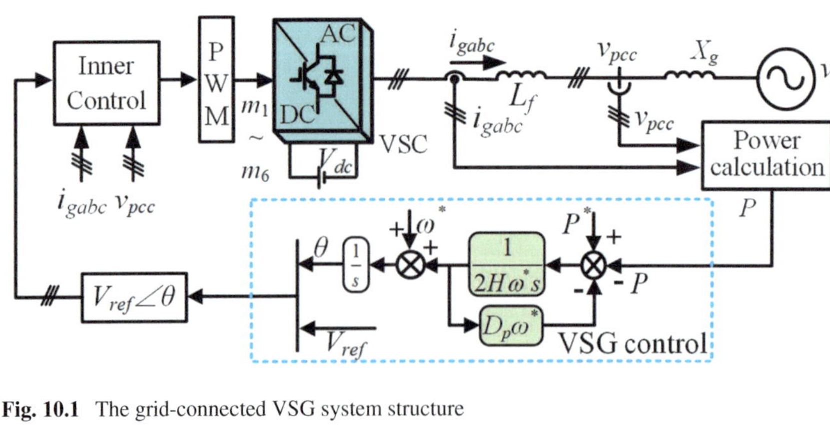

Fig. 10.1 The grid-connected VSG system structure

normally at least for 10 s when the grid voltage drop to 0.5–0.8 p.u. [8, 9]. The grid code requires that the following conditions should be satisfied: (1) The equilibrium point exists after the grid voltage drop to 0.5–0.8 p.u. (2) The deceleration area is larger than the acceleration area. Therefore, the transient synchronous stability issues discussed in this section are oriented to Type-I transient stability problems [10].

10.2.3 Parameters Design Requirements on Transient Stability

Figure 10.2 shows the damping effect on the design of the active power reference P^*, where δ_0 is the initial power angle and δ_s is the power angle at the new equilibrium point after grid voltage sag.

To ensure transient stability, the maximum deceleration area should be greater than the acceleration area. In Fig. 10.2a, P^* can only be less than 1 p.u. considering no damping effect. In Fig. 10.2b, the active power reference can be regarded as $P^* - D_p(\omega - \omega^*)$. With the damping effect, the deceleration area is larger and the acceleration area is smaller. Therefore, the value of P^* can be selected larger owing to the positive damping effect. However, as the damping effect lacks quantitative characterizing, how to determine the proper boundary of P^* to ensure transient stability remains to be studied.

As for the design of the virtual inertia constant H and the damping coefficient D_p, Fig. 10.3 shows that large D_p and small H are preferred for transient synchronization stability. However, D_p is often designed not to exceed its upper limit according to frequency regulation requirements and small H cannot provide enough inertia support. It is urgent to co-design H and D_p to ensure transient synchronization stability and meanwhile satisfy other grid code requirements.

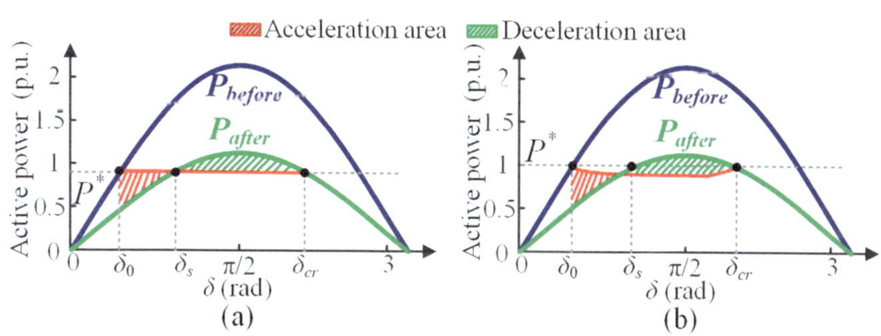

Fig. 10.2 Power angle curves under normal and grid voltage sag conditions: **a** Without considering the damping effect; **b** Considering the damping effect

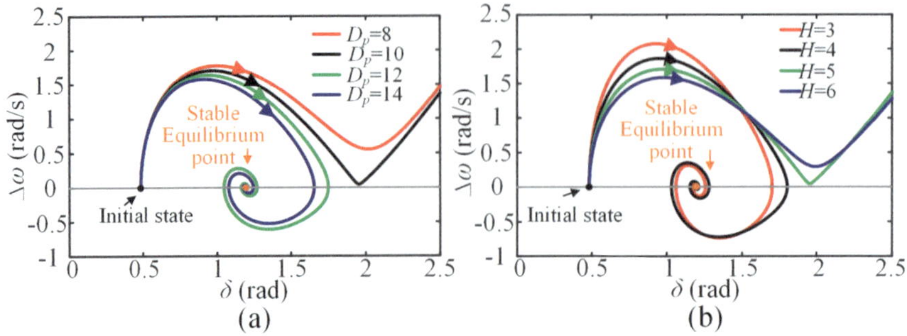

Fig. 10.3 Trajectories $\delta - \Delta\omega$ of the VSG when the grid voltage drops from 1 p.u. to 0.6 p.u.: **a** With different D_p (H = 5); **b** With different H (D_p = 10)

10.3 Parameters Design Guideline

10.3.1 Sufficient Conditions For VSG Parameter Design

To investigate the parameters design considering transient stability requirements, the Lyapunov function $V(\Delta\delta, \Delta\omega)$ of the system is employed with the form [6]

$$V(\Delta\delta, \Delta\omega) = H\omega^*\Delta\omega^2 - \frac{P_1^{\max}}{\omega^*}\Delta\delta + D_p\omega^*\lambda\Delta\omega\Delta\delta - \frac{P_2^{\max}}{\omega^*}[\cos(\Delta\delta+\delta_s) - \cos\delta_s] + \frac{D_p^2\omega^*}{4H}\lambda(\Delta\delta)^2 \tag{10.2}$$

where $\Delta\delta = \delta - \delta_s$, $P_1^{\max} = P^* + \frac{3V_{g\_sag}^2}{2X_g}$, $P_2^{\max} = \frac{3V_{pcc}V_{g\_sag}}{2X_g}$, λ is a constant $0 < \lambda < 1$ .v_{g_sag} represents the grid voltage amplitude after grid voltage sag. And v_{g_sag} is selected as 0.5 p.u.. According to Lyapunov's stability criterion [11], the stable system should satisfy

$$V_0 < V_{cr} \tag{10.3}$$

where, V_0 represents the initial energy with ($\Delta\delta_0 = \delta_0 - \delta_s$, $\Delta\omega = 0$), and V_{cr} is the critical energy with the maximum critical point ($\Delta\delta_{cr} = \pi - 2\delta_s$, $\Delta\omega = 0$). V_0 and V_{cr} are shown as

$$\begin{cases} V_0 = \frac{D_p^2\lambda\omega^*(\Delta\delta_0)^2}{4H} - \frac{P_1^{\max}\Delta\delta_0}{\omega^*} - \frac{P_2^{\max}[\cos(\Delta\delta_0+\delta_s)-\cos\delta_s]}{\omega^*} \\ V_{cr} = \frac{D_p^2\lambda\omega^*(\Delta\delta_{cr})^2}{4H} - \frac{P_1^{\max}\Delta\delta_{cr}}{\omega^*} - \frac{P_2^{\max}[\cos(\Delta\delta_{cr}+\delta_s)-\cos\delta_s]}{\omega^*} \end{cases} \tag{10.4}$$

Combine (10.3–10.4), the transient stability boundary can be given

$$\frac{\lambda\omega^* D_p^2}{4H}(\delta_0 + \pi - 3\delta_s) > \frac{P_1^{\max}}{\omega^*} + \frac{P_2^{\max}[\cos\delta_0 + \cos\delta_s]}{\omega^*(\delta_0 + \delta_s - \pi)} \tag{10.5}$$

where δ_0 and δ_s can be calculated as

$$\delta_0 = \arcsin(\frac{P^*}{P_3^{\max}}), \delta_s = \arcsin(\frac{P^*}{P_2^{\max}}) \tag{10.6}$$

where $P_3^{\max} = \frac{3V_{pcc}V_g}{2X_g}$

The condition $(\delta_0 + \pi - 3\delta_s) > 0$ should be ensured as a prerequisite. The active power reference P^* should satisfy

$$\pi + \arcsin\left(\frac{P^*}{P_3^{\max}}\right) - 3\arcsin\left(\frac{P^*}{P_2^{\max}}\right) > 0 \tag{10.7}$$

When $v_{g\_sag} = 0.5$ p.u., the boundary of P^* is derived as

$$P^* < \frac{\sqrt{14}}{8} P_3^{\max} \tag{10.8}$$

Based on (10.5) and (10.8), the boundary of the control parameters H and D_p can be obtained as

$$\frac{D_p}{\sqrt{H}} > \frac{2}{\omega^* \sqrt{\lambda}} \sqrt{\frac{P_1^{\max}}{\delta_0 + \pi - 3\delta_s} + \frac{P_2^{\max}(\cos\delta_0 + \cos\delta_s)}{(\delta_0 + \delta_s - \pi)(\delta_0 + \pi - 3\delta_s)}} \tag{10.9}$$

The boundaries (10.8) and (10.9) determine sufficient conditions for P^*, H, and D_p to ensure the transient synchronization stability of the VSG.

10.3.2 RoCoF Requirement on Virtual Inertia

According to IEEE Standards 1547-2018, the RoCoF is required to be no more 3 Hz/s during grid voltage sag [9]. The RoCoF is approximately calculated as

$$\text{RoCoF}_{\max} = \frac{1}{2\pi}\left|\frac{d\omega}{dt}\right|_{\max} \approx \frac{\Delta P}{4\pi\omega^* H} \tag{10.10}$$

The change of active power ΔP can be calculated as

$$\Delta P = \frac{3v_{pcc}(v_g - v_{g\_sag})}{2X_g}\sin\delta_0 = \frac{3(v_g - v_{g\_sag})P^*}{2v_g} \tag{10.11}$$

According to (10.10) and (10.11), Hmin should be designed to satisfy (10.12) to avoid the over-limitation of RoCoF.

$$H_{\min} \geq (1.2 \sim 1.5) \cdot \frac{3(v_g - v_{g\_sag})P^*}{8\pi\, v_g \omega^* \mathrm{RoCoF}_{\max}} \tag{10.12}$$

where v_{g_sag} = 0.5 p.u. and $\mathrm{RoCoF}_{\max}$= 3 Hz/s.Without losing conservatism, the designed minimum virtual inertia constant should be (1.2–1.5) times larger than the calculated result.

10.3.3 Feasible Parameters Region

The parameters design flow chart is depicted in Fig. 10.4. The setting for P* should satisfy (10.8) as a prerequisite, Then, the H and D_p should satisfy (10.9) for transient synchronous stability and the minimum value of H should satisfy (10.12) for the RoCoF requirement. Finally, considering the existing standards (2 s $\leq H \leq$6 s [5], and $Dp \leq (P^*/2\pi\omega^*)$) [3–5]) as a supplement, the feasible parameters region (FPR1) can be given in Fig. 10.5a. And the corresponding descriptions are listed in Table 10.1.

As the damping ratio ζin the VSG small-signal model is expressed as [8]

$$\xi = \frac{D_p}{\sqrt{H}} \sqrt{\frac{\omega^*}{8P_3^{\max} \cos \delta_0}} \tag{10.13}$$

The sufficient conditions in (10.9) can also be expressed in the form of the damping ratio in (10.14), which can provide a more intuitive interpretation of the parameters design.

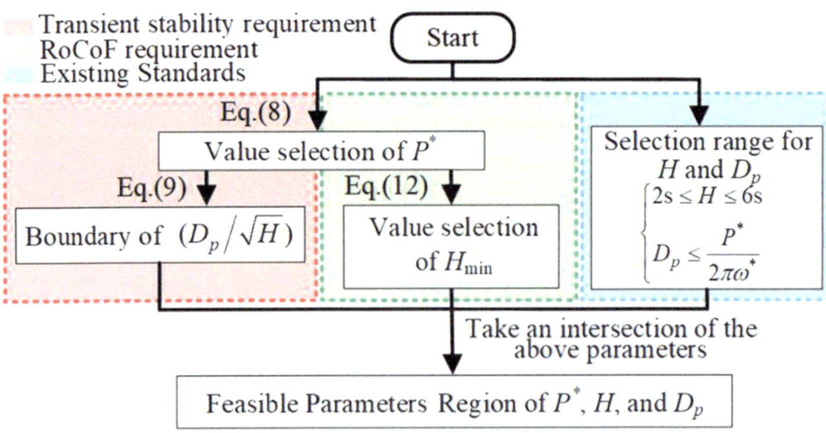

Fig. 10.4 Comparisons: **a** Introduced region; **b** The parameters range in [4]

Table 10.1 Stability region of control parameters

Parameter area	Description	
	Yes	Transient synchronization stability
FPR_1	Yes	RoCoF requirement
	Yes	Existing standards for H and D_n
Area A	No	RoCoF requirement cannot be ensured
Area B	No	Transient stability cannot be ensured
Area C	No	Frequency regulation cannot be satisfied
Area D	No	Slow dynamic response with large inertia

$$\xi > \sqrt{\frac{1}{2P_3^{\max}\cos\delta_0\omega^*\lambda}\left[\frac{P_1^{\max}}{\delta_0+\pi-3\delta_s}+\frac{P_2^{\max}(\cos\delta_0+\cos\delta_s)}{(\delta_0+\delta_s-\pi)(\delta_0+\pi-3\delta_s)}\right]} \tag{10.14}$$

Figure 10.5b also shows the FPR2 given by [4], which is aimed at the dynamic performance improvement near the equilibrium point and is merely the sufficient conditions for small-signal stability. Compared with our study, the FPR2 does not consider the RoCoF requirements on virtual inertia. As D_p has an upper limit, the value of H is restricted to a small value with the $\zeta = 0.4 - 0.8$. The allowable H is small and leads to exceeding the RoCoF limits.

Comparatively, when only transient synchronous stability condition (10.9) is considered, ζ is only required larger than 0.104. The stable area (A + FPR1 + C + D) in Fig. 10.5a satisfies the large-signal stability requirements, in which the small-signal stability is obviously satisfied. Also, this area (A + FPR1 + C + D) of Fig. 10.5a contains FPR2 of Fig. 10.5b. For practical implementation, the RoCoF requirement and other existing standards are further considered. H also satisfies (10.12), $H_{\max} \leq 6s$, and $D_pmax \leq (P^*/2\pi\omega^*)$. Then, area (A + C + D) should be removed and FPR1 can be obtained with systematic design among stability, RoCoF requirement, and

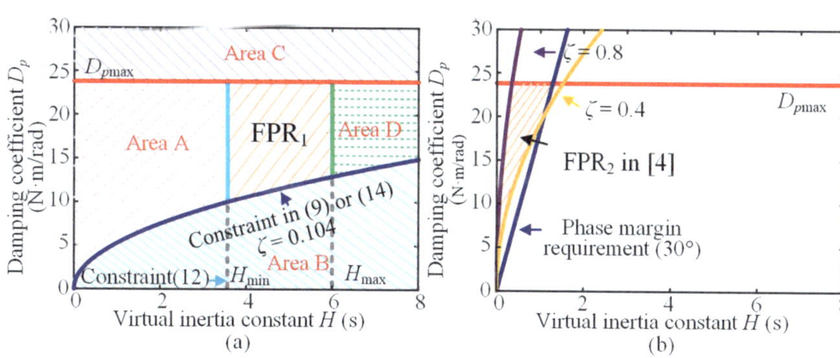

Fig. 10.5 Comparisons: **a** Introduced region; **b** The parameters range in [4]

existing standards. The introduced FPR1 may give engineers more freedom in the VSG parameters setting.

10.3.4 Discussions Under Different Current Limitation Strategies

The current limitation strategy is important since the power electronics converter can only handle 1.2-2 p.u. overcurrent [12]. The current saturation limiter and virtual impedance method are two typical current limitation strategies.

(1) Current saturation limiter: VSG triggering the current saturation limiter would switch to grid-following control mode. Under such circumstances to avoid transient instability, some effective measures, such as good anti-windup control for voltage loop and proper operating-mode switching method for recovery are needed [12], instead of the parameter design of P^*, H, and D_p.
(2) Virtual impedance method: The virtual impedance method can limit the current magnitude and maintain similar power-angle characteristics to normal VSG. However, the virtual impedance will deteriorate transient stability. It is necessary to redesign VSG parameters during overcurrent limitation. The virtual impedance method can be expressed as

$$X_v = \begin{cases} 0, I \leq I_{thres} \\ X_v, I > I_{thres} \end{cases} \tag{10.15}$$

where X_v presents the virtual impedance, I_{thres} represents the threshold of overcurrent and is selected as $(I_{thres} = I_0)$ here.

We choose I_0 as the rated current and the maximum overcurrent $(I_{\max} = 1.5 I_0)$. To avoid overcurrent, X_v can be approximately designed to satisfy $(X_{eq} \approx X_g + X_v > 1.5 \cdot X_g)$. Then, replace X_g with X_{eq} in $P_1^{\max}$ and $P_2^{\max}$. According to (10.7) and (10.9) the new reference $P_{new}{}^*$ and the new boundary of $(D_P/\sqrt{H})_{new}$ new during overcurrent limitation can be obtained. Therefore, the VSG needs to adjust P^* to $P^*{}_{new}$ and satisfy $(D_P/\sqrt{H})_{new}$ to realize both transient synchronous stability and overcurrent limitation.

10.4 Hardware-in-Loop Experimental Validation

The control hardware-in-loop (CHIL) experiments are carried out to verify the accuracy of the introduced feasible region, which are based on the OPAL-RT4510 with the system settings $v_{pcc} = 325$ V,v_g =311 V, $v_{g\_sag} = 155.5$ V(0.5 p.u. grid voltage

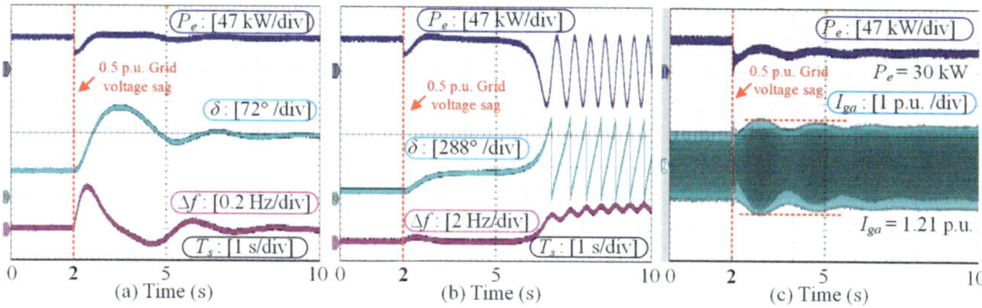

Fig. 10.6 CHIL experimental results. **a** Case 1; **b** Case 2; **c** Case 3

sag), and X_g =1.5 Ω. P^* is selected to be 47 kW through (10.8). Then, the feasible parameters region in the CHIL experiments can be obtained as

$$(D_p\sqrt{H}) > 5.28, 3.57 \leq \text{H} \leq 12{,}0 < D_p \leq 23.8 \tag{10.16}$$

Cases are carried out to verify the accuracy of the parameter design method and Fig. 10.6 shows the CHIL experimental results.

Case 1: Stable with $H = 5$ s, $D_p = 12\,\text{N} \cdot \text{m/rad}$, $\left(D_P/\sqrt{H}\right) \approx 5.36 > 5.28$. As is shown in Fig. 10.6a, transient stability is ensured when the parameters are located in the introduced feasible parameters region.

Case 2: Unstable with $H = 5$ s, $D_p = 10\,\text{N} \cdot \text{m/rad}$, $\left(D_P/\sqrt{H}\right) \approx 4.47 < 5.28$. Figure 10.6b shows that the VSG transient synchronization stability cannot be guaranteed when the parameters are out of range (10.16).

Case 3: Stable with virtual impedance $X_v = 0.75\,\Omega$. Parameters ($P_{new}{}^* < 30.756$ kW and $\left(D_p/\sqrt{H}\right) > 2.36$ can be determined . It is found that the requirement for $(D_P/\sqrt{H})_{new}$ new is lower after the derating operation of P^*. Therefore, only P^* is required to reduce to $P^*{}_{new}$, and the value of H and D does not need to be changed. The parameters can be selected as $P^*{}_{new}$ =30 kW and $H = 5\,\text{s}$, D_p=12 N · m/rad. As is shown in Fig. 10.5c , the output active power changes from 47 to 30 kW, and transient synchronous stability is ensured meanwhile the current is limited within 1.5 p.u. after 0.5 p.u. grid voltage sag.

10.5 Conclusion

This section provides an intuitive reference for VSG parameters design oriented to transient synchronization stability. The sufficient conditions of active power reference P^* , virtual inertia constant H, and damping coefficient D_p are derived to ensure transient stability. Furthermore, the feasible parameters region is given, considering grid-code-based transient stability and RoCoF requirement. This section provides a guideline for control parameters design and practical engineering application of VSG. Based on this work, the influence of the voltage dynamics and the reactive power loop, and the design oriented to the Type-II transient stability issues will be taken into consideration in the future.

References

1. H. Bevrani et al., Virtual synchronous generators: a survey and new perspectives. Int. J. Electr. Power Energy Syst. **54**, 244–254 (2014)
2. J. Chen, T. O'Donnell, Parameter constraints for virtual synchronous generator considering stability. IEEE Trans. Power Syst. **34**(3), 2479–2481 (2019). https://doi.org/10.1109/TPWRS.2019.2896853
3. Z. Lv, et al. Parameters analysis and operation area calculation of VSG applied to distribution network. CSEE JPES. https://doi.org/10.17775/CSEEJPES.2019.02330
4. H. Li et al., Parameters design in active power control of virtual synchronous generator considering power-angle characteristic nonlinearity. Int. J. Electr. Power Energy Syst. **130**, 106952 (2021)
5. J. Chen et al., Adaptive virtual synchronous generator considering converter and storage capacity limits. CSEE JPES. **8**(2), 580–590 (2022). https://doi.org/10.17775/CSEEJPES.2019.03360
6. Z. Shuai et al., Transient angle stability of virtual synchronous generators using Lyapunov's direct method. IEEE Trans. Smart Grid. **10**(4), 4648–4661 (2019)
7. D. Pan et al., Transient stability of voltage-source converters with grid-forming control: a design-oriented study. IEEE J. Emerg. Sel. Topics Power Electron. **8**(2), 1019–1033 (2020)
8. X. Xiong et al., Transient damping method for improving the synchronization stability of virtual synchronous generators. IEEE Trans. Power Electron. **36**(7), 7820–7831 (2021)
9. IEEE standard for interconnection and interoperability of distributed energy resources with associated electric power systems interfaces, in IEEE Std 1547-2018, vol., no., pp. 1–138 (2018)
10. H. Wu, X. Wang, A mode-adaptive power-angle control method for transient stability enhancement of virtual synchronous generators. IEEE. J. Emerg. Sel. Topics Power Electron. **8**(2), 1034–1049 (2020)
11. J. Machowski, *Power System Dynamics—Stability and Control* (Wiley, NewYork, NewYork, 2008), pp. 5–7, 207–209, 222–230
12. B. Fan et al., A review of current-limiting control of grid-forming inverters under symmetrical disturbances. IEEE Open J. Power Electron. **3**, 955–969 (2022)

Chapter 11
Modified VSG Combing Small-Signal and Synchronization Stability

Power grids are interfaced by more and more inverter-interfaced distribution generators (IIDG) [1, 2]. The Grid-forming converters control based on IIDG, among the various available control strategies, is deemed to be the most promising solution for low-inertia power systems [3–5]. Virtual synchronous generator (VSG) is one of the typical grid-forming control methods [6–8]. VSG control introduces the swing equation of SGs into the control algorithm to emulate the inertia and damping characteristics of the synchronous generator (SG) and provides grid-friendly virtual inertia to ensure frequency stability [9].

Virtual impedance or virtual resistance technology has been widely used on VSG-IIDGs [4, 10–16]. Substantial efforts have been devoted to the application of virtual impedance for various purposes, including reactive power-sharing enhancement in microgrids [11], current limitation [12], transient oscillations damping, and small-signal stability improvement [13–15]. In the aspect of small-signal stability improvement, virtual resistance is introduced to damp Synchronous resonance (SR) [16–18]. SR is one of the typical small-signal stability issues of grid-forming control converters [19]. A pair of conjugate open-loop poles at $-R/L \pm j\omega_g$ will be introduced and lead to potential instability, especially in a system with a low R/X ratio. Compared with adding a real resistor to the output of the IIDG, virtual resistance is effective for SR suppression without extra power loss.

Even though the above studies explored the benefits of virtual resistance in many aspects, virtual resistance jeopardizes the transient synchronization stability of VSG [20]. The transient stability represents the ability of the VSG-IIDG to maintain synchronization with the grid under large disturbances. In [21], the phase-portrait method is used to analyze the transient stability of the four typical grid-forming control schemes. And the conclusion has been drawn that smaller virtual inertia J and larger damping coefficient D conduces to the transient stability of VSG-IIDG. To further improve the transient stability of VSG, methods to modify the active power control loop, like the transient damping method [22–24], co-design control parameters method [25], and mode-adaptive-switching method [26, 27] have been

X. Hou et al., *Self-organized Operation and Stability Control of Electronized Power Networks*, https://doi.org/10.1007/978-981-96-9067-1_11

Table 11.1 The existing transient stability enhancement methods for VSG-IIDG

Methods	Proposals Parameter	Influence of R
Modify the control loop	Transient damping [22–24] Mode-switching [26]	Negative to transient stability
Change the power	Adaptive virtual inertia [27] Modify Q-V control loop [29, 30] Reducing P reference [20, 28, 33]	Negative to transient stability
Co-design control parameters	Increasing Q reference [34] Coordinate J and D [21]	Negative to transient stability
Output impedance optimization	Optimal Damping Design [25] Adaptive virtual resistance [35]	Negative to transient stability
	Transient virtual impedance cascading With HPF [16, 31, 32]	Limited improvement in transient stability

introduced. The influence of the reactive power control loop of VSG is also studied in [28]. It analyzed the deterioration of transient stability brought by the reactive power control loop with the Lyapunov direct method. In [29, 30], the VSG reactive control loop is modified to enhance transient stability. Besides, the virtual impedance cascading with HPF (High Pass Filter), also known as the transient virtual impedance [10, 16, 31, 32], can help improve transient stability, but its performance is limited. These methods can be summarized in Table 11.1. Although so many enhancement methods have been introduced recently, the conflict between small-signal stability and transient synchronization stability of VSG introduced by virtual resistance still needs to be solved.

This chapter focuses on the conflict caused by virtual resistance between SR suppression in small-signal stability and transient synchronization stability of VSG. The contribution is summarized as follows:

(1) A simple but effective modified VSG control scheme is introduced to address the conflict we studied. Virtual active power replaces real active power in the control loop and enhances both small-signal stability and transient synchronization stability with the existence of virtual resistance.

(2) Comprehensive and valid proof of the effectiveness of the introduced modified VSG control is presented. The introduced modified VSG control extends the VSG control structure and can be cooperated with other enhancement methods to further improve the transient synchronization stability while the required functions of grid-forming control remain.

The rest of this chapter is organized as follows. In Sect. 11.1, the conflict caused by virtual resistance between small-signal stability and transient synchronization stability of VSG is reviewed. Section 11.2 presents the modified VSG scheme in detail. CHIL experimental results are provided in Sect. 11.3. Conclusions are presented in Sect. 11.4.

11.1 Effects of Virtual Resistance

11.1.1 VSG System Description

Figure 11.1 depicts the general structure of the three-phase grid-connected VSG system. Virtual resistance R_v is introduced for SR suppression, and the voltage reference is changed to $\mathbf{V}_{\mathbf{ref}} = \mathbf{V}_{\mathbf{vref}} - R_v\mathbf{I}_{\mathbf{gabc}}$. R_g and L_g represent the grid resistance and inductance, respectively. $\mathbf{V}_{\mathbf{g}}$, $\mathbf{V}_{\mathbf{pcc}}$, $\mathbf{I}_{\mathbf{gab}}$, and $\mathbf{I}_{\mathbf{Labc}}$ are the space vectors to represent the grid voltage, the PCC voltage, the PCC current, and the inductor current, respectively. The transient stability of the VSG is mainly determined by the outer power control loops [21]. Besides, this chapter only focuses on the large disturbances which do not provoke the overcurrent limit control [26, 35, 36].

The outer power control loop of VSG can be expressed as follows

$$J\omega^*\frac{\mathrm{d}\omega}{\mathrm{d}t} = P_{ref} - P_e - D\omega^*(\omega - \omega^*) \tag{11.1}$$

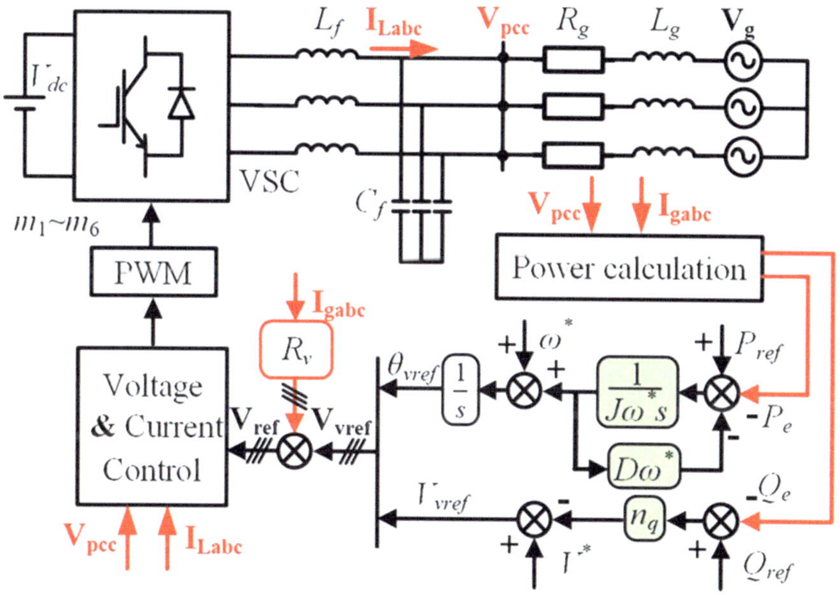

Fig. 11.1 The general structure of the three-phase VSG with virtual resistance

$$V_{vref} = V^* - n_q(Q_{ref} - Q_e) \tag{11.2}$$

where J, D, and n_q represent the virtual inertia constant, damping coefficient, and Q-V loop coefficient, respectively. P_{ref}, Q_{ref}, and P_e, Q_e are the power reference and output power at the PCC, respectively. ω^* and ω are the frequency reference and output voltage angular frequency, V^* and V_{vref} are the voltage normal amplitude and voltage amplitude.

11.1.2 Effects of Virtual Resistance on Small-Signal Stability

Figure 11.2 shows the equivalent circuit of the VSG with virtual resistance. Assuming the grid impedance is inductive, the dynamic small-signal model of P_e and Q_e can be expressed as

$$\Delta P_e = G_{P\delta} \Delta\delta_{vref} \tag{11.3}$$

$$\Delta Q_e = G_{QV} \Delta V_{vref} \tag{11.4}$$

where $G_{P\delta}$(s) and G_{QV}(s) are shown in Appendix.

The VSG open-loop transfer functions from output power to power reference can be derived as

$$G_{P\_ol}(s) = \frac{1}{s\omega^*(Js + D)} G_{P\delta}(s) \tag{11.5}$$

$$G_{Q\_ol}(s) = n_q G_{QV}(s) \tag{11.6}$$

where G_{P_ol}(s) and G_{Q_ol}(s) represent the open-loop transfer function of $\Delta P_e(s)/\Delta P_{ref}(s)$ and $\Delta Q_e(s)/\Delta Q_{ref}(s)$, respectively.

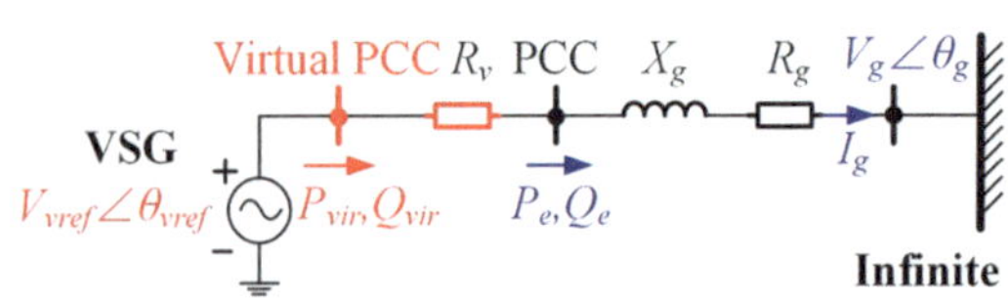

Fig. 11.2 The equivalent circuit of the VSG with virtual PCC and virtual resistance

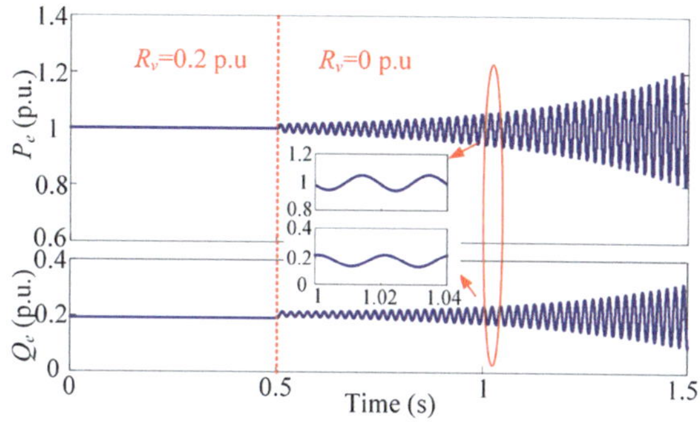

Fig. 11.3 Graphical diagram of the introduced communication network

It is reported that a pair of conjugate open-loop poles will be introduced and may lead to potential instability [16–18] and adding the virtual resistance can help to solve this issue [20]. Figure 11.3 shows the response of P_e and Q_e when Rv decreases from 0.2 p.u. to 0 at 0.5 s. Virtual resistance can effectively increase the system stability from the small-signal stability perspective.

11.1.3 Effects of Virtual Resistance on Transient Synchronization Stability

The transient stability issue is induced by the dynamic of the power angle under large disturbance [23]. Without loss of generality, we mainly analyze the transient stability problem with no equilibrium point [26]. The system can recover to be stable if the disturbance is cleared before the critical clear time (CCT). Critical clear time and critical clear angle (CCA) are utilized as the transient stability indicators.

According to EEAC (Expanded Equal Area Criterion) and the phase portrait, Fig. 11.4 is plotted to further illustrate the influence of virtual resistance R_v and grid resistance R_g on transient stability. An inverse time integral-based approach is applied to obtain a more accurate CCT and a less conservative transient stability boundary [5]. From Fig. 11.4, the effects of R_v and R_g on transient synchronization stability are opposite. CCT and CCA increase with the increase of R_g, while larger R_v results in the reduction of CCT and CCA. Therefore, the existence of R_v will jeopardize the transient synchronization stability of VSG while R_g is conducive to transient stability.

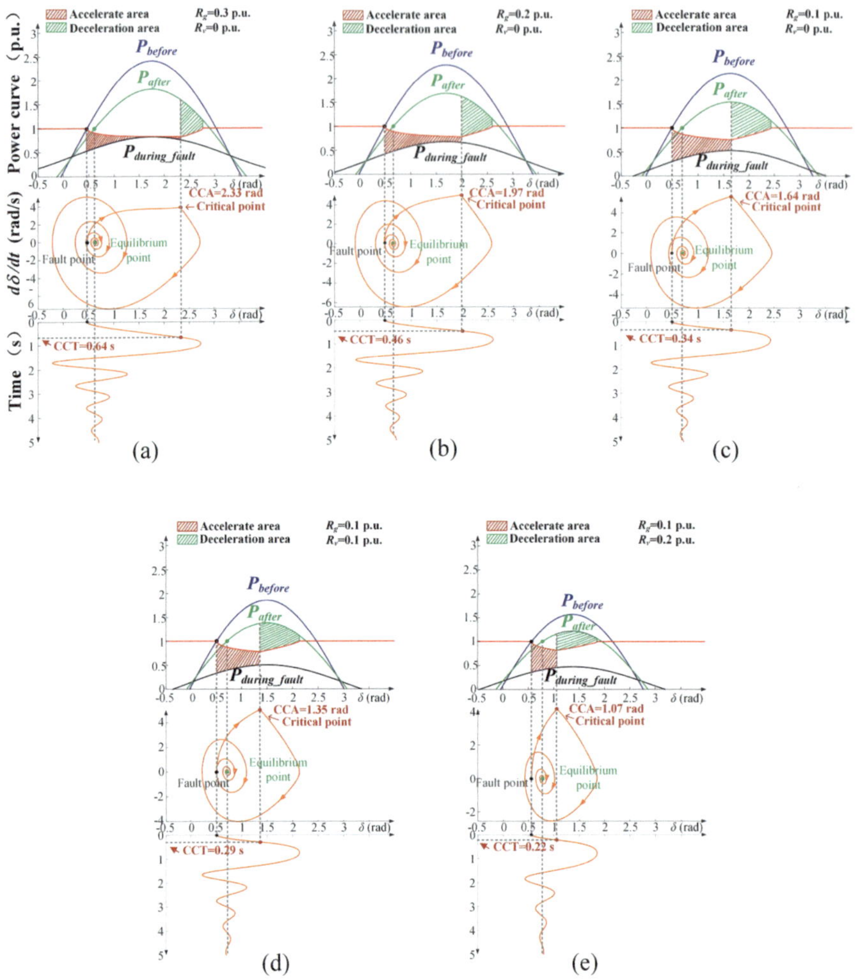

Fig. 11.4 Graphical diagram of the introduced communication network

11.1.4 Discussion

As discussed previously, the virtual resistance method can be adopted for SR suppression from a small-signal stability perspective. However, the existence of virtual resistance will inevitably lead to the deterioration of transient stability. Therefore, the conflict between small-signal stability and transient synchronization stability is urgent to be settled down.

11.2 Modified VSG Control Scheme with Virtual Resistance

11.2.1 Modified VSG Control Scheme

To address the conflict caused by virtual resistance, a modified VSG control scheme is introduced. The modified control scheme is shown in Fig. 11.5 and can be expressed in (11.7).

$$\begin{cases} J\omega^* \frac{d\omega}{dt} = P_m - P_{vir} + k \cdot [k_p(P_m - P_e) + k_i \int (P_m - P_e)dt] \\ P_m = P_{ref} - D\omega^*(\omega - \omega^*) \end{cases} \tag{11.7}$$

where, P_{vir} represents the virtual active power at virtual PCC. P_e represents the real active power. P_m represents the input power reference after frequency feedback. k_p and k_i are the proportional coefficient and the integral coefficient of the PI (Proportion Integral) controller. k takes the value of 0 or 1, which depends on the converter operating mode. The PI controller is designed to make the real active power p_e track the power reference P_{ref} in normal conditions (k=1). In the fault conditions (k=0), the PI controller is disconnected. The fault detection methods to trigger the value of k are not discussed in detail in this chapte, more detailed information can be referred to [23, 37, 38].

The virtual active power P_{vir} is expressed as

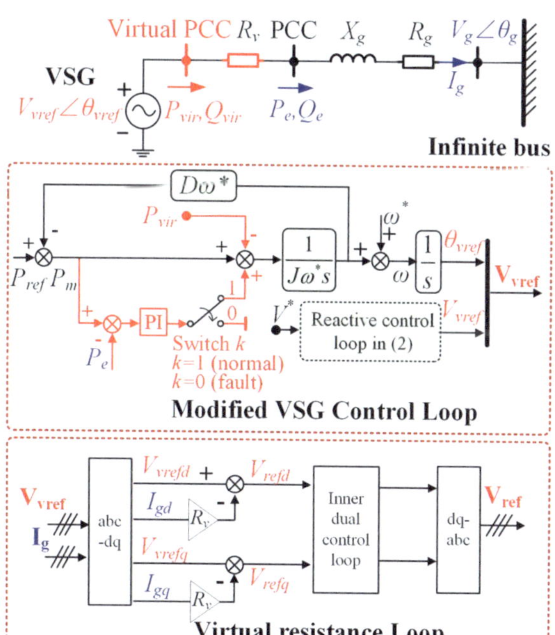

Fig. 11.5 The introduced modified VSG control with virtual resistance R_v

$$P_{vir} = \frac{3}{2}(V_{vrefd} I_{gd} + V_{vrefq} I_{gq}) \tag{11.8}$$

where, V_{vrefd}, V_{vrefd}, V_{vrefq}, I_{gd}, and I_{gq} are the voltage V_{vref} components and current I_g components in a rotating reference frame d_q with the d-axis aligned with the grid voltage.

Compared with other methods [20–30, 33–35], the key is that the virtual power P_{vir} is introduced to the active power control loop in (11.7) instead of the real power P_e. Then, the effect of virtual resistance R_v on transient stability can be equivalent to that of the grid resistance R_g. The small-signal stability and transient stability enhancement of the introduced modified VSG control scheme are analyzed as follows.

11.2.2 Small-Signal Stability Analysis

The small-signal control block diagram of the modified VSG with virtual resistance can be derived in Fig. 11.6. This analysis is based on the normal condition (the switch k =1) which means the PI controller is connected. $G_{Pvir\delta}(s)$ is the small-signal transfer function of $P_{vir}(s)/\delta(s)$, which is shown in the Appendix.

The bode diagram of the open-loop transfer function $\Delta P(s)/\Delta P_{ref}(s)$ is shown in Fig. 11.7. With the increase of R_v, the resonance peak value is reduced effectively. Thus, the effectiveness of virtual resistance on SR suppression remains under the introduced modified VSG control. The small-signal stability is improved.

To further analyze the influence of the rest control parameters on the small-signal stability, the root locus method is applied. Figure 11.8a and b show the root locus when J and D vary, which is similar to the conventional VSG control [8]. Figure 11.8c and d show the root locus when k_p and k_i vary.The existence of the integral term is unfavorable for small-signal stability. When the integral coefficient k_i is too large, a pair of closed-loop poles will move to the right region of the imaginary axis and lead to instability. Considering the requirement for dynamic response of real active power tracking and the system stability jointly, a comprehensive parameter design is required. The value of k_i is selected to the maximum under meeting the damping ratio requirement.

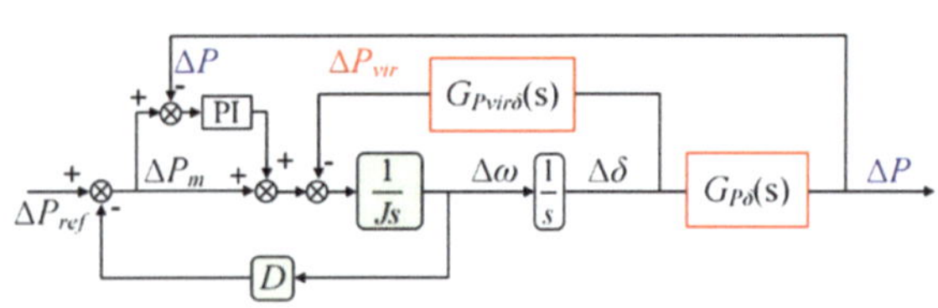

Fig. 11.6 The small-signal control block diagram of the modified VSG method with virtual resistance R_v

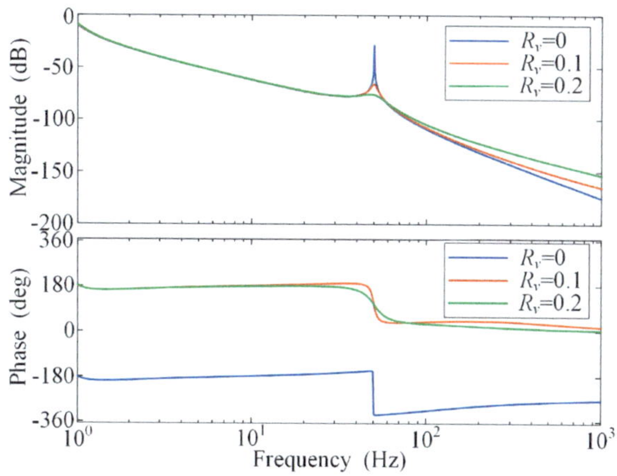

Fig. 11.7 Bode diagram of $P(s)/\Delta P_{ref}(s)$ with R_g=0.05 p.u, X_g=1.5 p.u

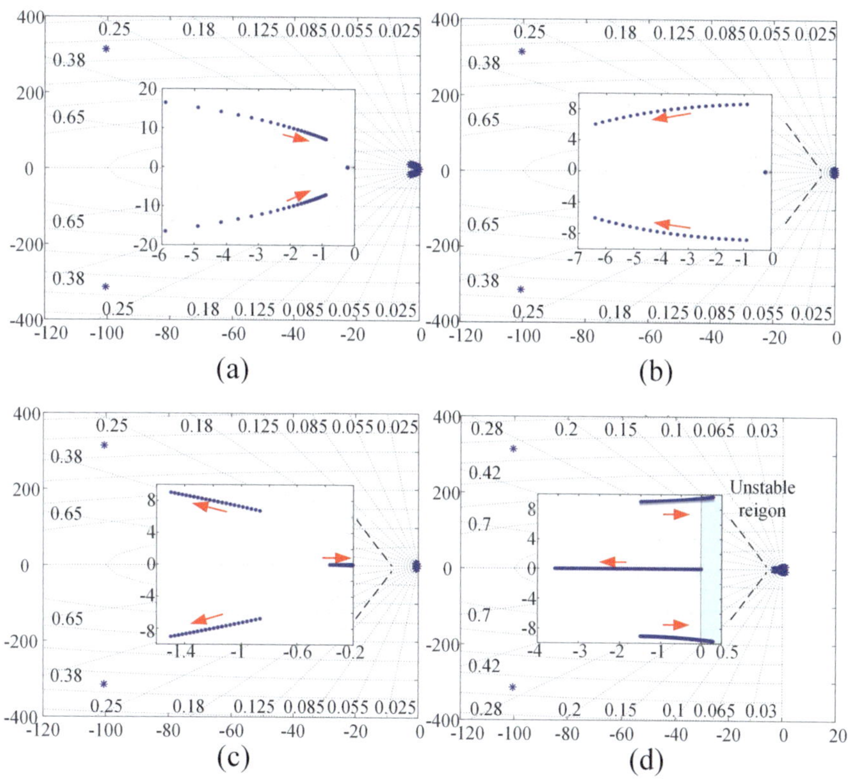

Fig. 11.8 Root locus: **a** when J varies from 1 to 10; **b** when D varies from 1 to 10; **c** when k_p varies from 0.01 to 0.5; **d** when k_i varies from 0.1 to 10

11.2.3 Transient Synchronization Stability Assessment

According to Fig. 11.5, the PI controller is disabled when the fault is detected. The virtual power angle is defined as $\delta = \theta_{vref} - \theta_g$. In this case, the swing equation in the introduced modified VSG control can be derived as

$$\begin{cases} \dot{\delta} = \omega - \omega_g \\ J\omega^* \dot{\omega} = P_{ref} - P_{vir} - D\omega^*(\omega - \omega^*) \end{cases} \tag{11.9}$$

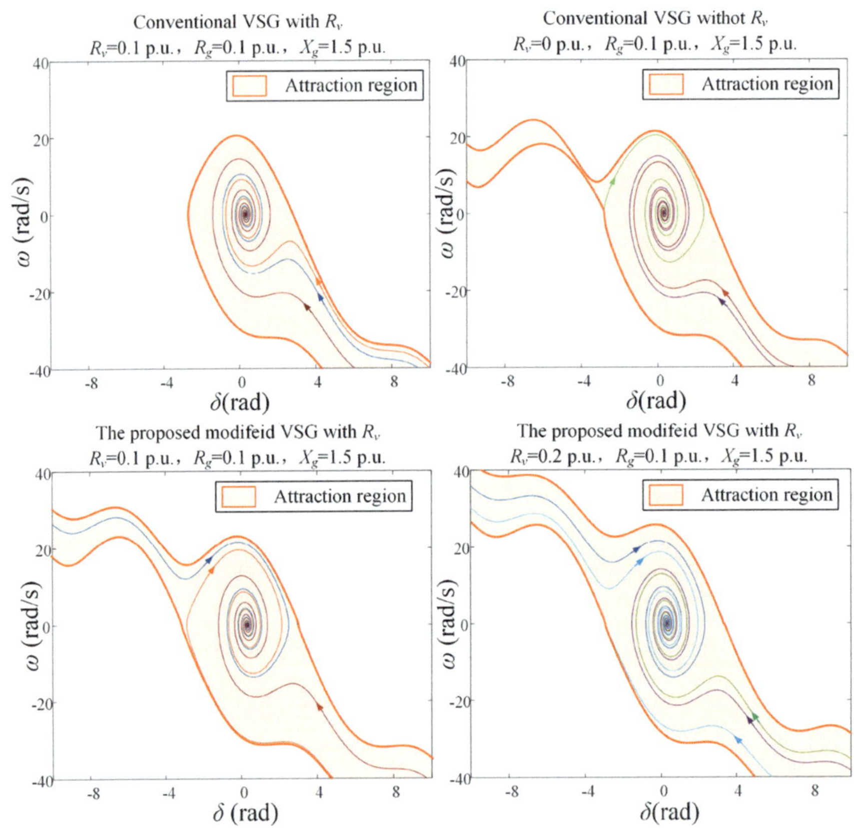

Fig. 11.9 The comparison of the attraction region between the basic VSG method with R_v and the introduced modified VSG control scheme with R_v

where P_{vir} is expressed as

$$P_{vir} = \frac{3}{2}\frac{V_{vref}^2\left(R_v + R_g\right) + V_{vref}V_g\left[X_g \sin\delta - \left(R_v + R_g\right)\cos\delta\right]}{\left(R_v + R_g\right)^2 + X_g^2} \tag{11.10}$$

According to the swing (11.9), the attraction region can be obtained for transient stability assessment. Figure 11.9 depicts the comparison of the attraction region between the conventional VSG method with R_v and the introduced modified VSG control. The existence of R_v in conventional VSG narrows the attraction region, which means that R_v harms transient stability. The introduced modified VSG control can greatly broaden the attraction region with the increase of virtual resistance R_v. The transient stability is greatly enhanced.

From the analysis mentioned above, the conflict between small-signal stability and transient stability due to virtual resistance is effectively solved under the introduced modified VSG control. The existence of virtual resistance benefits both small-signal stability and transient synchronization stability under the modified VSG control scheme.

11.2.4 Qualitative Analysis on Frequency Regulation and Virtual Inertia Support

As one of the grid-forming control schemes, the introduced modified VSG control is required to have the ability for frequency regulation and virtual inertia support.

Participating in frequency regulation refers to the ability that the grid-connected converter can adjust the output active power according to the grid frequency variation. Assuming that there is a grid frequency drop $\Delta\omega_g = 2\pi\Delta f_g$. According to (11.7), we have (11.11) in steady-state

$$P_e = P_{ref} - D\omega^*(\omega - \omega^*) \tag{11.11}$$

$$\omega = \omega_g - \Delta\omega_g \tag{11.12}$$

where ω^* is the same value as the nominal grid frequency ω_g.

Thus, the introduced modified VSG control scheme still has the $P - f$ droop relationship and can achieve the primary frequency regulation.

Providing virtual inertia support refers to the ability that the grid-connected converter can suppress its maximum RoCoF (Rate of Change of Frequency) to avoid triggering operation protection based on grid code during disturbances. Assuming that there is a step change of the power reference ΔPref. Usually, the maximum RoCoF is achieved when t=0.

Thus, according to (11.7), the expression of maximum RoCoF can be derived as (11.13) approximately.

$$\left.\frac{d\omega}{dt}\right|_{max} = \frac{(1 + k_p)\Delta P_{ref}}{J\omega^*} \tag{11.13}$$

From (11.13), the maximum RoCoF is determined by k_p and J. To avoid the negative impacts on frequency stability introduced by the proportional coefficient k_p, the value of k_p is selected to be no more than 0.2. Therefore, the introduced VSG control scheme can provide similar virtual inertia support to conventional VSG.

11.2.5 Compared with the Transient Virtual Resistance Method

To better illustrate the advantages of the introduced modified VSG control, comparisons with the transient virtual resistance method [16, 31, 32] are carried out. The control structure of the transient virtual resistance method is shown in Fig. 11.10.

Figure 11.11 shows the bode diagrams $\Delta P(s)/\Delta P_{ref}(s)$ under different methods. As indicated, the transient virtual resistance $R_v(s)$ is equivalent to R_v in the high-frequency region (nearly > 100 Hz). $R_v(s)$ suppresses the resonance peak value and improves the small-signal stability. However, its performance of reducing resonance peak value is weaker than that of the introduced modified VSG control.

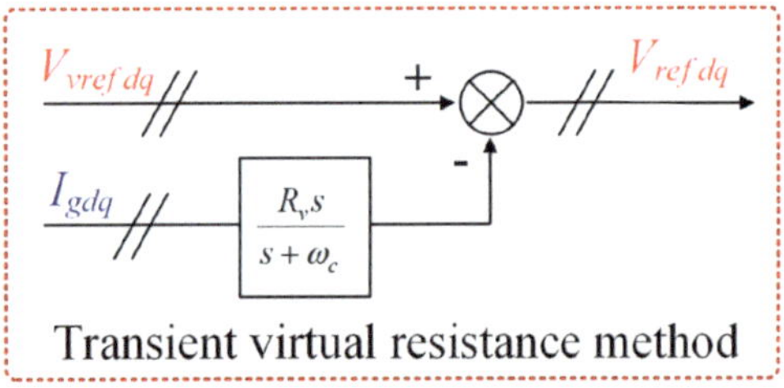

Fig. 11.10 The structure of the transient virtual resistance controller

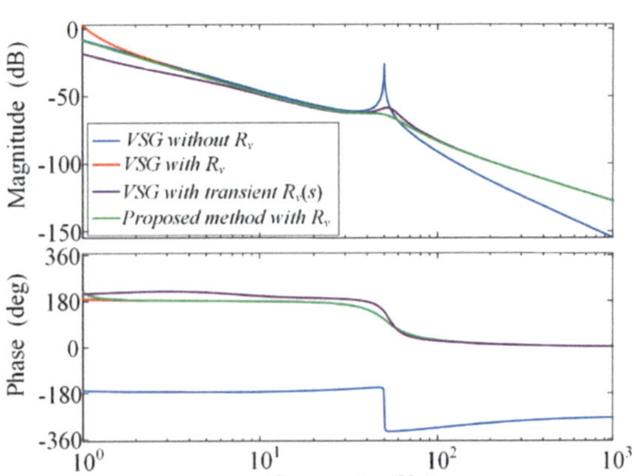

Fig. 11.11 The bode diagrams of the open-loop transfer function $\Delta P(s)/\Delta P_{ref}(s)$ under different control schemes with R_v=0.2 p.u, R_g=0.05 p.u, X_g=1.5 p.u

Usually, the quasi-steady-state approximation is used to analyze transient stability [39]. The dynamics of the inner control loops are neglected due to their higher bandwidth [21]. As the transient virtual resistance $R_v(s)$ only affects the dynamics of the inner control loop, the function of $R_v(s)$ is equivalent to (11.14).

$$\lim_{s\to 0} R_v(s) = \frac{R_v s}{s + \omega_c} = 0 \tag{11.14}$$

In other words, the transient virtual resistance method is equivalent to the virtual resistance method with R_v=0 from the perspective of transient synchronization stability. As discussed in [20], larger virtual resistance R_v will deteriorate the transient stability of VSG. Thus, the negative effect of $R_v(s)$ on the transient synchronization stability is trivial according to the above analysis, which also explains the reason why the transient stability of the transient virtual resistance method outperforms that of the virtual resistance method.

As indicated in (11.9), the effect of virtual resistance R_v in the introduced modified VSG control on transient stability is equivalent to that of the grid resistance R_g. Because the grid resistance R_g could effectively enhance the transient stability of VSG [20], it is safe to say that the introduced modified VSG control outperforms the transient virtual impedance method in terms of improving transient stability.

11.2.6 Compared with the Existing Advanced VSG Methods

To further illustrate the advantages of the introduced modified VSG control scheme, the comparisons between the introduced modified VSG control and the existing methods are shown in Table 11.2. The SR phenomena can be suppressed by adding virtual resistance. All the methods can improve the transient synchronization stability of VSG. However, only the introduced modified VSG control and the transient virtual impedance method [10, 16, 31] can solve the conflict caused by virtual resistance between small-signal stability and transient stability. And the introduced modified VSG control can greatly enhance transient stability. Besides, the implementation complexity of the introduced scheme is low and it has high expansibility to cooperate with other methods [19–30, 33, 34]. Moreover, the ability for virtual inertia support and power-sharing regulation also remains.

11.3 Control-Hardware-in-Loop Experimental Verification

To verify the effectiveness of the introduced modified VSG control scheme, experiments based on control hardware-in-loop (CHIL) are carried out. The CHIL platform is shown in Fig. 11.12. The three-phase grid-connected converters, the LC filter, the line impedance, and the grid of the main circuit are emulated in the OPAL-RT4510

Table 11.2 Comparison with the existing advanced VSG methods

Methods	SR suppression	Transient stability	Influence of R_v	Complexity	Expansibility
Introduced scheme with R_v	Yes	√	√ Greatly enhances transient stability	Low	High
Conventional VSG with R_v	Yes	×	× Deteriorates transient stability	Low	High
Transient virtual impedance [16, 31, 32]	Yes	√	– Limitedly improves transient stability	Low	High
Modify the control loop [21–23, 25, 26, 28, 29]	Needs R_v	√	× Deteriorates transient stability	Medium	Medium
Change the power reference [19, 27, 30, 33]	Needs R_v	√	× Deteriorates transient stability	Low	Low
Co-design control parameters [20, 24]	Needs R_v	√	× Deteriorates transient stability	High	High
Output impedance optimization [34]	Yes R_v	√	× Deteriorates transient stability	High	Medium

Table 11.3 Stability region of control parameters

Parameters	Values	Parameters	Values
Grid voltage V_g	311 V	Rated active power P_{ref}	45 kW
Rated voltage V^*	325 V	Rated reactive power Q_{ref}	0
Grid frequency f_g	50 Hz	Virtual inertia J	6
Rated frequency f^*	50 Hz	Damping coefficient D	8
Line resistance R_g	0.05 Ω	Q-V loop coefficient n_q	0.001
Line inductance L_g	1.5 mH	Proportional coefficient k_p	0.05
C_f of LC filter	50 μF	Integral coefficient k_i	0.2
L_f of LC filter	1.2 mH	Virtual resistance R_v	0.4 Ω

simulator. The controller of the converter is implemented in the DSP-TMS320F28335 control board, and the sampling frequency is set to be 10 kHz. The experimental data is recorded by the host machine of OPAL-RT4510 and can be displayed through an oscilloscope. The experimental parameters are listed in Table 11.3.

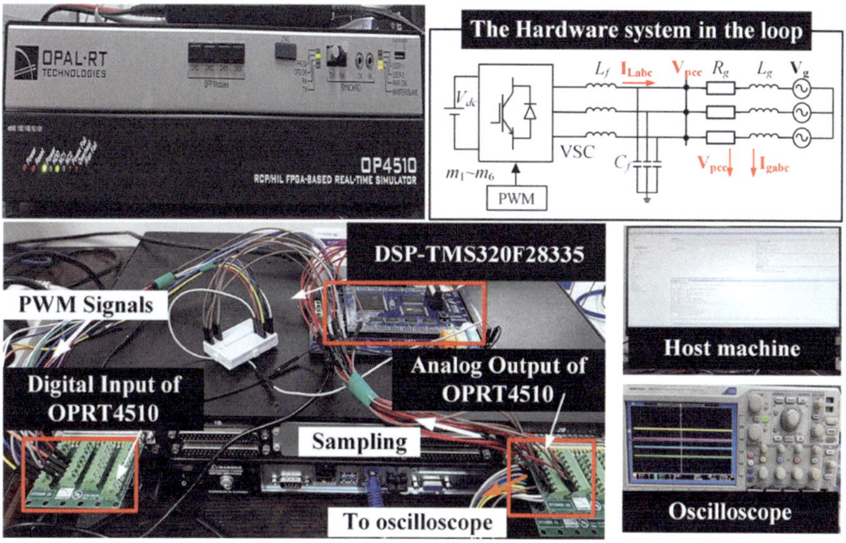

Fig. 11.12 The configuration of the control hardware in the loop system

11.3.1 Synchronous Resonance Suppression

To verify the effectiveness of SR suppression, a small disturbance of grid resistance change is considered in this case. The initial value of R_g is set as 0.2 Ω. After the system reached the steady state, R_g is changed to 0.05 Ω. Comparison CHIL experiments between conventional VSG and the introduced modified VSG control with R_v are carried out.

Figure 11.14a shows the CHIL experiment results underf conventional VSG. When R_g is reduced from 0.2 Ω to 0.05 Ω, SR phenomena occur in the curves of current, active power, and frequency. It may lead to potential instability for grid-forming control converters [18, 19]. Figure 11.13b shows the results under the introduced modified VSG control with R_v =0.4 Ω. SR phenomena have been suppressed effectively. Therefore, the introduced modified VSG control with virtual resistance can improve small-signal stability.

11.3.2 Frequency Regulation and Inertia Supportability

The modified VSG control scheme should have the ability for frequency regulation and virtual inertia support under normal conditions. In this case, a 0.5 Hz drop in grid frequency has been considered. It is required that the output power should be

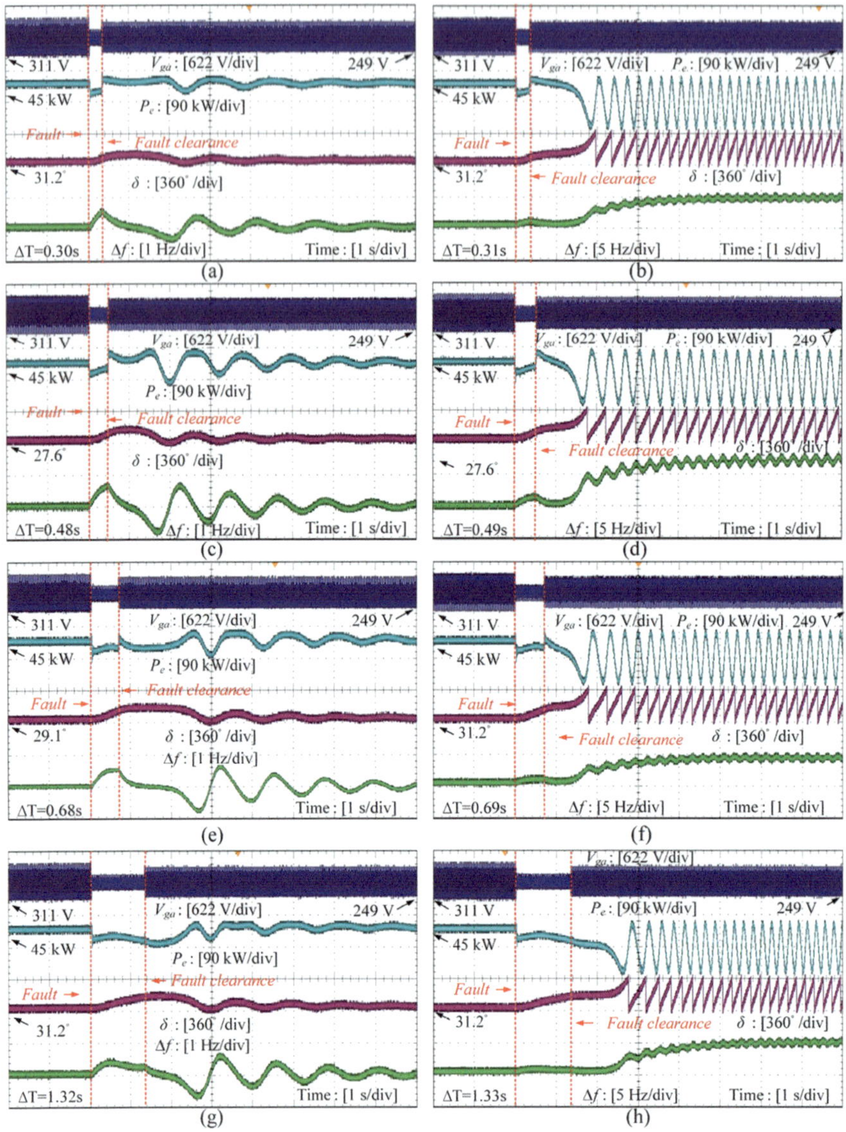

Fig. 11.13 CHIL experimental results under voltage sag: **a** The conventional VSG control with R_v=0.4 Ω under fault clearance time $\Delta T = 0.30$ s $= CCT_1$; **b** The conventional VSG control with R_v=0.4 Ω under fault clearance time $\Delta T = 0.31$ s $> CCT_1$; **c** The conventional VSG control with transient virtual impedance method, which integrates a HPF cascading with R_v=0.4 Ω under fault clearance time $\Delta T = 0.48$ s $= CCT_2$; **d** The conventional VSG control with transient virtual impedance method, which integrates a HPF cascading with $R_v = 0.4$ Ω under fault clearance time $\Delta T = 0.49$ s $> CCT_2$; **e** The introduced modified VSG control scheme with $R_v = 0.2$ Ω under fault clearance time $\Delta T = 0.68$ s $= CCT_2$; **f** The introduced modified VSG control scheme with $R_v = 0.2$ Ω. under fault clearance time $\Delta T = 0.69$ s $> CCT_2$; **g** The introduced modified VSG control scheme with $R_v = 0.4$ Ω under fault clearance time $\Delta T = 1.32$ s $= CCT_3$; **h** The introduced modified VSG control scheme with $R_v = 0.4$ Ω under fault clearance time $\Delta T = 1.33$ s $> CCT_3$

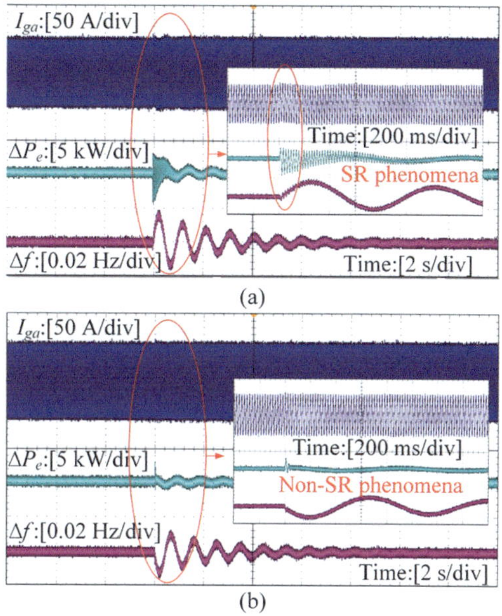

Fig. 11.14 CHIL experimental results for small-signal stability verification: **a** The conventional VSG control; **b** The introduced modified VSG control scheme with R_v

regulated to increase to participate in the frequency regulation when a 0.5 Hz drop occurs. Figure 11.15a shows the CHIL experimental results with the conventional VSG control and Fig. 11.15b shows the results with the introduced VSG control scheme. The introduced modified VSG control scheme can achieve similar control effects to conventional VSG in frequency regulation. Besides, the maximum RoCoF under the modified VSG control is about 2.21 Hz/s, which is close to it under the conventional VSG control. The modified VSG control hardly deteriorates the frequency stability.

11.3.3 Power-Sharing in Parallel Islanded Mode

In this case, two grid-forming converters are paralleled in islanded mode. The parameters are J_1=6, J_2=10, D_1=D_2=8, R_{g1}=R_{g2}=0.05 Ω, L_{g1}=1.5mH, L_{g2}=2.5 mH, R_{v1}=R_{v2}=0.4Ω, k_{p1}=k_{p2}=0.05, k_{i1}=k_{i2}=0.2. The load increase of 30 kW is considered in this case. The result is shown in Fig. 11.16. Under the introduced modified VSG control, the power-sharing of the two paralleled converters in islanded mode can still be realized.

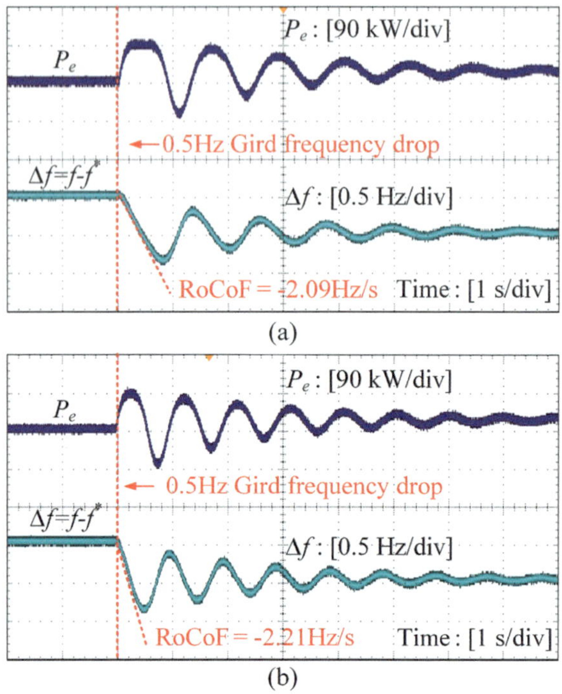

Fig. 11.15 CHIL experimental results under grid frequency drop: **a** The conventional VSG control; **b** The introduced modified VSG control scheme

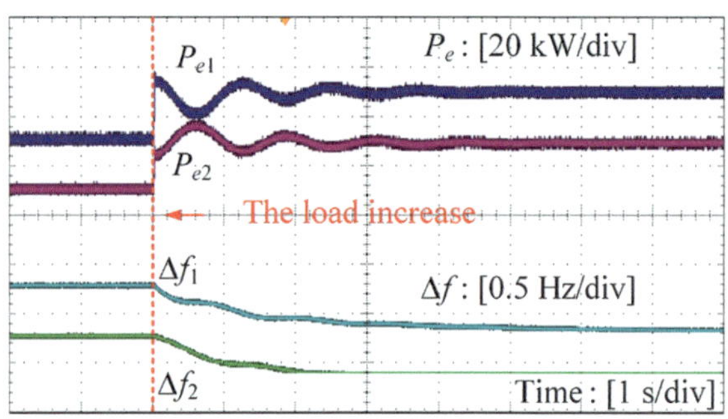

Fig. 11.16 CHIL experimental results of the introduced modified VSG control in parallel islanded mode

11.4 Conclusion

This chapter proposes a modified VSG control scheme with virtual resistance. Under this method, the conflict caused by virtual resistance between small-signal stability and transient stability is well solved. Besides, the Power-Sharing in Parallel Islanded Mode still has the expansibility to cooperate with other enhanced methods. The ability for frequency regulation, virtual inertia support, and power-sharing regulation remains. Finally, CHIL experimental results verify the effectiveness of the introduced modified VSG control scheme.

$$G_{P\delta}(s) = \frac{V_{vref}V_g(R_v+R_g+sL_g)\sin\delta_0+V_{vref}V_gX_g\cos\delta_0}{(R_v+R_g+sL_g)^2+X_g{}^2} - \frac{2V_{vref}V_gR_v[(R_v+R_g+sL_g)^2+X_g{}^2]\sin\delta_0}{(R_v+R_g+sL_g)^2+X_g{}^2} \tag{11.15}$$

$$G_{QV}(s) = \frac{V_{vref}V_gX_g\cos\delta_0-V_{vref}V_g(R_v+R_g+sL_g)\sin\delta_0}{(R_v+R_g+sL_g)^2+X_g{}^2} \tag{11.16}$$

$$G_{P_{vir}\delta}(s) = \frac{V_{vref}V_g(R_v+R_g+sL_g)\sin\delta_0+V_{vref}V_gX_g\cos\delta_0}{(R_v+R_g+sL_g)^2+X_g{}^2} \tag{11.17}$$

References

1. F. Blaabjerg, R. Teodorescu, M. Liserre, A.V. Timbus, Overview of control and grid synchronization for distributed power generation systems. IEEE Trans. Ind. Electron. **53**(5), 1398–1409 (2006)
2. F. Milano, F. Dorfler, G. Hug, D. J. Hill, G. Verbic, Foundations and challenges of low-inertia systems (Invited paper), in *2018 Power Systems Computation Conference (PSCC), Dublin, Ireland* (2018), pp. 1–25
3. X. Fu, J. Sun, M. Huang, Z. Tian, H. Yan, H.H.-C. Iu, P. Hu, X. Zha, Large-signal stability of grid-forming and grid-following controls in voltage source converter: a comparative study. IEEE Trans. Power Electron. **36**(7), 7832–7840 (2021)
4. J. Rocabert, A. Luna, F. Blaabjerg, P. Rodríguez, Control of power converters in AC microgrids. IEEE Trans. Power Electron. **27**(11), 4734–4749 (2012)
5. X. He, S. Pan, H. Geng, Transient stability of hybrid power systems dominated by different types of grid-forming devices. IEEE Trans Energy Convers. **37**(2), 868–879 (2022)
6. Q.C. Zhong, G. Weiss, Synchronverters: inverters that mimic synchronous generators. IEEE Trans. Ind. Electron. **58**(4), 1259–1267 (2011)
7. H. Bevrani, T. Ise, Y. Miura, Virtual synchronous generators: a survey and new perspectives. Int. J. Electr. Power Energy Syst. **54**, 244–254 (2014)
8. J. Liu, Y.S. Miura, H. Bevrani, T. Ise, Enhanced virtual synchronous generator control for parallel inverters in microgrids. IEEE Trans. Smart Grid. **8**(5), 2268–2277 (2017)
9. S. Chen, H. Han, X. Chen, Y. Sun, X. Hou, Reviews on inertia emulation technology with power electronics, in *2020 IEEE Energy Conversion Congress and Exposition (ECCE), Detroit, MI, USA* (2020), pp. 2101–2107
10. X. Wang, Y.W. Li, F. Blaabjerg, P.C. Loh, Virtual-impedance-based control for voltage-source and current source converters. IEEE Trans. Power Electron. **30**(12), 7019–7037 (2015)
11. A.S. Vijay, N. Parth, S. Doolla et al., An adaptive virtual impedance control for improving power sharing among inverters in islanded AC microgrids. IEEE Trans. Smart Grid. **12**(4), 2991–3003 (2021)
12. M.G. Taul, X. Wang, P. Davari, F. Blaabjerg, Current limiting control with enhanced dynamics of grid-forming converters during fault conditions. IEEE J. Emerg. Sel. Topics Power Electron. **8**(2), 1062–1073 (2020)
13. X. Lin, Y. Zheng, Z. Liang, Y. Kang, The suppression of voltage overshoot and oscillation during the fast recovery process from load shortcircuit fault for three-phase stand-alone inverter. IEEE J. Emerg. Sel. Topics Power Electron. **9**(1), 858–871 (2021)
14. A. Rodríguez-Cabero, J. Roldán-Pérez, M. Prodanovic, Virtual impedance design considerations for virtual synchronous machines in weak grids. IEEE J. Emerg. Sel. Topics Power Electron. **8**(2), 1477–1489 (2020)
15. G. Li et al., Analysis and mitigation of subsynchronous resonance in series-compensated grid-connected system controlled by a virtual synchronous generator. IEEE Trans. Power Electron. **35**(10), 11096–11107 (2020)

16. L. Zhang, L. Harnefors, H. Nee, Power-synchronization control of grid-connected voltage-source converters. IEEE Trans. Power Syst. **25**(2), 809–820 (2010)
17. L. Harnefors, M. Hinkkanen, U. Riaz, F.M.M. Rahman, L. Zhang, Robust analytic design of power-synchronization control. IEEE Trans. Ind. Electron. **66**(8), 5810–5819 (2019)
18. J. Wang, Y. Wang, Y. Gu, W. Li, X. He, Synchronous frequency resonance of virtual synchronous generators and damping control, in *Proceedings of 9th International Conference on Power Electronics, Asia* (2015), pp. 1011–1016
19. X. Wang, M.G. Taul, H. Wu, Y. Liao, F. Blaabjerg, L. Harnefors, Grid synchronization stability of converter-based resources—an overview. IEEE Open J. Ind. Appl. **1**, 115–134 (2020)
20. X. Xiong, C. Wu, F. Blaabjerg, Effects of virtual resistance on transient stability of virtual synchronous generators under grid voltage sag. IEEE Trans. Ind. Electron. **69**(5), 4754–4764 (2022)
21. D. Pan, X. Wang, F. Liu, R. Shi, Transient stability of voltage-source converters with grid-forming control: a design-oriented study. IEEE J. Emerg. Sel. Topics Power Electron. **8**(2), 1019–1033 (2020)
22. X. Xiong, C. Wu, B. Hu, D. Pan, F. Blaabjerg, Transient damping method for improving the synchronization stability of virtual synchronous generators. IEEE Trans. Power Electron. **36**(7), 7820–7831 (2021)
23. R. Sun, J. Ma, W. Yang, S. Wang, T. Liu, Transient synchronization stability control for LVRT with power angle estimation. IEEE Trans. Power Electron. **36**(10), 10981–10985 (2021)
24. M. Choopani, S.H. Hosseinian, B. Vahidi, New transient stability and LVRT improvement of multi-VSG grids using the frequency of the center of inertia. IEEE Trans. Power Syst. **35**(1), 527–538 (2020)
25. X. Xiong, C. Wu, P. Cheng, F. Blaabjerg, An optimal damping design of virtual synchronous generators for transient stability enhancement. IEEE Trans. Power Electron. **36**(10), 11026–11030 (2021)
26. H. Wu, X. Wang, A mode-adaptive power-angle control method for transient stability enhancement of virtual synchronous generators. IEEE J. Emerg. Sel. Topics Power Electron. **8**(2), 1034–1049 (2020)
27. P. Ge, C. Tu, F. Xiao, Q. Guo, Transient stability enhancement of a VSG based on flexible switching of control. Proc. CSEE. **42**(6), 2109–2124 (2022)
28. Z. Shuai, C. Shen, X. Liu, Z. Li, Z. Shen, Transient angle stability of virtual synchronous generators using Lyapunov's direct method. IEEE Trans. Smart Grid **10**(4), 4648–4661 (2019)
29. X. Xiong, C. Wu, F. Blaabjerg, An improved synchronization stability method of virtual synchronous generators based on frequency feedforward on reactive power control loop. IEEE Trans. Power Electron. **36**(8), 9136–9148 (2021)
30. M. Chen, D. Zhou, F. Blaabjerg, Enhanced transient angle stability control of grid-forming converter based on virtual synchronous generator. IEEE Trans. Ind. Electron. **69**(9), 9133–9144 (2022)
31. A.D. Paquette, D.M. Divan, Virtual impedance current limiting for inverters in microgrids with synchronous generators. IEEE Trans. Ind. Appl. **51**(2), 1–9 (2014)
32. X. Zhang et al., Analysis of dynamic power angle oscillation and its suppression strategy for the droop-controlled grid-connected inverter. IEEE J. Emerg. Sel. Topics in Power Electron. **9**(5), 5718–5731 (2021). https://doi.org/10.1109/JESTPE.2021.3078290
33. K. Sun, W. Yao, J. Wen, L. Jiang, A two-stage simultaneous control scheme for the transient angle stability of VSG considering current limitation and voltage support. IEEE Trans Power Syst. **37**(3), 2137–2150 (2022)
34. D. Pan, X. Wang, F. Liu, R. Shi, Transient stability impact of reactive power control on grid-connected converters, in *Proceedings of IEEE Energy Conversion Congress Exposition* (2019), pp. 4311–4316
35. S.P. Me, S. Zabihi, F. Blaabjerg, B. Bahrani, Adaptive virtual resistance for postfault oscillation damping in grid-forming inverters. IEEE Trans. Power Electron. **37**(4), 3813–3824 (2022)
36. L. Huang, H. Xin, Z. Wang, L. Zhang, K. Wu, J. Hu, Transient stability analysis and control design of droop-controlled voltage source converters considering current limitation. IEEE Trans. Smart Grid **10**(1), 578–591 (2019)

37. S.F. Zarei, H. Mokhtari, F. Blaabjerg, Fault detection and protection strategy for islanded inverter-based microgrids. IEEE J. Emerg. Sel. Topics Power Electron. **9**(1), 472–484 (2021)
38. E. Casagrande, W.L. Woon, H.H. Zeineldin, D. Svetinovic, A differential sequence component protection scheme for microgrids with inverter–based distributed generators. IEEE Trans. Smart Grid. **5**(1), 29–37 (2014)
39. X. Wang, H.-D. Chiang, Quasi steady-state model for power system stability: limitations, analysis and a remedy. Power Syst. Comput. Conf. **2014**, 1–7 (2014). https://doi.org/10.1109/PSCC.2014.7038362

Chapter 12
Adaptive Inertia and Damping Coordination Control for Grid-Forming VSG

12.1 AIDC Control Method of VSG

12.1.1 Classical VSG System Configuration

Figure 12.1 shows the typical VSG control diagram. The power control loop of VSG can be expressed as [1]:

$$J\frac{\mathrm{d}\omega}{\mathrm{d}t} = P^* - P - D\left(\omega - \omega^*\right) \tag{12.1}$$

$$V_{ref} = V^* - n\left(Q^* - Q\right) \tag{12.2}$$

where (12.1) and (12.2) represent the $P - \omega$ loop and $Q - V$ loop, respectively. P^* and Q^* are the reference of active power and reactive power reference, respectively. P and Q are the output power of active power and reactive power at the PCC. The virtual inertia coefficient and virtual damping coefficient are expressed as J and D, respectively. The $Q - V$ loop droop coefficient is expressed as n. V is the VSG output voltage at PCC. ω^* and V^* are the angular speed reference and the amplitude reference voltage. ω and V are the angular speed and amplitude of the VSG output voltage.

The active power transmitted from VSG to the grid is expressed as [2]:

$$P = \frac{3VV_g}{2X_g}\sin\delta \tag{12.3}$$

where V_g is the grid voltage amplitude and δ represents the power angle between VSG and the grid. X_g represents the line impedance, and the line resistance is ignored in the highly inductive transmission line.

X. Hou et al., *Self-organized Operation and Stability Control of Electronized Power Networks*, https://doi.org/10.1007/978-981-96-9067-1_12

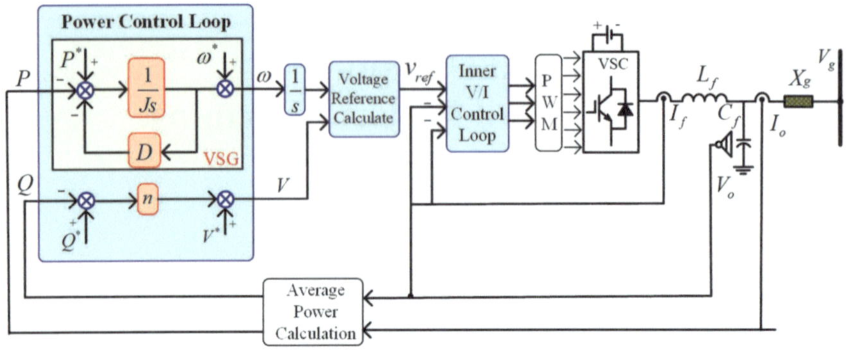

Fig. 12.1 Typical VSG control diagram of an inverter-based DG

12.1.2 Universal AIDC Control Principle Oriented to Transient Stability

As is shown in Fig. 12.2a, the maximum deceleration area should be larger than the maximum acceleration area to remain in synchronization with the grid under large distances according to the equal area criterion (EAC) [3]. Figure 12.2b shows the time-domain response of power angle/frequency. It can be mainly divided into four segments. The grid fault occurs at point B and time t_1. During the first segment from point B to point C (from t_1 to t_2), the frequency will accelerate, and the transient power angle is increased. This area is called the accelerated area. At this moment, a large virtual inertia is needed to slow down the deviation trend and reduce the maximum frequency deviation. At point C, the fault is cleared. During the second segment from point C to point D (from t_2 to t_3), the frequency will decelerate, but $\Delta\omega$ is greater than zero, and then the power angle is still increased. This area is called the decelerated area. In this process, small inertia is needed to make VSG return to the nominal frequency quickly. For example, if the virtual inertia is changed to 0 at point C, the frequency can return to the nominal frequency instantaneously. Then, the power angle of VSG will reduce, return back to point A, and remain stable in the steady state. The decelerated area for transient stability is no longer needed, which is totally different from the SG.

Therefore, a large amount of inertia is needed in the accelerated area, and a small amount of inertia is needed in the decelerated area. So, $\triangle J$ should be positive during the segments of $t_1 - t_2$ and $t_3 - t_4$, and it should be negative during the segments of $t_2 - t_3$ and $t_4 - t_5$. For the virtual damping control, the virtual damping coefficient D should be always greater than 0 during all segments. In addition, the dynamic adaptive compensation should only affect the dynamic response ($\triangle J = 0$, $\triangle D = 0$ in steady-state), and $\triangle J$ and $\triangle D$ should be designed by sampling available signals. The corresponding design guidelines are summarized in Table 12.1.

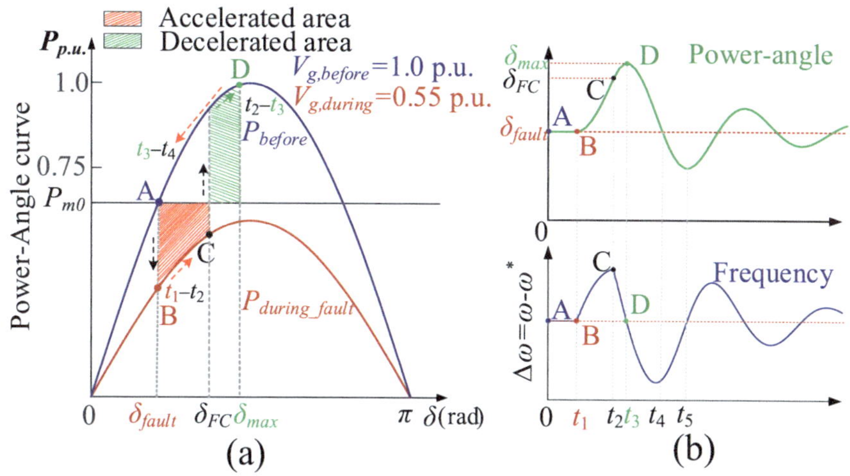

Fig. 12.2 Power angle curve (**a**); the time-domain response of power angle/frequency (**b**)

Table 12.1 Design principles of virtual inertia and damping coefficient oriented to transient synchronization stability at different operation states

Segment	$\Delta\omega$	$\mathrm{d}\omega/\mathrm{d}t$	ω State	Power angle operation area	ΔJ	ΔD
$t_1 - t_2$	> 0	> 0	Deviation	Accelerated aera	> 0	> 0
$t_2 - t_3$	> 0	< 0	Return	Decelerated aera	< 0	> 0
$t_3 - t_4$	< 0	< 0	Deviation	Accelerated aera	> 0	> 0
$t_4 - t_5$	< 0	> 0	Return	Decelerated aera	< 0	> 0

12.1.3 Arctan-Function-Based AIDC Control Method

The introduced AIDC control method can be expressed as:

$$(J_0 + \Delta J)\frac{\mathrm{d}\omega}{\mathrm{d}t} = P^* - P - (D_0 + \Delta D)(\omega - \omega^*) \tag{12.4}$$

where J_0 and D_0 represent the static control variables for the virtual inertia constant and damping coefficient, while ΔJ and ΔD represent the designed dynamic control variables. According to the design guidelines, the adaptive inertia and damping control items of AIDC can be designed based on real-time conditions of operation frequency deviation and its rate of change of frequency (RoCoF):

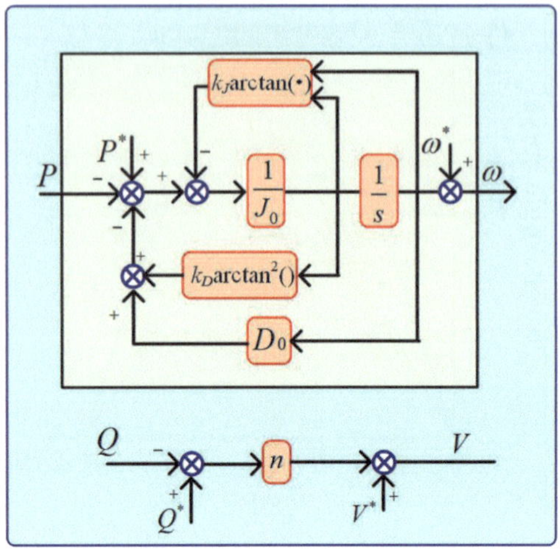

Fig. 12.3 The schematic diagram of the introduced AIDC control method

$$\begin{cases} \Delta J = k_J \arctan\left(\Delta\omega \cdot \frac{d\omega}{dt}\right) \\ \Delta D = k_D \arctan^2\left(\frac{d\omega}{dt}\right) \end{cases} \tag{12.5}$$

where K_J and K_D represent the adaptive constants of virtual inertia and the damping constant, respectively.

The introduced method contains two adaptive items, which can be seen in Fig. 12.3.

12.1.3.1 The Adaptive Virtual Inertia According to $\Delta\omega$ and $\frac{d\omega}{dt}$

The virtual inertia coefficient $(J_0 + \triangle J)$ is required to increase during the accelerated area and decrease in the decelerated area. It can adaptively change according to the combination of $\Delta\omega$ and $\frac{d\omega}{dt}$. The adaptive virtual inertia is also required to have the upper and lower boundaries. The maximum virtual inertia J_{max} is relevant to the instantaneous power capacity limitation of the inverter [4], and the minimum virtual inertia J_{min} is required to satisfy the minimum frequency stability requirement. Therefore, the upper and the lower limits of virtual inertia can be designed as [4, 5, 7]:

$$J_{min} \begin{cases} PO\left(J_{\max}\right)\Delta\omega_{\max} \leq P_{\max} - P^* \\ J_{\min} > \frac{\Delta P_{dis}}{|\,d\omega/dt|_{\max}} \end{cases} \tag{12.6}$$

where $PO\left(J_{\max}\right)$ represents the VSG output power overshoot with J_{max} caused by a step disturbance in $\Delta\omega$. P_{max} represents the instantaneous power capacity limitation of the inverter, and P^* represents the rated power. ΔP_{dis} represents the output power change due to disturbance, and $|\,d\omega/dt|_{\max}$ is the required maximum frequency change according to the grid code. To realize the required virtual inertia control, the mathematical function $\Delta J = \arctan(\Delta\omega \cdot \mathrm{d}\omega/\mathrm{d}t)$ is introduced in this chapte. The adaptive virtual inertia curve is shown in Fig. 12.4a. Then, the reasonable value of the coefficient K_J can also be determined as shown in (12.7) and (12.8). It guarantees the boundaries for the adaptive virtual inertia, which is easy for engineers to design.

$$\begin{cases} J = J_0 + \Delta J = J_0 + k_J \arctan\left(\Delta\omega \cdot \frac{\mathrm{d}\omega}{\mathrm{d}t}\right) \\ J_{\min} \le J_0 + k_J \arctan\left(\Delta\omega \cdot \frac{\mathrm{d}\omega}{\mathrm{d}t}\right) \le J_{\max} \end{cases} \tag{12.7}$$

As the boundary of the arctan function is $[-\pi/2, \pi/2]$, the adaptive inertia control coefficient K_J should meet the following equation.

$$0 \le k_J \le \min\left\{\frac{2}{\pi}(J_{\max} - J_0), \frac{2}{\pi}(J_0 - J_{\min})\right\} \tag{12.8}$$

12.1.3.2 The Adaptive Virtual Damping According to $\frac{\mathrm{d}\omega}{\mathrm{d}t}$

The increase in the virtual damping coefficient can increase the system positive damping effect for transient synchronization stability enhancement. Therefore, the virtual damping coefficient $(D_0 + \Delta D)$ is designed to be adaptively increased according to $(d\omega/dt)$ under disturbance. ΔD can also be understood as the transient damping coefficient, which is positive during transient stability issues and returns to zero in the steady state. Since large transient virtual damping may cause a large power fluctuation when the grid frequency disturbances occur, the upper limit of the virtual damping coefficient can be designed as [5]:

$$D_{\max} \le \frac{P_{\max} - P^*}{|\Delta\omega|} \tag{12.9}$$

In addition, the mathematical function $\Delta D = \arctan^2(\mathrm{d}\omega/\mathrm{d}t)$ provides an upper limit and guarantees that ΔD remains positive during the whole accelerated/decelerated transient process. The adaptive virtual damping curve is shown in Fig. 12.4b. According to (12.10) and (12.11), a reasonable value of the coefficient k_D can also be determined. The arctan function facilitates the parameters design.

$$\begin{cases} D = D_0 + \Delta D = D_0 + k_D \arctan^2\left(\frac{\mathrm{d}\omega}{\mathrm{d}t}\right) \\ D_{\min} \le D_0 + k_D \arctan^2\left(\frac{\mathrm{d}\omega}{\mathrm{d}t}\right) \le D_{\max} \end{cases} \tag{12.10}$$

As the boundary of the arctan function is $[-\pi/2, \pi/2]$, the adaptive virtual damping control coefficient k_D should meet the following equation.

$$\frac{4}{\pi^2}(D_{\min} - D_0) \le k_D \le \frac{4}{\pi^2}(D_{\max} - D_0) \tag{12.11}$$

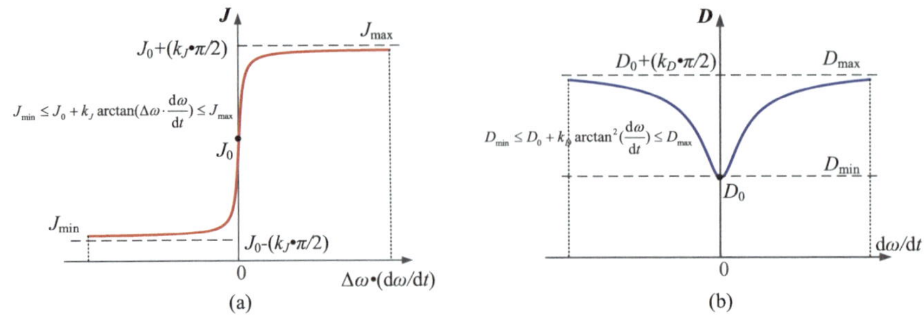

Fig. 12.4 Changing curves of adaptive virtual inertia (**a**) and adaptive virtual damping (**b**)

12.2 Transient Stability Analysis of the AIDC Control

Before analyzing the transient synchronization stability of the introduced AIDC control, some assumptions are made: (1) Due to the fast inner voltage/current control loop, the transient stability is mainly determined by the outer power control loop. (2) The influence of the $Q - V$ loop is approximately neglected. (3) The line impedance is highly inductive and the dynamics of line impedance is ignored. Therefore, the power angle swing equation of the VSG can be expressed as:

$$\begin{cases} \delta = \omega \\ \delta = \frac{1}{J}P^* - \frac{3VV_g}{2JX_g}\sin\delta - \frac{D}{J}\delta \end{cases} \tag{12.12}$$

where $J = J_0 + \triangle J$, and $D = D_0 + \triangle D$. To facilitate transient stability analysis, this chapte focuses on faults that will not trigger current limitation protection.

The attraction region analysis can represent the transient synchronization stability margin. According to the equivalent dynamic (12.12), the attraction regions among the traditional VSG control ($k_J = 0, k_D = 0; \rightarrow \triangle J = 0, \triangle D = 0$), adaptive virtual inertia control ($k_D = 0; \rightarrow \triangle D = 0$), and the introduced AIDC control method can be drawn and seen in Fig. 12.5. Through the comparison of the above attraction regions, the introduced AIDC method has a broader attraction region. Therefore, transient synchronization stability under AIDC control is greatly enhanced.

To further verify the effectiveness of the introduced adaptive inertia and damping coordination (AIDC) control, a comparison on transient performances of VSG without/with dynamic items is shown in Fig. 12.6. In this comparison, the second transient stability issue is considered, where the system has no equilibrium point under the severe disturbance. This kind of disturbance is required to be cleared within the Critical Clear Time (CCT) to make the system recover to sTable In this case, the CCT and Critical Clear Angle (CCA) are used as indicators to represent the strength of transient synchronization stability with larger values indicating stronger stability. To obtain a more accurate CCT and less conservative transient stability boundary, the inverse time integral-based approach is employed here [6].

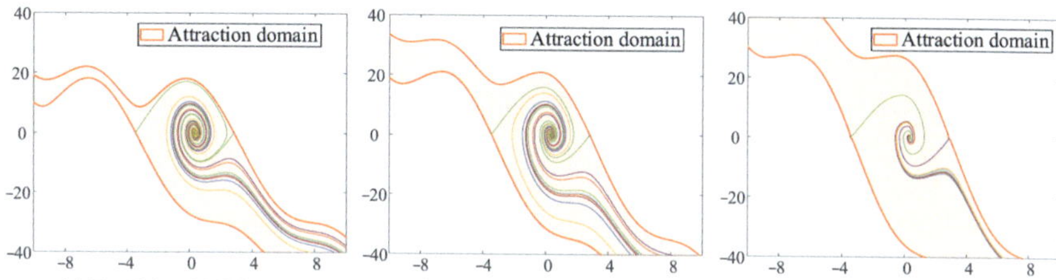

(a) Traditional VSG Control (b) Adaptive Inertia Control (c) Proposed AIDC Control

Fig. 12.5 The comparison of the attraction region between the traditional VSG control ($\triangle J = 0$, $\triangle D = 0$), adaptive virtual inertia control ($\triangle D = 0$) and the introduced AIDC control method

As shown in Fig. 12.6, the power oscillation under the introduced Adaptive Inertia and Damping Coordination (AIDC) control is greatly suppressed during the transient process, and the settling time is reduced from 19.249 to 3.288 s, indicating a faster dynamic response speed. The CCA is increased from 2.23 to 2.65 rad, while the CCT is increased from 1.219 to 1.936 s. Thus, we can see that AIDC control items can improve both transient stability and dynamic performance.

12.3 Control-Hardware-In-Loop Tests

CHIL tests are carried out to further verify the feasibility of the introduced Adaptive Inertia and Damping Coordination (AIDC) control in transient synchronization stability enhancement. The HIL system consists of two parts: the hardware circuits in the loop and the controller outside. The hardware physical circuits in the loop are simulated by the OP5600 real-time simulator with a time-step of 20 μs to accurately simulate the dynamic characteristics, which includes the inverter switches, gate-driven circuit, LC filters, line impedance, and the utility grid. The dSPACE 1202 Microlab-Box device is applied as the system controller with a practical sampling frequency of 20 kHz, and the detailed introduced control algorithm is shown in Fig. 12.3.

12.3.1 Transient Synchronization Stability Enhancement

Figure 12.7 shows the model of the introduced adaptive inertia and damping coordination (AIDC) control-based DGs, which includes the average power calculation, the introduced AIDC control, the dual voltageâŁ"current closed loops, and Pulse Width Modulation (PWM). The CHIL parameters are listed in Table 12.2. In this case, the grid voltage drops to 0.55 p.u. when the disturbance occurs, and after the

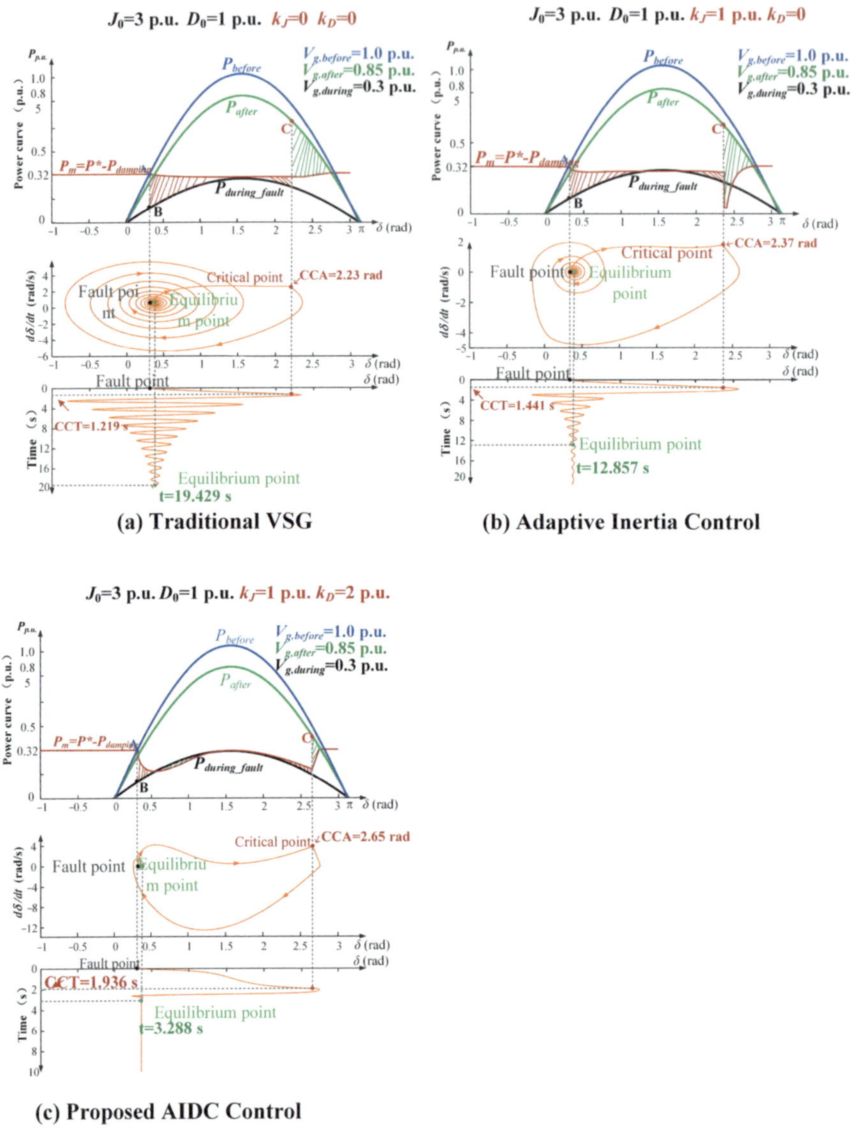

Fig. 12.6 Power angle curve, phase portrait, and time-domain response between the traditional VSG control ($\triangle J = 0$, $\triangle D = 0$), adaptive virtual inertia control ($\triangle D = 0$), and the introduced AIDC control method

fault is cleared, the grid voltage recovers to 1 p.u. of the normal condition. The CHIL experimental results are shown in Fig. 12.8.

Figure 12.8(a1, a2) shows the waveforms of the traditional VSG control under different fault clearance times (FCTs). The critical fault clearance time CCT_1 under

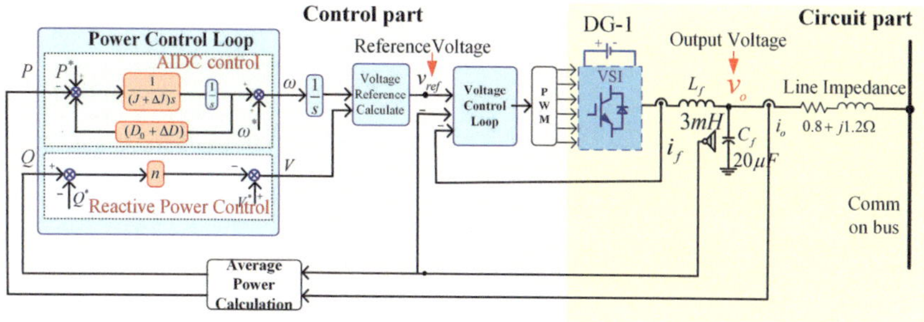

Fig. 12.7 Schematic diagram of improved power outer loop based on introduced AIDC control algorithm

Table 12.2 Control-Hardware-In-Loop (CHIL) test parameters

Parameter	Symbol	Value
System parameters		
Nominal frequency	f^*	50 Hz
Nominal voltage	V^*	311 V
Rated active power	P^*	20 kW
Rated reactive power	Q^*	0 kvar
Control parameters		
Power filter time constant	τ	1/60
Q-V droop coefficient	n	1 p.u.
Virtual inertia constant	J_0	4 p.u.
Virtual damping coefficient	D_0	2 p.u.
Adaptive inertia coefficient	k_J	1 p.u.
Adaptive damping coefficient	k_D	2 p.u.

this condition is 1.23 s. In Fig. 12.8a1, FCT = 1.23 s $\leq$ CCT_1, and VSG can recover stable. In Fig. 12.8a2, FCT = 1.24 s > CCT_1, the VSG loses synchronization stability. Figure 12.8a3 shows the value curves of virtual inertia J and damping coefficient Dm of the traditional VSG control. J and Dm remain fixed.

Figure 12.8(b1, b2) shows the waveforms of the adaptive virtual inertia control under different fault clearance times (FCTs). Under such conditions, the critical fault clearance time CCT_2 is 1.55 s. When FCT = 1.56 s, the system converges to another equilibrium point, which is deemed to be unstable in engineering [6]. Figure 12.8b3 shows the value curves of virtual inertia J and damping coefficient D of the adaptive virtual inertia control, and J is adaptively altered to enlarge the CCT and enhance transient stability during the transient process.

Figure 12.8(c1, c2) shows the waveforms of the AIDC control under different fault clearance times (FCTs). Under such conditions, the critical fault clearance time CCT_3 is 2.00 s, which is greatly longer than the two previous methods. When

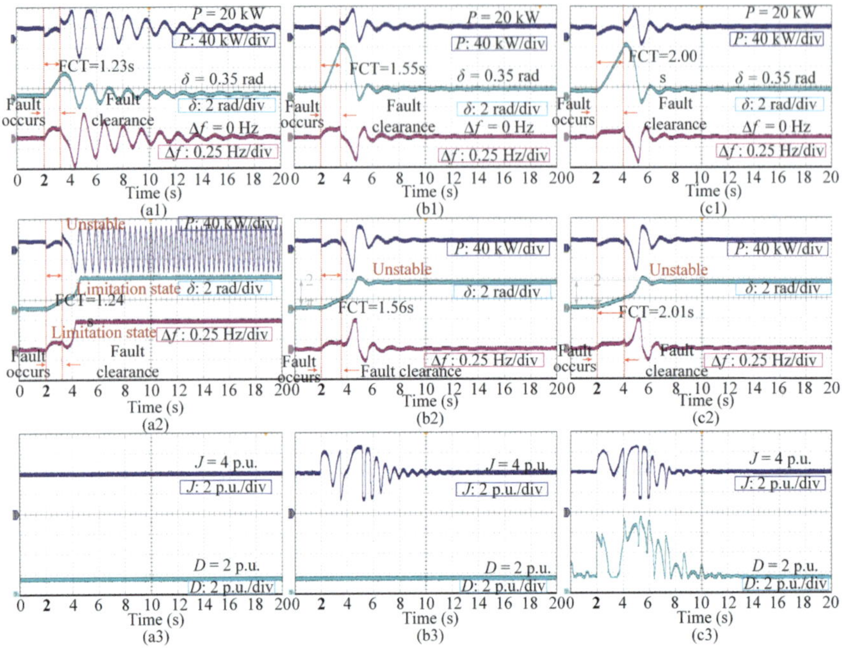

Fig. 12.8 CHIL experimental results under grid voltage sag. **(a1)** The traditional VSG control with fault clearance time FCT $= 1.23\,\text{s} \leq \text{CCT}_1$; **(a2)** The traditional VSG control with fault clearance time FCT $= 1.24\,\text{s} > \text{CCT}_1$; **(a3)** The value curves of virtual inertia J and damping coefficient D of VSG control; **(b1)** The adaptive virtual inertia control with fault clearance time FCT $= 1.55\,\text{s} \leq \text{CCT}_2$; **(b2)** The adaptive virtual inertia control with fault clearance time FCT $= 1.56\,\text{s} > \text{CCT}_2$; **(b3)** The value curves of virtual inertia J and damping coefficient D of the adaptive virtual inertia control; **(c1)** The introduced AIDC control with fault clearance time FCT $= 2.00\,\text{s} \leq \text{CCT}_3$; **(c2)** The introduced AIDC control with fault clearance time FCT $= 2.01\,\text{s} > \text{CCT}_3$; **(c3)** The value curves of virtual inertia J and damping coefficient D of the introduced AIDC control

FCT $= 2.01\,\text{s}$, the system converges to another equilibrium point to be unsTable Fig. 12.8c3 shows the value curves of virtual inertia J and damping coefficient D of AIDC control. Both J and D are adaptively altered to enlarge the CCT and enhance transient stability during the transient process. As there is some small disturbance in $d\omega/dt$, it is reasonable that the virtual damping coefficient fluctuates before the grid voltage sag, and the positive fluctuating of the virtual damping is preferred for system stability.

To further illustrate the advantages of the introduced adaptive inertia and damping coordination (AIDC) control in transient stability enhancement, Table 12.3 displays the comparisons among the above three methods according to the CHIL test results. It can be seen that the introduced AIDC control can greatly enhance the transient synchronization stability by enlarging of CCT.

Table 12.3 Comparisons among the above three methods according to the CHIL test results

Methods	ΔJ	ΔD	Critical clearing time (CCT) (s)	Transient synchronization stability
Traditional VSG control	0	0	1.23	/
Only adaptive virtual inertia control	$\Delta J = k_J \arctan(\Delta\omega \cdot \frac{d\omega}{dt})$	0	1.55	+
introduced AIDC control	$\Delta J = k_J \arctan(\Delta\omega \cdot \frac{d\omega}{dt})$	$\Delta D = k_D \arctan^2(\frac{d\omega}{dt})$	2.00	++

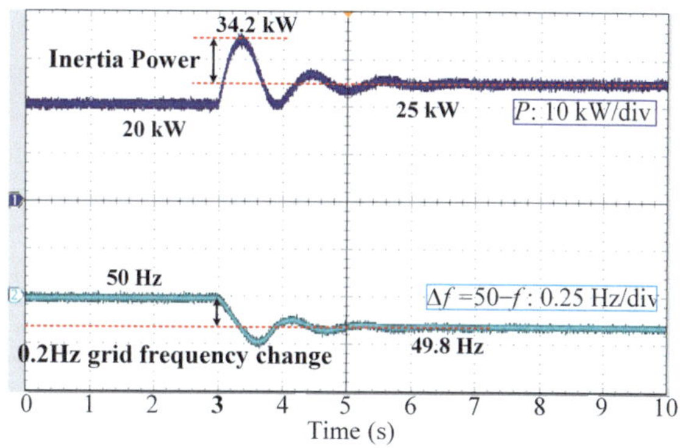

Fig. 12.9 CHIL experimental results of the introduced AIDC control under grid frequency change

12.3.2 Inertia Support and Frequency Regulation Ability

In this case, 0.2 Hz grid frequency change occurs at 3 s. The output active power and frequency of grid-connected inverter are shown in Fig. 12.9. It can be seen that when the disturbance occurs at 3 s, the introduced AIDC control can provide the instantaneous inertia power for frequency support. The steady-state active power rises from 20 to 25 kW to provide the primary frequency regulation. The maximum output power is 34.2 kW to provide the inertia support. In addition, the AIDC control can realize frequency regulation according to the value of virtual damping coefficient. In addition, the converter under the Adaptive Inertia and Damping Coordination (AIDC) control can quickly reach the new steady state within about 2 s.

12.4 Conclusion

This chapte proposes an Adaptive Inertia and Damping Coordination (AIDC) control method to improve the transient synchronization stability of VSG rather than the frequency support viewpoint. The introduced method contains two items: the adaptive virtual inertia and the adaptive damping coefficient. According to the designed principle $\Delta J = \arctan(\Delta\omega \cdot \mathrm{d}\omega/\mathrm{d}t)$, the virtual inertia adaptively increases in the acceleration area and decreases in the deceleration area based on real-time conditions of operation frequency deviation and its rate of change of frequency (RoCoF). In addition, the virtual damping coefficient increases to enlarge the positive virtual damping effect during the transient process with the principle $\Delta D_m = \arctan^2(\mathrm{d}\omega/\mathrm{d}t)$. Compared with traditional VSG control and only adaptive virtual inertia control, the introduced AIDC control has a significant advantage in transient synchronization stability enhancement, resulting in a larger attraction area, CCA, and CCT. The effectiveness of the introduced AIDC control is verified through HIL experiments under grid voltage sag. From the large-signal attraction area analysis and time-domain experimental results, the introduced coordinating design for virtual inertia and damping can significantly improve the system attraction area, dynamic performance, and transient stability.

In future work, we will consider a more practical implementation of the introduced virtual and damping control method. It is noted that the instantaneous power limit of overload capacity for a normal inverter is 1.2–1.8 p.u., which is very weak compared with the synchronous generator (2.5–6 p.u.). For a normal inverter, the buffer power and over-load capacity is limited, and thus, the inertia support is very limited. To apply the inertia support control of the special inverter, there are two measures: (1) from the physical hardware, we will focus on designing a large-margin power capacity of the special inverter by considering overload heat dissipation; (2) from the control viewpoint, we should design the proper virtual inertia and damping moments, which is given in (12.6)–(12.11) in this study.

References

1. H. Bevrani, T. Ise, Y. Miura, Virtual synchronous generators: a survey and new perspectives. Int. J. Electr. Power Energy Syst. **54**, 244–254 (2014)
2. J. Guerrero, L.G. de Vicuna, J. Matas, M. Castilla, J. Miret, Output impedance design of parallel-connected ups inverters with wireless load-sharing control. IEEE Trans. Ind. Electron. **52**, 1126–1135 (2005)
3. Z. Yuan, X. Shen, Y. Tan, Z. Tan, J. Lei, Interactive power oscillation and its suppression strategy for VSG-DSG paralleled system in islanded microgrid. Chin. J. Electr. Eng. **8**, 113–120 (2022). https://doi.org/10.23919/CJEE.2022.000043
4. Adaptive virtual synchronous generator considering converter and storage capacity limits. CSEE J. Power Energy Syst. **8**, 580–590 (2020). https://doi.org/10.17775/CSEEJPES.2019.03360
5. P.C. Loh, D. Li, Y.K. Chai, F. Blaabjerg, Autonomous operation of hybrid microgrid with ac and dc subgrids. IEEE Trans. Power Electron. **28**, 2214–2223 (2013)

6. X. Hou, Y. Sun, X. Zhang, J. Lu, P. Wang, J.M. Guerrero, Improvement of frequency regulation in VSG-based AC microgrid via adaptive virtual inertia. IEEE Trans. Power Electron. **35**, 1589–1602 (2019)
7. L. Sun, P. Wang, J. Han, Y. Wang, C. Yu, M. Chen, Adaptive inertia control of virtual synchronous generator based on power feedback, in *Proceedings of the 2021 IEEE 4th International Electrical and Energy Conference (CIEEC), Wuhan, China*, 28–30 May 2021. https://doi.org/10.1109/CIEEC50170.2021.9510442

Chapter 13
COI-Frequency Consensus Control for Distributed Multi-VSG Grids

13.1 Dynamic Frequency Characteristic Analysis of Multi-VSG Grids

13.1.1 Mathematical Modeling for the Multi-VSG Grids

Figure 13.1 shows the topology and control structure of VSG. The frequency-active power control loop of the basic VSG control is expressed as (13.1).

$$\omega = \omega^* + \frac{1}{Js + D}\left(P_{ref} - P_e\right) \tag{13.1}$$

where P_{ref}, and P_e are the reference active power and output active power of the VSG, respectively. J and D represent the virtual inertia and damping coefficient, respectively. ω^* and ω are the reference frequency and output frequency of the VSG, respectively. As reactive power control is not the focus of this chapter, the reactive power control loop is not considered.

The equivalent circuit of the multi-VSG gird is shown in Fig. 13.2. P_{ei} and $U_i \angle\delta_i$ represent the output active power and voltage of i-th VSG, for i varies from 1 to n. $R_i + jX_i$ is the line impedance, while R_i can be ignored due to the application of virtual impedance [1]. P_L is the load power is the load voltage.

According to [1], the closed-loop transfer function of $\Delta P_i(s)/\Delta P_L(s)$ and $\Delta\omega_i(s)/\Delta P_L(s)$ in the multi-VSG grid can be expressed as:

$$\frac{\Delta P_i(s)}{\Delta P_L(s)} = \frac{G_i(s)\left(J_i s + D_i\right)}{\sum_{m=1}^{n} G_m(s)\left(J_m s + D_m\right)} \tag{13.2}$$

$$\frac{\Delta\omega_i(s)}{\Delta P_L(s)} = -\frac{G_i(s)}{\sum_{m=1}^{n} G_m(s)\left(J_m s + D_m\right)} \tag{13.3}$$

X. Hou et al., *Self-organized Operation and Stability Control of Electronized Power Networks*, https://doi.org/10.1007/978-981-96-9067-1_13

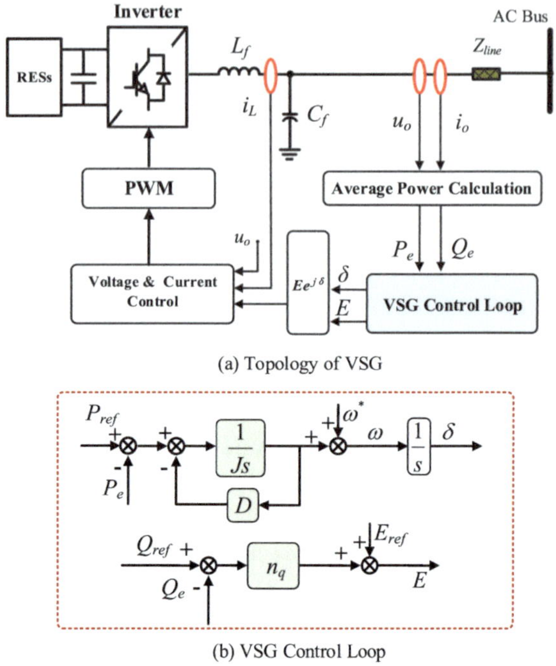

Fig. 13.1 Topology and control structure of VSG **a** Main circuit and control structure **b** Basic VSG control loop

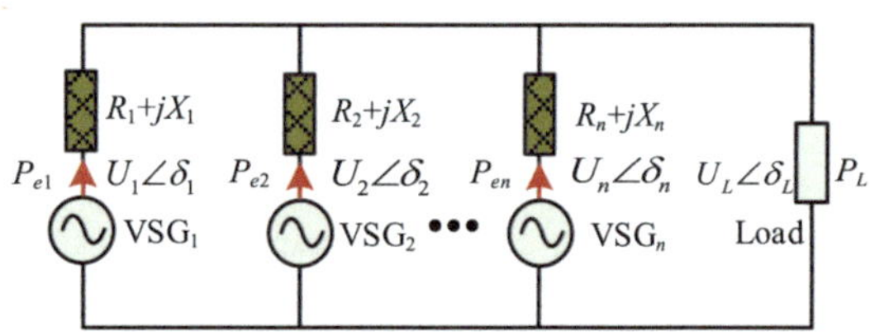

Fig. 13.2 Equivalent circuit of the multi-VSG gird

where, $G_i(s) = \frac{k_{iL}}{J_i s^2 + D_i s + k_{iL}}$, $k_{iL} = \frac{U_i U_L}{X_i} \cos\left(\delta_i^\circ - \delta_L^\circ\right)$, $G_m(s) = \frac{k_{mL}}{J_m s^2 + D_m s + k_{mL}}$, δ_i° and δ_L° are the voltage phase angle of i-th VSG and the load in the steady-state equilibrium, respectively. k_{iL} is the power coefficient of i-th VSG.

Through (13.2), the active power response characteristic of each VSG is interconnected with each other in the multi-VSG grid. The interactions among virtual inertia, damping coefficient, and line impedance are the main cause of active power oscillation and inappropriate transient active power-sharing, which have been fully discussed in previous studies [2–4]. In this chapte, we do not repeat the above analysis but focus on the frequency response characteristic analysis of multi-VSG grids.

13.1.2 Dynamic Frequency Response Characteristics

The transfer function (13.3) can reflect the frequency response characteristic in multi-VSG grids. To further explore the difference in dynamic frequency response characteristics, we decompose the frequency into two parts: the common-mode component and the differential-mode component. The common-mode component reflects the dynamic frequency consistency of all VSGs, while the differential-mode component can reflect the difference in dynamic frequency characteristics. In this chapte, the common-mode component of frequency can be represented by COI frequency. Therefore, the dynamic frequency response can be expressed as:

$$\Delta\omega_i = \Delta\omega_{coi} + \Delta\omega_{ie} \tag{13.4}$$

$$\Delta\omega_{coi} = \frac{H_1\Delta\omega_1 + \cdots + H_n\Delta\omega_n}{H_1 + \cdots + H_n} \tag{13.5}$$

$$\Delta\omega_{ie} = \Delta\omega_{coi} + \Delta\omega_{ie} \tag{13.6}$$

where $\Delta\omega_{coi}$ represents the common-mode component based on COI and $\Delta\omega_{ie}$ represents the differential-mode component. All frequencies will converge to the COI frequency finally.

The step response of $\Delta\omega_{coi}$,ω_1, ω_2 is depicted in Fig. 13.3 in the two-VSG system. ω_{coi} is the common-mode component of ω_1 and ω_2, which performs no oscillation, reflecting the dynamic frequency characteristic of the overall system. And $\omega_{ie} = \omega_i - \omega_{coi}$ represents the dynamic frequency deviation towards ω_{coi}.

The 2-norm $\mathcal{H}_2$ of the transfer function $\Delta\omega_i(s)/\Delta P_L(s)$ is used as the assessment to indicate the robustness in the dynamic frequency response of the multi-VSG grids. The smaller the value of $\mathcal{H}_2$, the stronger to suppress the dynamic frequency deviation. Taking the two-VSG system as an example, the line impedance parameter of the system is k_{1L} : k_{2L} =2:1. Figure 13.4 can be calculated to analyze the influence of J_1, J_2,D_1, D_2.

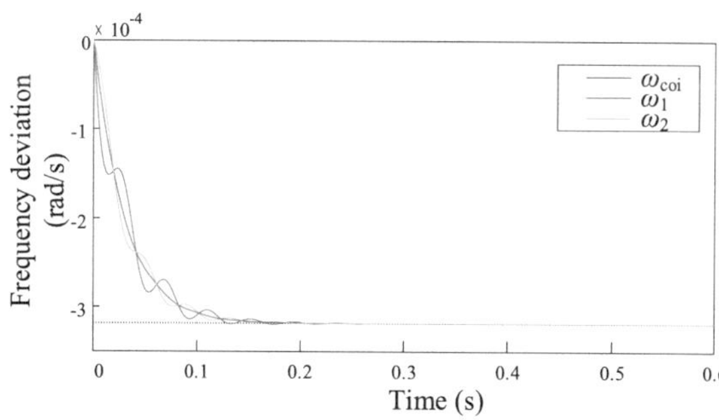

Fig. 13.3 The step load change response of two VSG systems

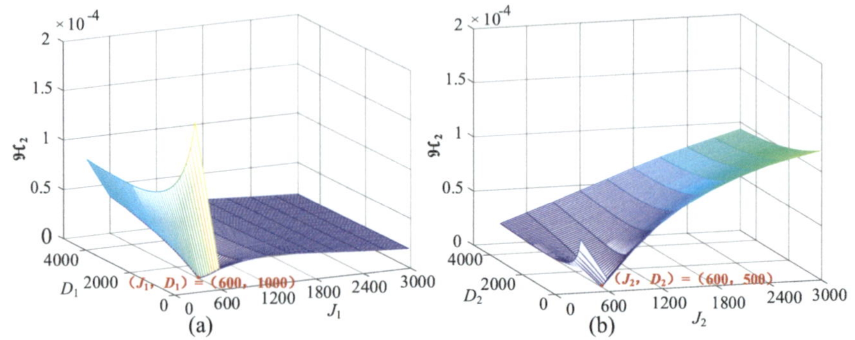

Fig. 13.4 **a** The relationship among J_1, D_1, and $\mathcal{H}_2$ of $\Delta\omega_{1e}(s)/\Delta P_L(s)$, when $J_2 = 300$, $D_2 = 500$. **b** The relationship among J_2, D_2 and $\mathcal{H}_2$ of $\Delta\omega_{1e}(s)/\Delta P_L(s)$, when $J_1 = 1200$, $D_1 = 1000$

When $D_1 : D_2 = J_1 : J_2 = k_{1L} : k_{2L}$ =2:1, the value of $\mathcal{H}_2$ reaches the minimum and the dynamic response is optimal under this condition.

13.1.3 Discussion

In the multi-VSG grids, each VSG shows significantly different and interacted responses under the mismatched system parameters, which will deteriorate the dynamic performance. The system parameters should satisfy $J_1 : J_2 : ... : J_n = D_1 : D_2 : ... : D_n = k_{1L} : k_{2L} : ... : k_{nL}$ to reach the optimal, which gives a reliable reference for locating and sizing VSGs (virtual inertia coefficient J_i and damping coefficient D_i are usually designed based on the capacity of the converters). However, the line impedance x_i in multi-VSG grids is hardly obtained in practical application. The optimization principle is hardly applied in practice due to these physical constraints. More practical methods need to be explored to reduce the frequency deviation towards the COI frequency and achieve a better dynamic performance of the multi-VSG grid under severe disturbances.

13.2 COI-Frequency Consensus Control

13.2.1 COI-Frequency Control

As discussed previously, it is effective to realize better dynamic performance that all the VSG track the ω_{coi}. Therefore, the modified VSG control methods based on COI frequency are introduced to improve the dynamic performance and enhance frequency stability in multi-VSG grids [5–7]. However, the calculation for ω_{coi} relies

Fig. 13.5 The introduced control method

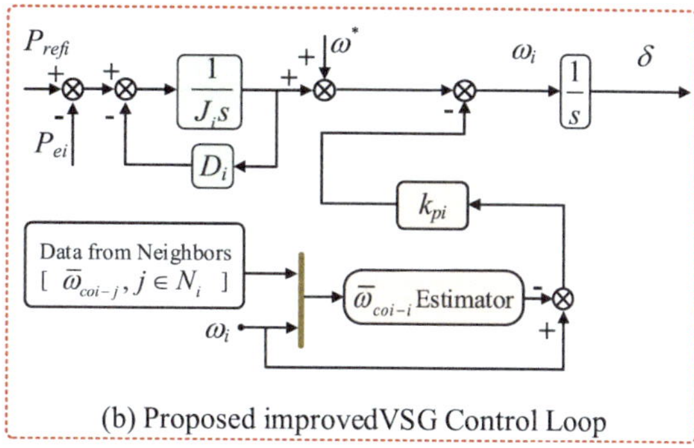

on a centralized controller and global communication. To overcome the disadvantages of the centralized controller and global communication, this chapter designs a COI frequency estimator based on dynamic consensus control. The introduced control structure is shown in Fig. 13.5. And the method is designed as:

$$\begin{cases} \omega_i = \omega^* + \frac{1}{J_i s + D_i}\left(P_{refi} - P_{ei}\right) - \frac{k_{pi}}{(J_i s + D_i)}\bar{\omega}_{ie} \\ \bar{\omega}_{ie} = \omega_i - \bar{\omega}_{coi_i i} \end{cases} \tag{13.7}$$

where k_{pi} is defined as the additional control parameters.$\bar{\omega}_{coi-i}$ denotes the COI frequency calculated by the local estimator-i.

13.2.2 Local Estimator Design for COI Frequency

(1) The communication Topology
To facilitate the COI frequency estimation in the multi-VSG grid, the distributed communication network is designed as shown in Fig. 13.6a. The communication graph consists of n nodes, and the communication links are bidirectional to connect with neighbor nodes. N_i denotes the set of all neighbors of node i. The system based on centralized control with global communication in [5–7] is shown in Fig. 13.6b. Compared with Fig. 13.6b, the introduced communication network in Fig. 13.6a does not require a centralized controller and reduces the cost of communication. Both the sparsity and reliability of the communication network can be ensured.
(2) Local estimator for COI frequency based on the distributed communication
The local estimator based on distributed communication in Fig. 13.6a to estimate the COI frequency is designed as follows:

$$\bar{\omega}_{coi-i} = \left(n J_i \omega_i + \int_0^t \sum_{j \in N_i} a_{ij}\left(\bar{\omega}_{coi-j} - \bar{\omega}_{coi-i}\right) dt \right) / \bar{J}_{\text{total-i}} \tag{13.8}$$

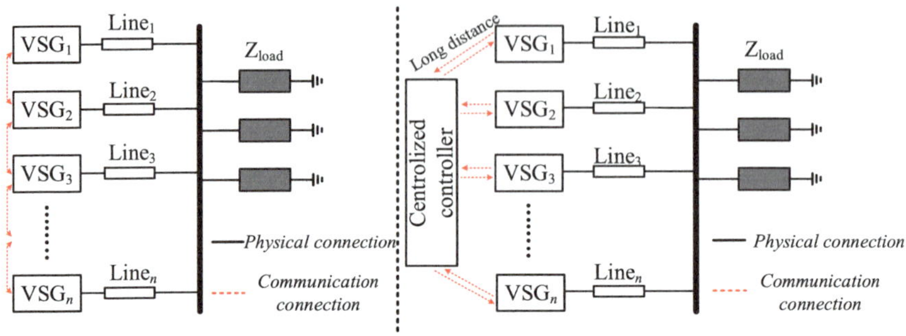

Fig. 13.6 **a** Distributed cyber-physical system; **b** System with global communication

$$\bar{J}_{\text{total-i}} = nJ_i + \int_0^t \sum_{j \in N_i, j \neq i} a_{ij} \left(\bar{J}_{\text{total-}ij} - \bar{J}_{\text{total-i}\,i} \right) dt \tag{13.9}$$

where $\bar{\omega}_{coi-i}$, and $\bar{\omega}_{coi-j}$ are the COI angular speed that node i and node j estimate, respectively. $\bar{J}_{\text{total-i}}$ is the sum of all inertia coefficients that node i estimates, and the estimation of $\bar{J}_{\text{total-i}}$ can be realized by (13.9). n represents the number of VSGs in the microgrid, N_i represents the collection of neighboring nodes of node i, and a_{ij} represents the communication weight of node i receiving the data of node j, where $a_{ij} = 1$ if there is an edge connecting node i to node j, and $a_{ij} = 0$, otherwise. can be proved to converge to the sum of all virtual inertia in previous research [8, 9] and we are not going to prove it anymore in this chapter. Each node updates the output of the estimator in real time by processing its local output and the estimation information of neighboring nodes.

(3) Validity analysis for the consensus protocol for COI frequency estimation

Taking the derivative of both sides of (13.8), we can obtain (13.10).

$$\dot{\bar{\omega}}_{coi-i} = \frac{\left(nJ_i\dot{\omega}_i - d_i\omega_{coi-i} + \sum_{j \in N_i} a_{ij}\bar{\omega}_{coi-j} \right)}{\bar{J}_{\text{total-i}}} \tag{13.10}$$

where $d_i = \sum_{j \in N_i} a_{ij}$

Then the COI angular speed estimator dynamic of node i can be written in matrix form.

$$\begin{cases} \dot{\bar{\boldsymbol{\omega}}}_{\text{coi}} = (n\mathbf{J}\dot{\boldsymbol{\omega}} - \text{L}\bar{\boldsymbol{\omega}}_{\mathbf{coi}}) \,/\, \sum_{i=1}^{n} J_i \\ \mathbf{L} = \mathbf{d} - \mathbf{a} \end{cases} \tag{13.11}$$

where, $\bar{\boldsymbol{\omega}}_{\text{coi}} = \left[\bar{\omega}_{coi-1} \cdots \bar{\omega}_{coi-n} \right]^T$ and $\boldsymbol{\omega} = [\omega_1 \cdots \omega_n]^T$ are estimated COI frequency phasor and instantaneous angular speed phasor. $\mathbf{J} = \text{diag}\{J_i\}$ represents the inertia coefficient matrix, $\mathbf{d} = \text{diag}\{d_i\}$ and $\mathbf{a} = \left[a_{ij} \right] \in \mathbb{R}^{n \times n}$ are the intrinsic matrix and adjacency matrix.$\mathbf{L} = \mathbf{d} - \mathbf{a}$is the Laplace matrix that carries communication information of the system.

From (13.11), the initial conditions can be deduced and substituted into (13.12).

$$\overline{\boldsymbol{\omega}}_{\text{coi}} = \mathbf{M}s(s\mathbf{I} + \mathbf{N})^{-1}\omega \tag{13.12}$$

where $\mathbf{I} \in \mathbb{R}^{n \times n}$ represents a unit matrix,$\mathbf{M} = \frac{n}{\sum_{i=1}^{n} J_i}\mathbf{J}$,$\mathbf{N} = \frac{1}{\sum_{i=1}^{n} J_i}\mathbf{L}$. If $\mathbf{L}$ is balanced, all elements of the matrix $\overline{\boldsymbol{\omega}}_{\text{coi}}$ will converge collectively to the weighted average of all nodes weighted according to COI [8, 9].

$$\bar{\omega}_{coi-1} = \bar{\omega}_{coi-2} = \cdots = \bar{\omega}_{coi-i} = \frac{\sum_{i=1}^{n} (J_i \omega_i)}{\sum_{i=1}^{n} J_i} \tag{13.13}$$

13.2.3 Comparisons with Other Improved VSG Control Methods

To further highlight the advantages of the introduced method, a comprehensive comparison with the aforementioned control methods [10–17] is presented in Table 13.1. Being superior to methods [10–15], the introduced method considers the dynamic interaction among other VSGs. Therefore, it is suitable for multi-VSG grids and has a better control effect on dynamic performance improvement and active power oscillation suppression. The consensus control based on distributed communication does not depend on global information [16] and overcomes the disadvantages of centralized methods [11, 17], exhibiting robustness and high communication reliability. Besides, the introduced control method is based on the basic VSG scheme, and the

Table 13.1 Comparisons with other improved VSG control methods

Methods	Dynamic performance improvement	Active power oscillatio suppression	Communication requirement	Implementation complexity	Suitable for multi-VSG grids
Adaptive virtual inertia methods [10, 12–15]	Medium	Low	Low	Low	×
VSG with LQR controller [16]	High	Medium	Medium	Medium	✓
Centralized COI-based VSG control [11, 17]	High	High	High	Low	✓
Intelligent control [14]	High	Medium	Low	High	×
Introduced method	High	High	Medium	Low	✓

control parameters can ensure the effectiveness of the control effect in a wide range. Therefore, the implementation complicity is low and it is easy to be mastered by the engineers.

13.3 Stability Analysis

13.3.1 Small Signal Modeling

The small-signal model and the analysis of the eigenvalues are carried out to verify the stability of the introduced method.

Assume the line impedance is inductive to simplify the analysis. Combining the network model of the multi-VSG grid and the introduced method, we can obtain the small-signal model after linearization, which is shown as (13.14).

$$\mathrm{J}\Delta\dot{\omega} = -\mathrm{K}_{\mathrm{line}}\,\Delta\boldsymbol{\delta} - \mathbf{D}\Delta\omega - \mathbf{k_p}\left(\Delta\omega - \Delta\bar{\omega}_{\mathrm{coi}}\right) \tag{13.14}$$

where, $\mathrm{K}_{\mathrm{line}}$ is the power transmission matrix, which is shown in the Appendix. $\mathbf{J} = \mathrm{diag}\{J_i\}$ and $\mathbf{D} = \mathrm{diag}\{D_i\}$ represent the virtual inertia matrix and the damping coefficient matrix, respectively. $\mathbf{k_p} = \mathrm{diag}\left\{k_p\right\}$ is the proportional coefficient matrix.

To describe the estimated COI frequency dynamic in (13.11), we define a new state variable vector $\mathbf{x} = \left[x_1\ x_2\ \cdots\ x_n\right]^T$. Then we have

$$\dot{\mathbf{x}} = \mathbf{N}\bar{\boldsymbol{\omega}}_{\mathrm{coi}} \tag{13.15}$$

$$\dot{\mathbf{x}} = \mathbf{N}(\mathbf{M}\omega - \mathbf{x}) \tag{13.16}$$

Therefore, we can obtain the state-space model of the introduced control method.

$$\begin{bmatrix} \Delta\dot{\delta} \\ \Delta\dot{\omega} \\ \Delta\dot{\mathbf{x}} \end{bmatrix} = \begin{bmatrix} \mathbf{0} & \mathbf{I} & \mathbf{0} \\ -\mathbf{J}^{-1}\mathbf{K}_{\mathrm{line}} & -\mathbf{J}^{-1}\mathbf{G} & \mathbf{J}^{-1}\mathbf{H} \\ \mathbf{0} & \mathbf{N}\mathbf{M}^2 & \mathbf{N} \end{bmatrix} \begin{bmatrix} \Delta\delta \\ \Delta\omega \\ \Delta\mathrm{x} \end{bmatrix} \tag{13.17}$$

where, **F** ,**G** ,**H** are the sub-matrixes of the state space model which are shown in Appendix, respectively.

13.3.2 Eigenvalues Analysis

Taking the multi-VSG grid in Fig. 13.8 as the study case, the root locus method is used to analyze the matrix eigenvalues in (13.17). The matrix eigenvalues are given according to different control parameters k_p. The system parameters are shown in Table 13.3.

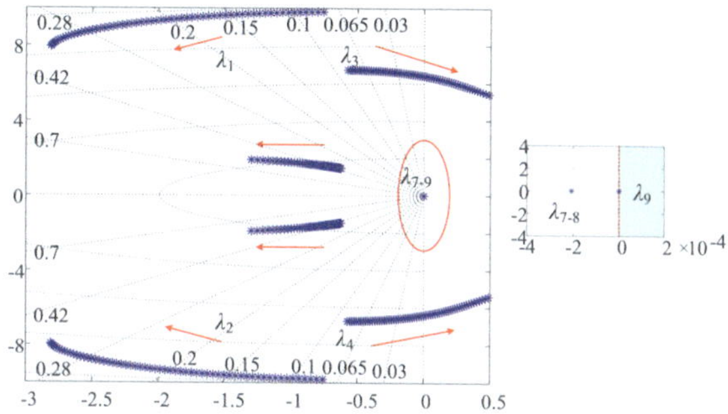

Fig. 13.7 Root locus when kp varies in [0,1.5e5]

Figure 13.7 shows the root locus diagram when k_p varies in $[0, 1.5 \times 10^5]$. It is worth noting that matrix A has an eigenvalue of 0 due to the rotational symmetry of the system, but it does not affect the dynamic stability of the system [18]. When k_p is between 0 and 1×10^5, all eigenvalues locate on the left of the imaginary axis, the system is stable. The imaginary value of the eigenvalue decreases as k_p increases. But when k_p increases, the system is prone to be unstable as the dominated eigenvalues move towards the right. When k_p is gradually greater than 1×10^5, λ_3 and λ_4 enter the right of the imaginary axis, and the system is unstable.

13.3.3 Modal Analysis

The modes whose damping ratio is lower than 0.1 and oscillatory frequency is lower 3 Hz are chosen as the critical ones [19, 20], as shown in Table 13.2.

According to Fig. 13.7 and Table 13.2, the damping ratio of λ_1&λ_2 is enhanced as k_p is increased, and the damping ratio of λ_3&λ_4 is decreased because of the migration

Table 13.2 Parameters of the simulation test system

	Mode number	Eigenvalues	Damping ratio	Oscillatory frequency (Hz)
Basic VSG	λ_1&λ_2	$-0.759 + 9.851j$	0.0772	1.57
		$-0.759 - 9.851j$		
	λ_3&λ_4	$-0.574 + 6.716j$	0.0852	1.07
		$-0.574 - 6.716j$		
Introduced method	λ_1&λ_2	$-1.199 + 9.621j$	0.1112	1.53
		$-1.199 - 9.621j$		
	λ_3&λ_4	$-0.401 + 6.689j$	0.0662	1.06
		$-0.401 + 6.689j$		

to the right-half plane (RHP). The too-large k_p is undesirable for the stability of the system. The proper selection of k_p can ensure the stability of the introduced control method and optimize the dynamic response.

13.4 Simulation Verification

To verify the feasibility and effectiveness of the introduced method in this section, simulations in Matlab/Simulink have been carried out. The three-VSG test system for simulation verification is shown in Fig. 13.8. And system parameters are listed in Table 13.3.

13.4.1 Comparisons with VSG Control Under Load Step Change

Comparison simulations between the introduced method and the basic VSG control which is shown in Fig. 13.8 are carried out. When k_p is set as 0, the introduced method is equal to the basic VSG control. In this case, a load step change is simulated. Active power load increases at 15 s.

It can be observed from Fig. 13.9a that serious active power oscillation and undesired frequency dynamic performance occur. The degraded active power oscillation and deteriorated frequency response will greatly increase the probability of

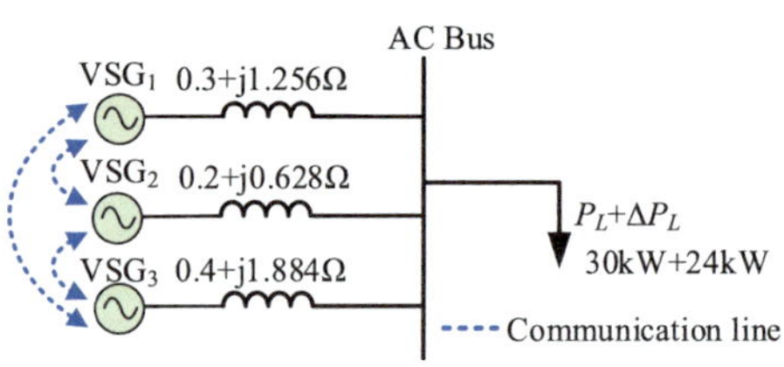

Fig. 13.8 The simulation test system

Table 13.3 Parameters of the simulation test system

Parameter	Symbol	Value
Nominal frequency	f^*	50 Hz
Nominal voltage	E^*	311 V
Virtual inertia	$J_1/J_2/J_3$	1720/3640/960
Damping coefficient	$D_1/D_2/D_3$	4000/4000/4000r
Damping coefficient	$k_{p1}/kp2/k_{p3}$	8000/8000/8000

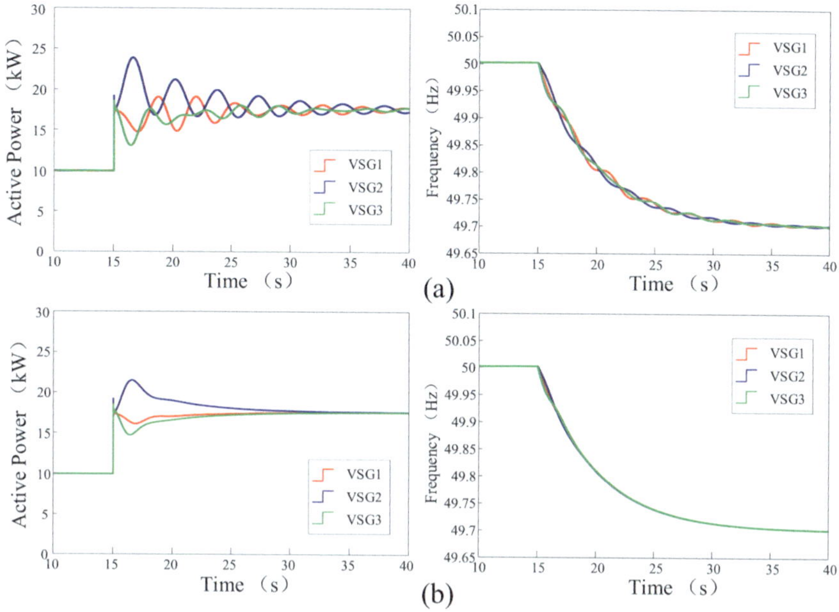

Fig. 13.9 Simulation results: **a** with the basic VSG; **b** with the introduced improved VSG control

irreversible damage to power electronic devices. Figure 13.9b shows the simulation results with the introduced improved VSG method. The settling time is about 6.8s, while the settling time is about 14.5s with the basic VSG. Power oscillation is suppressed effectively and the oscillation periods are reduced. The active power overshoot is reduced from 36.9 to 17.8%.

13.4.2 Effects of Communication Time Delay

The effect of communication delay on the introduced method should be taken into consideration, which is critically important to control performance and stability.

Simulations are carried out considering time delays of 30 ms, 50 ms, and 100 ms. The simulation results are shown in Fig. 13.10a–c. The system under the introduced method is proven to remain stable within the communication time delay range of 0–100 ms. Different communication time delay hardly influences the active power overshoot. Although the control effect for power oscillation suppression is attenuated with the existence of communication time delay, it is still acceptable if the communication time delay is less than 50 ms. And the time delay results in a small deviation of the COI estimated by the distributed consistency algorithm, which leads to the upward shift of the equilibrium point of the frequency. Power sharing among VSGs in steady-state can still be realized.

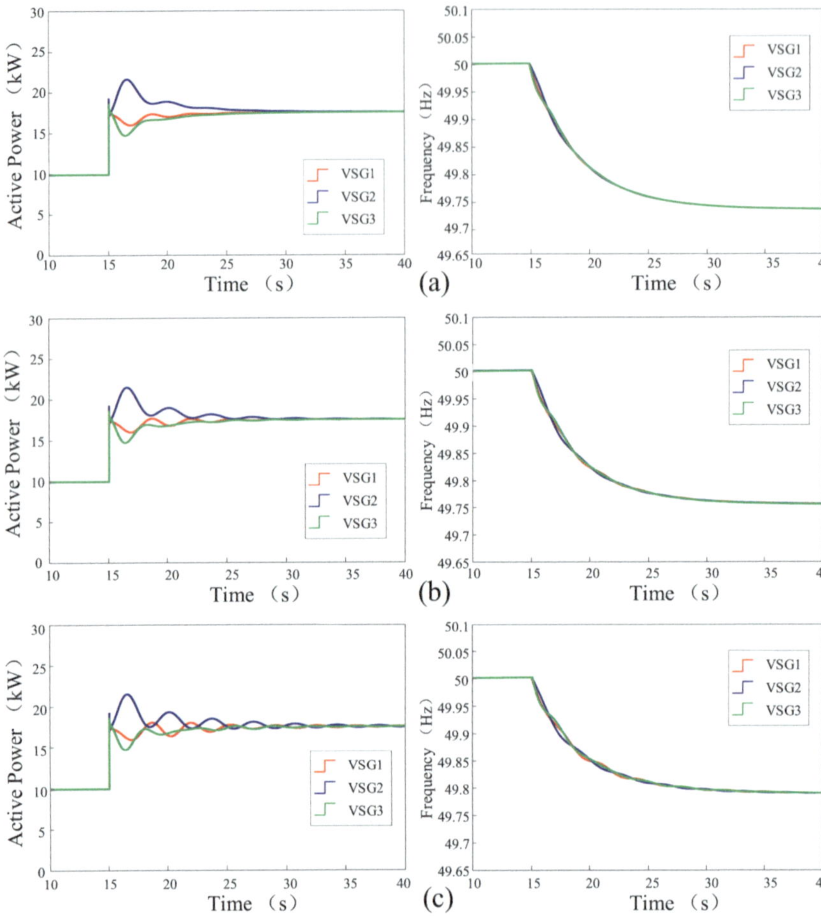

Fig. 13.10 Simulation results under the introduced improved VSG control: **a** with 30 ms communication time delay; **b** with 50 ms communication time delay; **c** with 100 ms communication time delay

13.4.3 Comparisons Under Frequent-Variation Load

To further test the performances of the introduced improved VSG control, frequent load fluctuations are considered. In this case, frequent-variation load is connected to the system at 10 s. The simulation results are shown in Fig. 13.11a, b. Compared with basic VSG control, the introduced method can realize smooth active power regulation and reduce the frequency oscillation and dynamic deviation as much as possible under variable loads.

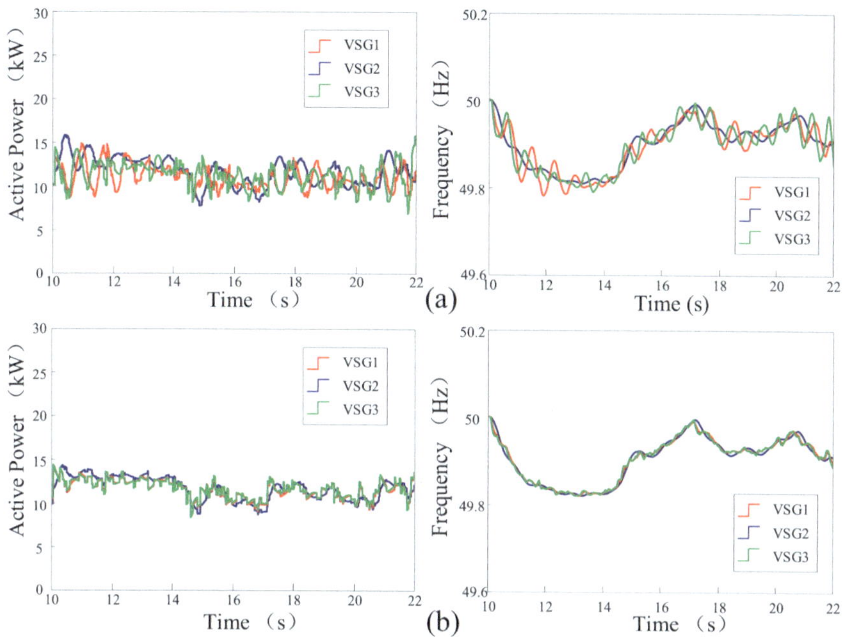

Fig. 13.11 Simulation results: **a** with the basic VSG; **b** with the introduced improved VSG control

13.4.4 Simulation Verification in IEEE 9-Bus System

The effectiveness of the introduced method is further verified in the IEEE 9-bus system [21, 22] in Fig. 13.12. Two VSGs replace the synchronous generators in the IEEE 9-Bus System. Figure 13.13a shows the simulation results with the basic VSG control after a load change (500 MW load increases) at 14 s. Figure 13.13b shows the comparative results with the introduced improved VSG control method. At 14 s, the basic VSG is switched to the introduced method. Severe power oscillation and frequency oscillation are suppressed with the introduced method effectively. The simulation results verify the effectiveness of the introduced method in a system with a mixture of VSGs and conventional synchronous generators.

13.5 Experimental Verification

Subject to the experimental conditions, the experiment consists of three DC power supplies, three parallel single-stage voltage source inverters, AC loads, and DSP28335+FPGA control board. It is set up to confirm the viability of the method introduced in this chapte. Parameters used in the experiment are listed in Table 13.4.

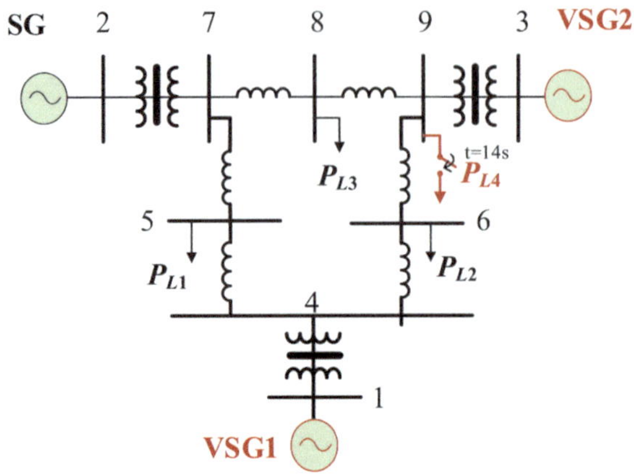

Fig. 13.12 IEEE 9 bus system [20, 21]

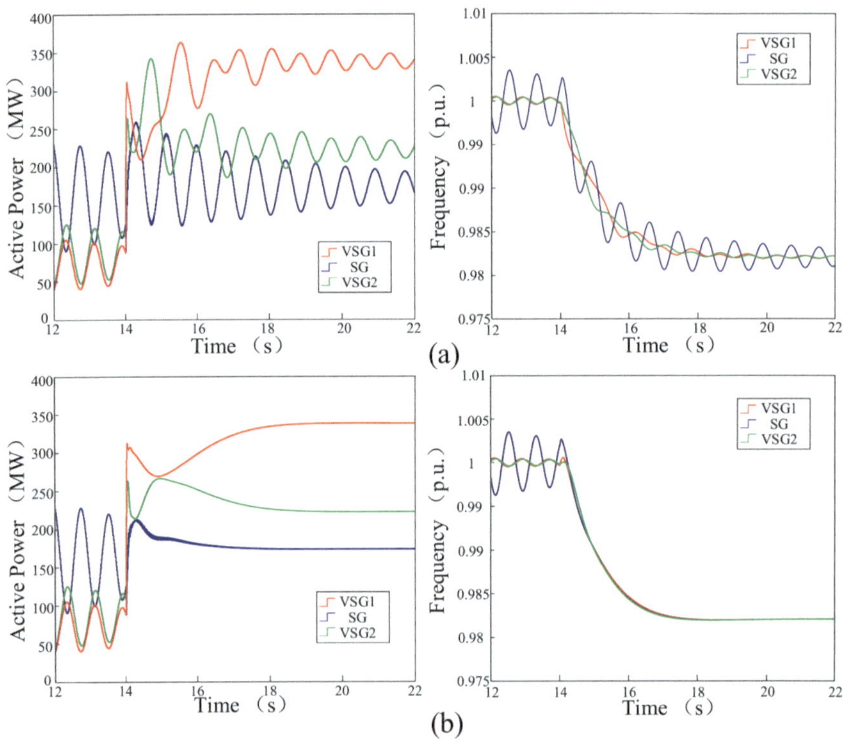

Fig. 13.13 Simulation results in IEEE 9-Bus System **a** with basic VSG control; **b** with the introduced improved VSG control

Table 13.4 Parameters of the experimental test system

Parameter	Symbol	Value
Nominal frequency	f^*	50 Hz
Nominal voltage	E^*	48 V
Filter inductance	L_f	1.5 mH
Filter capacitor	C_f	20 μF
Impedance	$L_{line1}/L_{line2}/L_{line3}$	4.4 mH/4.2 mH/4.5 mH
Load	Z_{load}	25Ω
Load change	ΔZ_{load}	25/Ω
Virtual inertia	$J_1/J_2/J_3$	16/32/48
Damping coefficient	$D_1/D_2/D_3$	80/80/80
Damping coefficient	$k_{p1}/kp2/k_{p3}$	200/200/200

13.5.1 Comparisons with the Basic VSG Control

The waveforms of the current of three VSG are shown in Fig. 13.15. The load demand has increased around 5s. Figure 13.15a shows that with the basic VSG control, the system current oscillates after the load change. While Fig. 13.15b shows that with the introduced improved VSG control method, the system current oscillations have been damped effectively. Figure 13.14a, b show the waveforms of active power and frequency with different control methods. The performance quantitative indicators are summarized in Table 13.5. Compared with the basic VSG control, the introduced improved VSG control reduces the active power overshoot, increases the response speed, and suppresses the power/frequency oscillations effectively.

13.5.2 Experimental Under Communication Link Failure

To further verify the advantages of the introduced control strategy, the experiment, in this case, is carried out under a single communication link failure. The experimental results are shown in Figs. 13.16 and 13.17. Although the existence of communication link failure will reduce the control effect, the introduced control strategy can still suppress the power/frequency oscillation and improve the dynamic response.

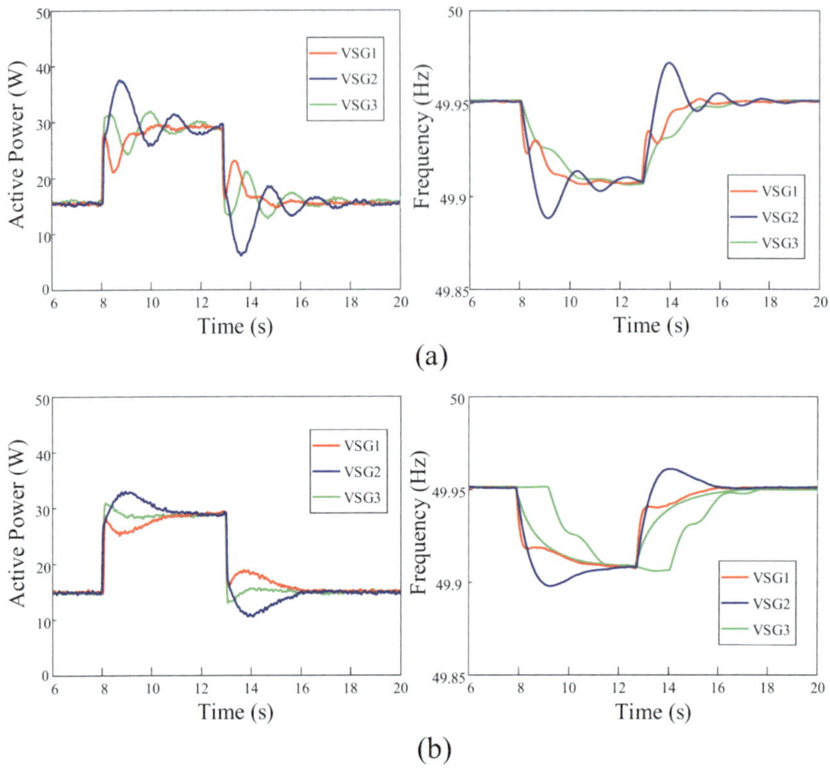

Fig. 13.14 Experimental results **a** with basic VSG control; **b** with the introduced improved VSG control

13.6 Conclusion

This chapter proposes an improved VSG control method to suppress power/frequency oscillation and improve the dynamic performance of multi-VSG grids. To overcome the disadvantages of the centralized COI frequency calculation method, a local real-time COI frequency estimator based on dynamic consensus control is designed, which only relies on communication between adjacent neighbors. The dynamic performance of multi-VSG grids is greatly improved under the introduced method. Simulation and experimental results verify the effectiveness and superiority.

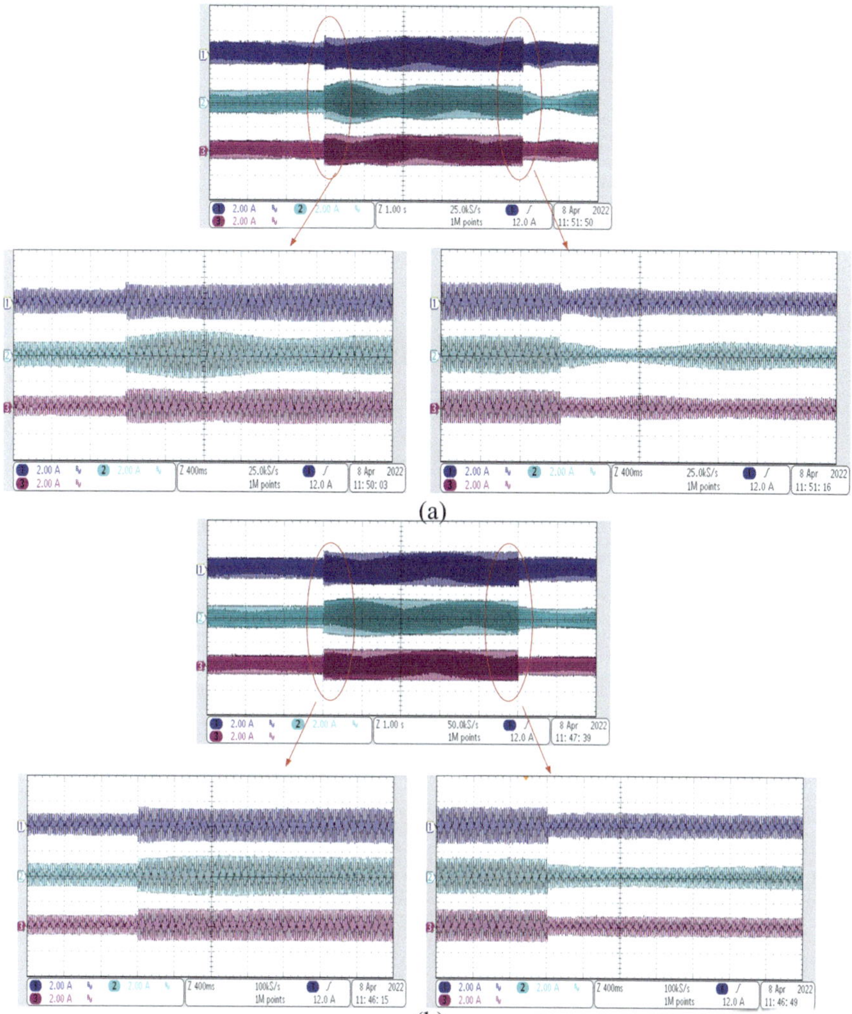

Fig. 13.15 Experimental current waveforms of three single VSGs **a** with basic VSG control; **b** with the introduced improved VSG control

Table 13.5 Quantitative indicators for experimental results

Indexes		
Methods	Basic VSG method	introduced Improved VSG method
Frequency nadir	0.07 Hz	0.052 Hz
Power overshoot	31.03%	13.7%
Settling time (5%)	4.1 s	2.2 s

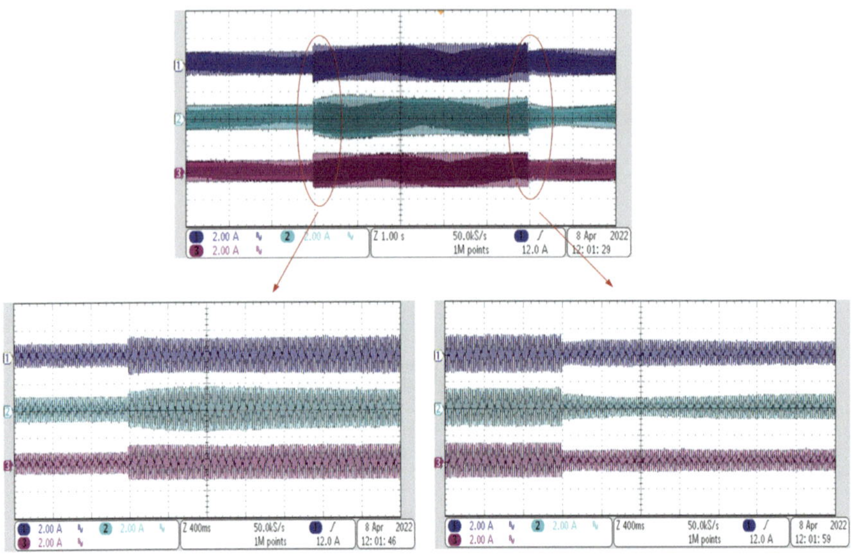

Fig. 13.16 Experimental current waveforms of three single VSGs with the introduced improved VSG control under single communication link failure

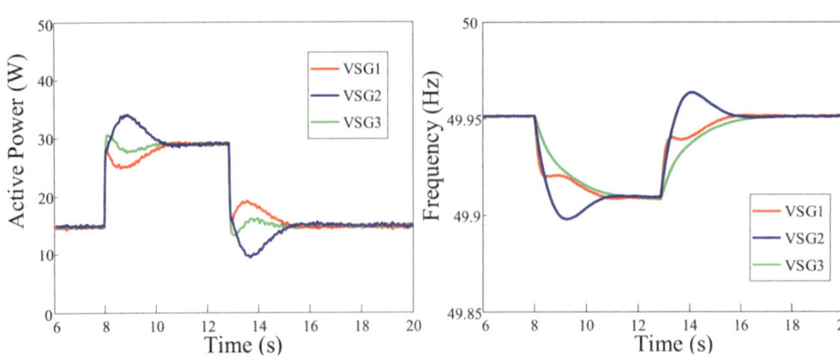

Fig. 13.17 Experimental results with the introduced improved VSG control under single communication link failure

Appendix

$$\mathbf{K}_{\text{line}} = \begin{bmatrix} \sum_{i=2,3,4} k_{1i} & -k_{12} & \dots & -k_{1n} \\ -k_{21} & \sum_{i=1,3,4} k_{2i} & \dots & -k_{2n} \\ \dots & \dots & \dots & \\ -k_{n1} & -k_{n2} & \dots & \sum_{i=1,2,3} k_{ni} \end{bmatrix} \tag{13.18}$$

$$k_{ij} = \frac{U_i U_j X_{ij} \cos\delta_{ij}}{X_{ij}^2}, (i, j = 1, 2, .., n; i \neq j) \tag{13.19}$$

$$\mathbf{G} = \begin{bmatrix} g_{11} & -n & \dots & -n^2 \\ -n & g_{22} & \dots & 0 \\ \dots & \dots & \dots & \dots \\ -n^2 & 0 & \cdots & g_{nn} \end{bmatrix}_{n\times n} \tag{13.20}$$

$$g_{ii} = \frac{2nk_d^2 J_i + D_i + k_p}{\left(\sum_{j=1}^{n} J_j\right)^2}, i = 1, 2, 3 \dots, n \tag{13.21}$$

$$\mathbf{H} = \begin{bmatrix} -k_p & 0 & \dots & 0 \\ 0 & -k_p & \dots & 0 \\ \dots & \dots & \dots & \dots \\ 0 & 0 & \dots & -k_p \end{bmatrix}_{n\times n} \tag{13.22}$$

References

1. S. Fu, Y. Sun, Z. Liu, X. Hou, H. Han, M. Su, Power oscillation suppression in multi-VSG grid with adaptive virtual inertia. Int. J. Electr. Power Energy Syst. **99**, 1–1 (2021)
2. M. Chen, D. Zhou, C. Wu, F. Blaabjerg, Characteristics of parallel inverters applying virtual synchronous generator control. IEEE Trans. Smart Grid. **12**(6), 4690–4701 (2021)
3. Z. Wang, H. Yi, F. Zhuo, J. Wu, C. Zhu, Analysis of parameter influence on transient active power circulation among different generation units in microgrid. IEEE Trans on Ind. Electron. **68**(1), 248–257 (2021)
4. Z. Shuai, W. Huang, Z.J. Shen, A. Luo, Z. Tian, Active power oscillation and suppression techniques between two parallel synchronverters during load fluctuations. IEEE Trans. Power Electron. **35**(4), 4127–4142 (2020)
5. J. Alipoor, Y. Miura, T. Ise, Stability assessment and optimization methods for microgrid with multiple VSG units. IEEE Trans. Smart Grid. **9**(2), 1462–1471 (2018)
6. M. Choopani, S.H. Hosseinain, B. Vahidi, A novel comprehensive method to enhance stability of multi-VSG grids. Int. J. Electr. Power Energy Syst. **104**, 502–514 (2019)
7. M. Choopani, S.H. Hosseinian, B. Vahidi, New transient stability and LVRT improvement of multi-VSG grids using the frequency of the center of inertia. IEEE Trans. Power Syst. **35**(1), 527–538 (2020)

8. V. Nasirian, Q. Shafiee, J.M. Guerrero et al., Droop-free distributed control for AC microgrids. IEEE Trans on Power Electron. **31**(2), 1600–1617 (2016)
9. H. Han, X. Chen, Z. Liu, Y. Liu, M. Su, S. Chen, A completely distributed economic dispatching strategy considering capacity constraints. J. Emerg. Sel. Top. Circuits Syst. **11**(1), 210–221 (2021)
10. M. Li, W. Huang, N. Tai, L. Yang, D. Duan, Z. Ma, A dual-adaptivity inertia control strategy for virtual synchronous generator. IEEE Trans. Power Syst. **35**(1), 594–604 (2020)
11. J. Alipoor, Y. Miura, T. Ise, Power system stabilization using virtual synchronous generator with alternating moment of inertia. IEEE. J. Emerg. Sel. Top. Power Electron. **3**(2), 451–458 (2015)
12. D. Li, Q. Zhu, S. Lin, X.Y. Bian, A self-adaptive inertia and damping combination control of VSG to support frequency stability. IEEE Trans. Energy Convers. **32**(1), 397–398 (2017)
13. X. Hou, Y. Sun, X. Zhang, J. Lu, P. Wang, J.M. Guerrero, Improvement of frequency regulation in VSG based AC microgrid via adaptive virtual inertia. IEEE Trans. Power Electron. **35**(2), 1589–1602 (2020)
14. Z. Shen J. Z et al., Variable-inertia emulation control scheme for VSC-HVDC transmission systems. IEEE Trans. Power Syst, Early Access
15. A. Karimi, Y. Khayat, M. Naderi et al., Inertia response improvement in AC microgrids: a fuzzy-based virtual synchronous generator control. IEEE Trans. Power Electron. **35**(4), 4321–4331 (2020)
16. U. Markovic, Z. Chu, P. Aristidou, G. Hug, LQR-based adaptive virtual synchronous machine for power systems with high inverter penetration. IEEE Trans. Sustain Energy. **10**(3), 1501–1512 (2019)
17. W. Wang, L. Jiang, Y. Cao, Y. Li, A parameter alternating VSG controller of VSC-MTDC systems for low frequency oscillation damping. IEEE Trans. Power Syst. **35**(6), 4609–4621 (2020)
18. W. Wang, X. Yin, Y. Cao, L. Jiang, Y. Li, A distributed cooperative control based on consensus protocol for VSC-MTDC systems. IEEE Trans. Power Syst. **36**(4), 2877–2890 (2021). https://doi.org/10.1109/TPWRS.2021.3051770
19. X. Ge, H. Han, W. Xiong, S. Mei, Z. Liu, Y. Sun, Locally-distributed and globally-decentralized control for hybrid series-parallel microgrids. Int. J. Electr. Power Energy Syst. **116**(99), 1–1 (2020)
20. M. Choopani, S.H. Hosseinain, B. Vahidi, A novel comprehensive method to enhance stability of multi-VSG grids. Int. J. Electr. Power Energy Syst. **104**, 502–514 (2019)
21. H. Wu, X. Wang, A mode-adaptive power-angle control method for transient stability enhancement of virtual synchronous generators. IEEE J. Emerg. Sel. Top. Power Electron. **8**(2), 1034–1049 (2020)
22. J. Liu, H. Golpira, H. Bevrani, et al. Grid integration evaluation of virtual synchronous generators using a disturbance-oriented unified modeling approach. IEEE Trans. Power Syst., **36**(5), 4660–4671

Chapter 14
PLL-Synchronized Voltage-Supporting Control Under Near-Zero Line Impedance

14.1 Issues for UPS Systems Under Near-Zero Line Impedance

Synchronization and power-sharing control are essential for the islanded operation of the parallel inverter system [1]. In islanded AC microgrids which contain parallel inverters, decentralized control methods, such as droop control [2] and Virtual Synchronous Generator (VSG) control [3], have received widespread attention due to the salient features of communication-free and plug-and-play capability. Figure 14.1 shows the architecture of the typical modular UPS system. When the UPS system becomes offline with the grid, the bypass switch is open, and the parallel inverters with batteries work in the islanded mode as the backup for the power supply.

However, the UPS system presents some differences from the islanded AC microgrid. In the application of the UPS system, the transmission distance is quite short between parallel inverters and the critical load, and the line impedance is nearly zero. When parallel voltage sources are electrically close to each other, voltage regulation tends to be more difficult. The droop-controlled inverters are prone to loss of synchronization [4]. Although adding an additional inductor can help increase the equivalent line impedance, it also increases the volume and costs of the UPS system.

Different from grid-forming droop control, the grid-following phase-lock-loop (PLL) method is suitable for strong grid strength with low grid impedance, which can exhibit stable operating capability under near-zero line impedance conditions.However, in the islanded operation mode, how to utilize PLL to realize frequency synchronization, voltage support, and power-sharing operation among parallel inverters with near-zero line impedances has not been investigated, which needs theoretical analysis.

X. Hou et al., *Self-organized Operation and Stability Control of Electronized Power Networks*, https://doi.org/10.1007/978-981-96-9067-1_14

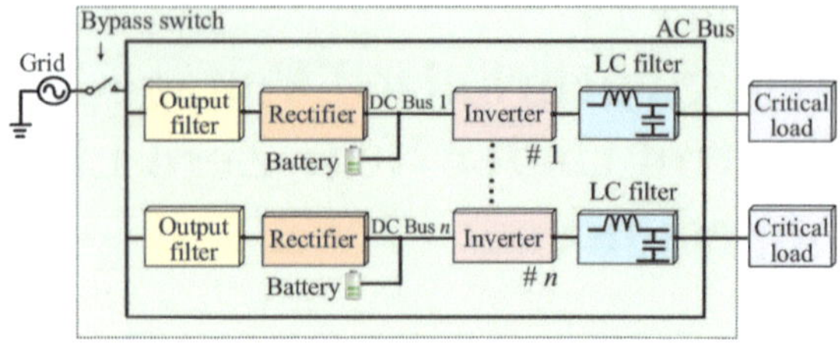

Fig. 14.1 Configuration of the parallel inverters in modular UPS system

14.2 First-Order PLL-Synchronized Voltage-Supporting Control

The introduced control scheme for i-th inverters is shown in Fig. 14.2. Compared with the traditional PLL-based control method [5], the main improvements lie in the first-order PLL and the modified current reference control. The inner current-tracking loop remains unchanged.

14.2.1 First-Order PLL with Basic Frequency Synchronization

The expression of the first-order PLL is expressed as:

$$\dot{\theta}_i = \int \left(\omega^* + k_p V_{iq}\right) \tag{14.1}$$

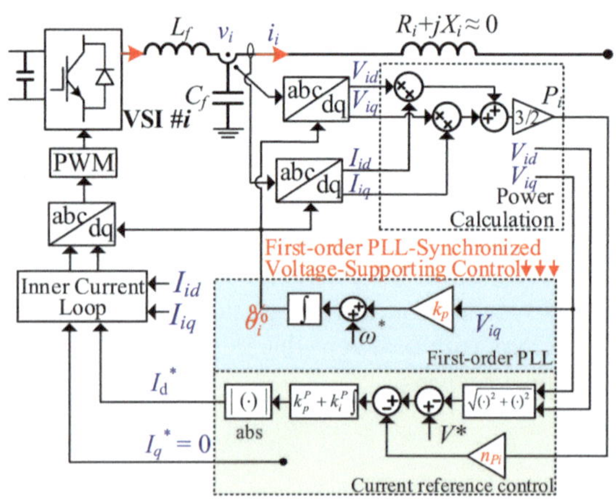

Fig. 14.2 The introduced first-order PLL-synchronized voltage-supporting control scheme for i-th inverter in the islanded UPS system

where k_p is the proportional coefficient of the first-order PLL. θ_i represents the estimated phase angle of the first-order PLL and V_{iq} is the q-axis component of the output voltage v_i.

Different from SRF-PLL, the integral item is removed, thus the traditional second-order PLL reduces to a first-order PLL, which generates the reference phase of the output current for frequency self-synchronization among parallel inverters. The inherent synchronization mechanism is analyzed later.

14.2.2 Modified Current Reference Control with Voltage Support

The modified current reference control is designed in dq frame, which is expressed as follows

$$\begin{cases} I_d^* = \left|\left(k_p^P + k_i^P \int\right)(V^* - V_i - n_{Pi} P_i)\right| \\ I_q^* = 0 \end{cases} \tag{14.2}$$

where I_d^*, and I_q^* are the desired output current reference. k_p^P and k_i^P are the proportional coefficient and the integral coefficient for I_d reference (I_d^*) control, respectively. V^* is the amplitude reference. P_i and V_i represent the output reactive power and the voltage amplitude of the i-th inverter, respectively. n_{Pi} is the active power-voltage control coefficient.

As for I_d^* control, the absolute value stage ensures that $I_d^* > 0$. The proportional-integral item guarantees the $P - V$ regulation characteristic, which generates the current amplitude reference for voltage support and power sharing.

I_{q^*} is set to 0 to orient the output current on the d-axis, thus $\left(\theta_i = \dot{\theta}_i, \omega_i = \dot{\theta}_i\right)$ can be ensured, where θ_i and ω_i are the output current phase and angular speed, respectively. Besides, controlling I_q^*=0 can facilitate power decoupling analysis.

14.3 Synchronization and Steady-State Analysis

14.3.1 Steady-State Analysis

As shown in Fig. 14.3, $V_i \angle\theta_{Vi}$ and $I_i \angle\theta_{Ii}$ are the output voltage and current of i-th inverter, respectively. The parallel inverters are treated as controlled current source inverters. $Z_L \angle\theta_Z$ represents the load impedance at the public point. $R_i + jX_i$ is the i-th line impedance, which is nearly zero due to the quite short connection distance to the public point.

According to the differential of (14.1), the $V_q - \omega$ output characteristics can be expressed as

Fig. 14.3 Schematic of parallel inverters in UPS system with a public load

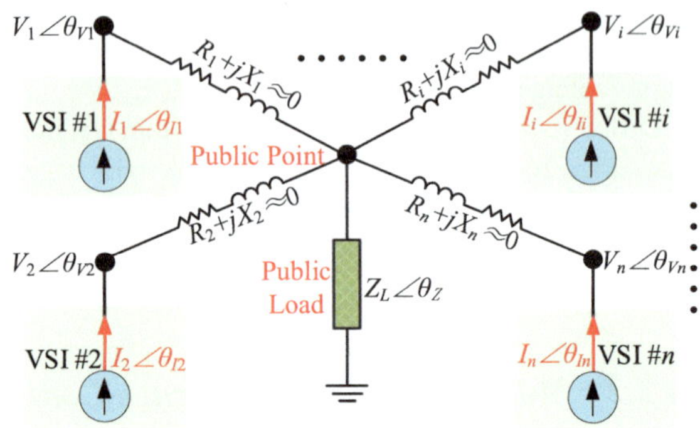

$$\omega_i = \omega^* + k_p V_{iq} \tag{14.3}$$

In the steady state, we have $(\omega_1 = \omega_2 = \ldots = \omega_n)$. If k_p of the parallel inverters are selected to be the same, $\left(V_{1q} = V_{2q} = \ldots = V_{nq}\right)$ can also be established.

As I_q is controlled to zero in (14.2) for decoupling power control, the output power of the i-th inverter can be expressed as

$$P_i = \frac{3}{2}\left(V_{id}I_{id} + V_{iq}I_{iq}\right), Q_i = \frac{3}{2}\left(V_{iq}I_{id} - V_{id}I_{iq}\right) \tag{14.4}$$

where the active power control mostly depends on V_d, and reactive power control depends on V_q.

Owing to the proportional-integral function of I_{d*} control in (14.2), the following equation is established in the steady state.

$$V^* - \sqrt{V_{id}^2 + V_{iq}^2} - n_{Pi}P_i = 0 \tag{14.5}$$

Due to the near-zero line impedance conditions, the output voltage of all inverters is roughly equal $(v_1 \approx v_2 \approx \ldots \approx v_n)$. According to (14.3), $\left(V_{1q} = V_{2q} = \ldots = V_{nq}\right)$. Thus, $(V_{1d} = V_{2d} = \ldots = V_{nd})$ is guaranteed in the steady state. Therefore, we have

$$\begin{cases} n_{P1}P_1 = n_{P2}P_2 = \mathrm{K} = n_{Pn}P_n \\ n_{P1}Q_1 = n_{P2}Q_2 = \mathrm{K} = n_{Pn}Q_n \\ \varphi_1 = \varphi_2 = \ldots = \varphi_n = \theta_Z \end{cases} \tag{14.6}$$

As a result, the output active power P and reactive power Q of parallel inverters can be proportionally shared according to $(1/n_{P1} : 1/n_{P2}:\ldots : 1/n_{Pn} = P_1 : P_2 : \ldots : P_n = Q_1 : Q_2 : \ldots : Q_n)$. To change the output power of inverter #i , it just needs to adjust the value of n_{Pi} without a centralized controller.

14.3.2 Synchronization Analysis of PLL-Based Inverters

Multiple parallel PLL-based inverters must synchronize with each other to guarantee stable islanded operation [6]. In this letter, frequency synchronization is realized through the first-order PLL-synchronized method. Herein, the dynamic model of the parallel inverters-based system with the first-order PLL is built, and the synchronization mechanism is clearly analyzed.

For the convenience of the analysis, the inner loop is ignored and the output current can be expressed as $I_i^* \angle\theta_i$. Under near-zero line impedance conditions ($R_i \approx 0$, $X_i \approx 0$), V_{iq} can be simplified as:

$$V_{iq} = Z_L \left[I_i^* \sin\theta_Z - \sum_{i \neq i}^{n} I_j^* \left(\sin\theta_{Iij} \cos\theta_Z - \cos\theta_{Iij} \sin\theta_Z \right) \right] \tag{14.7}$$

where $\theta_{Iij} = \theta_{Ii} - \theta_{Ij}$.

Due to the power-sharing operation, the current angle difference θ_{Iij} is always small among the parallel inverters, thus, $\cos\theta_{Iij} \approx 1$, $\sin\theta_{Iij} \approx \theta_{Iji}$. (14.7) can be further simplified as

$$V_{iq} = Z_L \left[\sin\theta_Z \sum_{j=1}^{n} I_j^* - \cos\theta_Z \sum_{j \neq i}^{n} I_j^* \left(\theta_{Ii} - \theta_{Ij} \right) \right] \tag{14.8}$$

Combing (14.3) and (14.9), the system dynamics are presented as:

$$\begin{cases} \omega_1 = \dot{\theta}_{I1} = \omega^* + k_p Z_L \sin\theta_z \sum_{j=1}^{n} I_j^* - k_p Z_L \cos\theta_z \sum_{j=1}^{n} I_j^* \left(\theta_{I1} - \theta_{Ij} \right) \\ \omega_2 = \dot{\theta}_{I2} = \omega^* + k_p Z_L \sin\theta_z \sum_{j=1}^{n} I_j^* - k_p Z_L \cos\theta_z \sum_{j \neq 2}^{n} I_j^* \left(\theta_{I2} - \theta_{Ij} \right) \\ \vdots \\ \omega_n = \dot{\theta}_{In} = \omega^* + k_p Z_L \sin\theta_z \sum_{j=1}^{n} I_j^* - k_p Z_L \cos\theta_z \sum_{j \neq n}^{n} I_j^* \left(\theta_{In} - \theta_{Ij} \right) \end{cases} \tag{14.9}$$

For the coupling parallel inverters, when the conditions in (14.10), are fulfilled, (14.9) meet the form of the commonly studied update rule (14.11), in multi-agent systems and nonlinear networked systems [6].

$$\left(k_p Z_L I_j^* \cos\theta_Z \right) > 0,\ j \in 1, 2 \ldots, n \tag{14.10}$$

$$\dot{x}_i = -\sum_{j \neq i}^{n} a_{ij} \left(x_i - x_j \right) \tag{14.11}$$

According to (14.10), frequency synchronization among parallel inverters with resistive (R) load, resistive-inductive (RL) load, and resistive-capacitive (RC) load can be guaranteed by the positive I_j^* through I_d reference control in (14.2). As a result, the current angles $\theta_{I1}, \theta_{I2}, \ldots, \theta_{In}$ will converge and synchronize with each other. Frequency synchronization is realized through the dynamics of current angle difference. Initially, $\theta_{Ii} > \theta_{Ij}$, then $\omega_i < \omega_j$ is obtained from (14.9) and finally, $(\theta_{Ii} - \theta_{Ij})$ decreases. The convergence process will continue until the system reaches the steady state $\theta_{Ii} = \theta_{Ij}, \omega_i = \omega_j$.

14.3.3 Experimental Verifications Under Load Change

As the parameters k_p and n_P influence the frequency and voltage amplitude variation in the steady state, the proper parameter design for k_p and n_P is crucial to the introduced method. According to some operation standards like German-VDE-AR-N 4110 and EN50438, the steady state voltage frequency/amplitude variation can be required less than 10% [7, 8]. The parameter design guideline considering the steady-state operation requirements can be derived as follows.

(1) Design for k_p

The parameter k_p is designed to meet the requirement that the steady state frequency variation is less than 10%.

V_{iq} can be expressed as

$$V_{iq} = \frac{P_i}{S_i} \cdot V_i \tag{14.12}$$

Considering that the range of the steadystate voltage amplitude is also required as [0.9 p.u., 1.1p.u.], k_p can be designed according to (14.3) and (14.12) as:

$$0 < k_p < 10\% \cdot \frac{S_i^*}{1.1 \cdot V^* P_i^*} \tag{14.13}$$

where S_i^* and P_i^* represent the rated apparent power and active power of i-th inverter, respectively.

Equation (14.13) gives a reference for k_p to meet the steady state requirement. Besides, a fast convergence rate of the frequency dynamic can be obtained by increasing k_p. Thus, a proper k_p should be designed by making a compromise between settling time and the steady-state operation requirement for frequency.

(2) Design for n_{Pi}

The parameter n_{Qi} is designed to meet the requirement that the steady-state voltage amplitude variation is less than 10%.

According to (14.5), n_{Qi} should be designed as:

$$0 < n_{P_i} \leq 10\% \cdot \frac{V^*}{P_i^*} \tag{14.14}$$

where P_i^* represents the rated active power of i-th inverter. The positive value of n_{Pi} can realize the $P - V$ regulation characteristic. Increasing n_{Pi} will enlarge the effect of the active power P on the voltage amplitude regulation. Thus, a proper n_{Pi} should be designed according to the steady-state operation requirement for voltage amplitude.

14.3.4 Compared with the Droop Control Method

As discussed previously, frequency synchronization under the introduced method is based on the dynamics of the current angle difference, which is different from that of the traditional voltage angle difference based on the droop control method.

Neglecting voltage dynamics, the small-signal model of the current angle dynamics in the paralleled inverters under the introduced method is expressed as:

$$\Delta\dot{\theta}_i = \kappa_i \Delta\theta_i \tag{14.15}$$

where κ_i is the synchronization matrix of the introduced method, which is shown in detail in the Appendix.

The small-signal model of the voltage angle dynamics under the droop control method is expressed as:

$$\begin{bmatrix} \Delta\dot{\boldsymbol{\omega}}_{\mathrm{v}} \\ \Delta\dot{\boldsymbol{\theta}}_{\mathrm{v}} \end{bmatrix} = \kappa_{\mathrm{v}} \begin{bmatrix} \Delta\omega_{\mathrm{v}} \\ \Delta\theta_{\mathrm{v}} \end{bmatrix} \tag{14.16}$$

where κ_{v} is the synchronization matrix of the droop control method, which is also shown in the appendix. It is noted that the voltage angle dynamics under the droop control method is a classical second-order system due to the low pass filter for average power calculation [9].

To demonstrate the advantages of the introduced method over droop control with near-zero impedance, the root loci of (14.15) and (14.16) in the two-paralleled-inverters system are plotted in Fig. 14.4a, b as the line impedances are varied.

Figure 14.4a shows that frequency synchronization under the introduced method is insensitive to the small change of low line impedances. However, Fig. 14.4b shows that reducing line impedance leads to the reduction of the system damping ratio under the droop control method, and further a worse dynamic. In other words, the droop-based parallel inverters are vulnerable to conditions with low line impedances. The corresponding comparative experiments between the introduced method and the droop control method are carried out later.

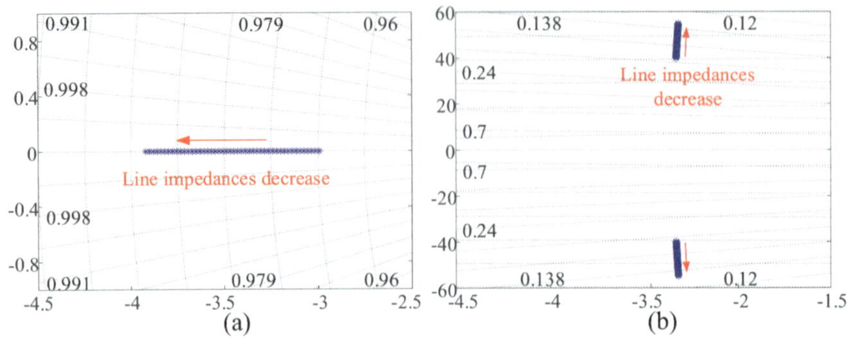

Fig. 14.4 Illustration of frequency synchronization stability with variations of line impedances. **a** Root loci under the introduced method with line impedances decrease from $j0.1\ \Omega$ to $j0.05\ \Omega$. **b** Root loci under the droop control method with line impedances decrease from $j0.1\ \Omega$ to $j0.05\ \Omega$

14.3.5 Compared with the PLL-Based Method

The comparisons between the introduced method and the PLL-based method in [10] are shown in Table 14.1. For the sake of analysis, the inertia support response in [10] is ignored and this comparison only analyzes the $\omega - P$ droop equation.

The introduced method employs only the first-order PLL to generate the reference phase of the output current. The $\omega - V_q$ equation of the first-order PLL enables frequency synchronization among parallel inverters in islanded mode. The frequency synchronization and voltage support are achieved through decoupling PLL and current reference control, separately, which facilitates controller design.

While the PLL-based method in [10] needs the SRF-PLL for voltage phase tracking and couples the current I_d reference control to construct the $\omega - P$ droop equation to generate the output current reference. Voltage support is realized through I_q reference control with a $Q - V$ droop equation.

Besides, the introduced method can operate in both RL and RC loads, while the method in [10] is only suitable for RL loads. To further explore the adaptability of load types under these two methods, the single inverter islanded system shown in Fig. 14.5 is considered.

Under the PLL-based method in [10], the output voltage amplitude variation is mainly influenced by the dynamic of i_q reference control, and the dynamic effects of PLL and I_d reference are too small to be readily neglected. Then linearizing at steady-state operating points yields

$$\Delta \dot{V} = -Z_L \sin\theta_Z k_i^Q \left(1 - n_Q I_q^S\right) \Delta V \tag{14.17}$$

where I_q^S is the steady-state value of I_q.

(14.17) shows that the stability of the system under the method in [10] is related to the load type. The system is unstable with RC load ($\sin\theta_z < 0$).

Table 14.1 Comparisons between the introcuced method and the PLL-based method in [10]

—	The introduced method	PLL-based method in [10]
Output Interfacing Characteristics	Voltage-support current-forming	Voltage-support current-forming
Synchronization Control	First-order PLL $\begin{cases} \theta_i = \theta_{PLL}, \theta_{PLL} = \int (\omega^* + k_p V_q) \\ \omega_i = \dot{\theta}_{PLL}, \underbrace{\omega_i = \omega^* + k_p V_q}_{\omega - V_q \text{ equation}} \end{cases}$	$SRF - PLL + I_d$ reference control $\begin{cases} \theta_i = \theta_{PLL} + \varphi, \omega_i = \dot{\theta}_{PLL} + \frac{d\varphi}{dt}, \varphi = \arctan \frac{I_q^*}{I_d^*} \\ \underbrace{\theta_{PLL} = \int \left[\omega^* + \left(k_p + k_i \int \right) V_q \right]}_{\text{Voltage phase tracking}}, \omega_{PLL} = \dot{\theta}_{PLL} \end{cases}$
Voltage-Supporting Control	$\begin{cases} I_d^* = \left| \left(k_p^P + k_i^P \int \right) (V^* - V - n_P P) \right| \\ I_q^* = 0 \end{cases}$	$\begin{cases} I_d^* = \left(k_p^P + k_i^P \int \right) \cdot \underbrace{[(\omega^* - \omega_{PLL}) - m_P P]}_{\omega - P \text{ equation}} \\ I_q^* = \left(k_p^Q + k_i^Q \int \right) [n_Q Q - (V^* - V)] \end{cases}$
Current Angle dynamics	Dominated by first-order PLL	Dominated by the coordination of second-order PLL and current reference control
Output Power Decoupling Analysis	Controlling $I_q = 0$ through I_q reference control $P_i = \frac{3}{2} V_{id} I_{id}, Q_i = \frac{3}{2} V_{iq} I_{id}$	Controlling $V_g = 0$ through SRF-PLL $P_i = \frac{3}{2} V_{id} I_{id}, Q_i = -\frac{3}{2} V_{id} I_{iq}$
Adaptability of Load types	Both RL and RC loads, seen in (18)	Only RL load, seen in (17)

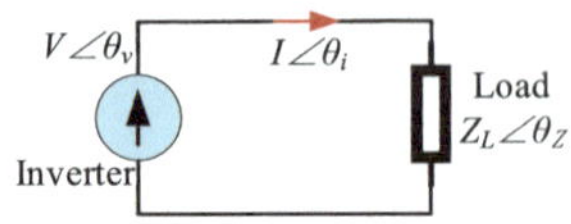

Fig. 14.5 Single inverter islanded system

Under the introduced method, the linear model of the voltage amplitude can be derived as

$$\Delta\dot{V}_d = -Z_L\cos\theta_Z \cdot k_i^P\left(\frac{1}{|\sin\theta_Z|} + n_P I_d^s\right)\Delta V_d \tag{14.18}$$

where I_d^s is the steady-state value of I_d.

(14.18) show that the system under the introduced method can operate in both RL load and RC load.

The comparative simulation results verify the superiority of the introduced method under all types of loads. The system parameters are consistent with those in the following experiments, the simulation results of the load change at 2s from RL to RC load with a single inverter islanded system are shown in Fig. 14.6. From Fig. 14.6a, it can be seen that the introduced method can work in both RL and RC load conditions while Fig. 14.6b shows that the system under the PLL-based method in [10] loses stability during the load change from RL load to RC load.

14.4 Experimental Verification

Experiments are carried out to verify the feasibility of the introduced method for parallel inverters. The configuration of the experimental setup is based on the RTU-BOX 204 experimental platform. Three inverters, three DC power supply sources, and loads are used. The DSP-based Rtunit-controller compiles programs and runs as the controller with a 10 kHz inverter switching frequency. Corresponding parameters are listed in Table 14.2.

14.4.1 Experimental Verifications Under Load Change

In this case, two tests with different loads are carried out.

Resistive-Inductive (RL) load change: three inverters are connected in parallel to the RL load. Transmission wires of different lengths are used without additional series RLC impedance to realize the near-zero line impedance conditions. A step change of the RL load occurs at t=5 s. The experimental results are shown in Fig. 14.7. The output voltage amplitude V_{PP} of the public load remains 100V after the RL load change. The active power P and reactive power Q of the three inverters are proportionally shared based on n_P.

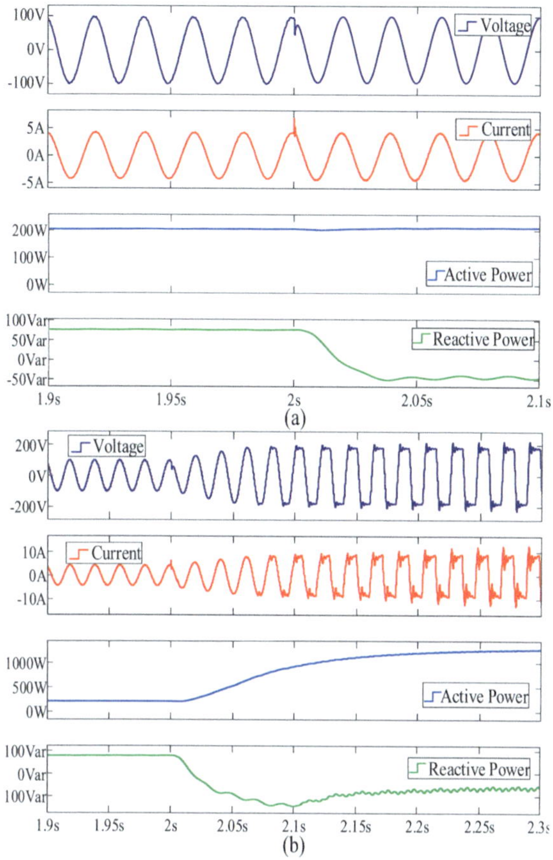

Fig. 14.6 Simulation results of the load change for RL load to RC load. **a** Under the introduced method; **b** Under the PLL-based method in [10]

Table 14.2 Parameters of the experimental test system

Parameters	Value	Parameters	Value
DC voltage	120 V	V_{ref}	100 V
Filter inductance L_f	1.5 mH	ω^*	100π
Filter capacitor C_f	30 μF	m	0
k_p	0.02	k_i	0.1
k_p^Q	0.005	k_i^Q	2
$n_{Q1}/n_{Q2}/n_{Q3}$	0.075/0.075/0.05	/	/

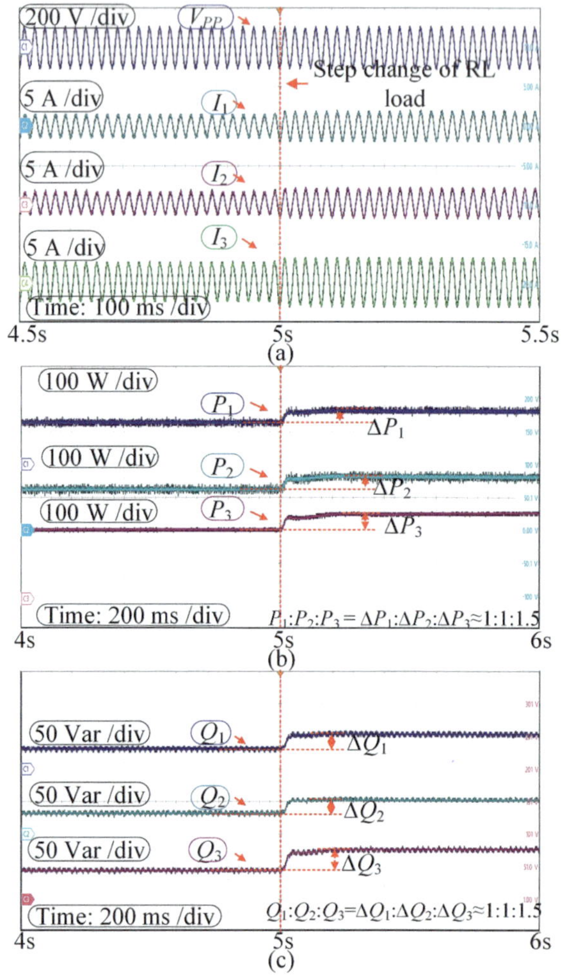

Fig. 14.7 Experimental results with RL load: **a** Voltage and current waveforms; **b** the output active power; **c** the output reactive power

Load Change from RL to RC load: three inverters are connected in parallel to the RL load, initially with the near-zero line impedance. A step change of the capacitive load occurs at t=5 s. The experimental results are shown in Fig. 14.8, which shows that the introduced control scheme is robust for the load change from RL to RC load.

The results in case A indicate that the introduced control realizes frequency synchronization, voltage support, and power-sharing operation under both RL and RC loads autonomously.

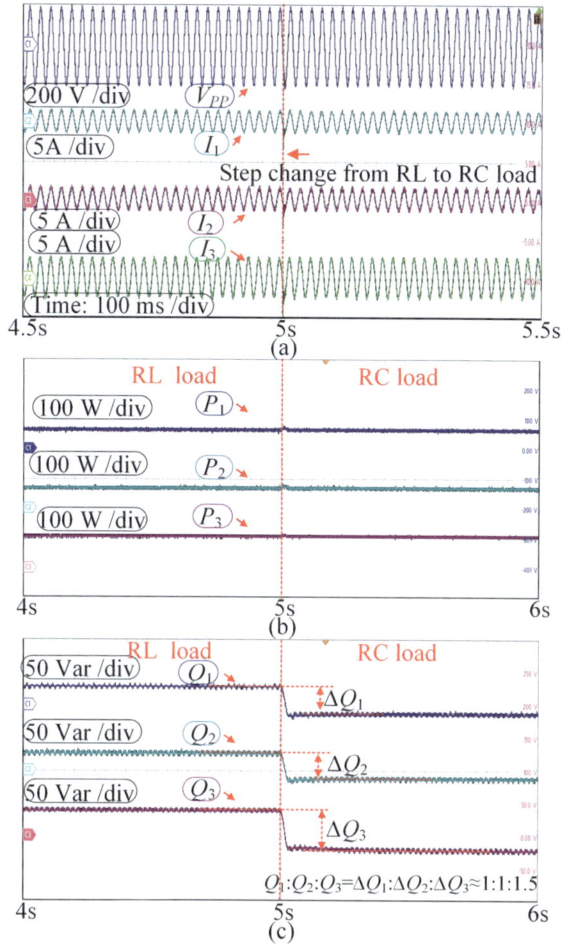

Fig. 14.8 Topology and control structure of VSG **a** main circuit and control structure **b** basic VSG control loop

14.4.2 Comparison with the Droop Control Method

To further explore the advantages of the introduced method under near-zero line impedance conditions, comparative experiments between the introduced method and droop control [11] are carried out.

The control structure diagram of the droop control is shown in Fig. 14.9, where ω^* is the rated angular speed (frequency), and V^* is the rated voltage amplitude. m_i represents the $P - \omega$ droop coefficient, and n_i represents the $Q - V$ droop coefficient, respectively. The parameters are selected as $V^* = 100$ V , $\omega^* = 100\pi$, $m_1 = m_2 = 0.001$, and $n_1 = n_2 = 0.005$ in the experiment.

Two parallel inverters are connected in this case. The line impedances of the two inverters are switched from 1.8 mH to 0 mH at 5 s. Figure 14.10 shows the results. It is shown in Fig. 14.10a that the parallel inverter system under the introduced method

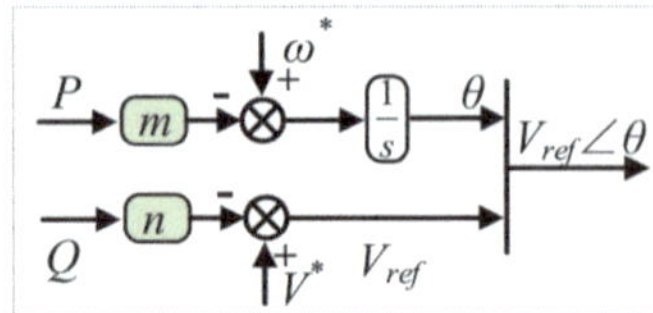

Fig. 14.9 Control diagram of the droop control method in case B

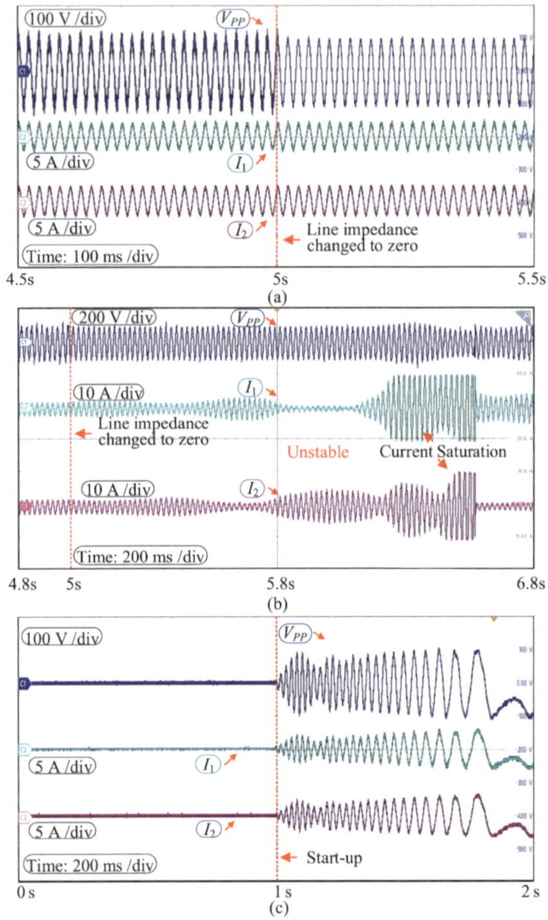

Fig. 14.10 Experimental results in case B: **a** Voltage/current waveforms under the introduced method; **b** voltage/current waveforms under droop control [11]; **c** voltage/current waveforms of the droop control method startup with the near zero line-impedance

remains stable even when the line impedance changes from 1.8 mH to 0 mH. The output voltage amplitude V_{PP} of the public load remains 100V. Figure 14.10b shows that the droopbased parallel inverters lose synchronization when the line impedance changes to 0 mH. Figure 14.10c shows that when the two parallel inverters start at 1 s without any additional series RLC impedance, the parallel inverters under droop control still lose synchronization. The experiment results indicate the superiority of the introduced method under the near zero line impedance conditions.

14.5 Conclusion

This chapter proposes a first-order PLL-synchronized voltage-supporting control method for parallel inverters in UPS systems with near-zero line impedance. A first-order PLL is employed for frequency synchronization among parallel inverters in islanded mode. The modified current control guarantees voltage support and power-sharing operation. The inverters are controlled as the controlled current source rather than the traditional controlled voltage source. It is revealed that under the introduced method, frequency synchronization is based on the dynamics of current angle difference instead of the traditional voltage angle difference. The introduced method can provide a new perspective to explore more possibilities for the operation of parallel inverters without using droop control.

Appendix

In the tested two-paralleled-inverters system in Section 14.3.4, the matrix in (14.15) and (14.16) can be expressed as:

$$\Delta\boldsymbol{\theta}_{\mathbf{i}} = \Delta\theta_{i1} - \Delta\theta_{i2} \tag{14.19}$$

$$\boldsymbol{\kappa}_{\mathbf{i}} = -k_p \left(I_1^s + I_2^s\right) Z_L \cos\theta_Z \cos\theta_{i12}^s \tag{14.20}$$

$$\begin{bmatrix} \boldsymbol{\Delta\omega}_{\mathbf{v}} \\ \boldsymbol{\Delta\theta}_{\mathbf{v}} \end{bmatrix} = \begin{bmatrix} \Delta\omega_1 - \Delta\omega_2 \\ \Delta\theta_{v1} - \Delta\theta_{v2} \end{bmatrix} \tag{14.21}$$

$$\kappa_{\mathbf{v}} = \begin{bmatrix} -\frac{1}{\tau} & -\frac{(m_1-m_2)V_1^s V_2^s \cos\theta_{v12}^s}{\tau X_{12}} \\ 1 & 0 \end{bmatrix} \tag{14.22}$$

where I_1^s, I_2^s, V_1^s, V_2^s, θ_{i12^s}, θ_{v12^s} are steady-state values of current, voltage, and current/voltage angle differences, respectively. τ is the time constant of the low pass filter in the droop control and X_{12} is the mutual line impedance between inverters #1 and #2. The values of $\cos\theta_{i12^s}$ and $\cos\theta_{v12^s}$ are related to the line impedance.

References

1. J.M. Guerrero et al., Control strategy for flexible microgrid based on parallel line-interactive UPS systems. IEEE Trans. Ind. Electron. **56**(3), 726–736 (2009)
2. J.M. Guerrero et al., Output impedance design of parallel-connected UPS inverters with wireless load-sharing control. IEEE Trans. Ind. Electron. **52**(4), 1126–1135 (2005)
3. Q.C. Zhong, G. Weiss, Synchronverters: inverters that mimic synchronous generators. IEEE Trans. Ind. Electron. **58**(4), 1259–1267 (2011)

4. X. Wang et al., Grid-synchronization stability of converter-based resources—an overview. IEEE Open J. Ind. Appl. **1**, 115–134 (2020)
5. C. Li et al., Dominant instability mechanism of VSI connecting to a very weak grid. IEEE Trans. Power Syst. **37**(1), 828–831 (2022)
6. X. Hou et al., Improvement of frequency regulation in VSG-based AC microgrid via adaptive virtual inertia. IEEE Trans. Power Electron. **35**(2), 1589–1602 (2020)
7. Rauch, et al. New revisions of FGW power quality guidelines, in *15th International Workshop on Large-Scale Integration of Wind Power into Power Systems as well as on Transmission Networks for Offshore Wind Power Plants* (2016)
8. B. O'Connor, Discussion - EN50438, ed. Dundalk: ESBN (2010)
9. S. D'Arco, J.A. Suul, Equivalence of virtual synchronous machines and frequency-droops for converter-based microGrids. IEEE Trans. Smart Grid. **5**(1), 394–395 (2014)
10. M. Zhang et al., Inertia and primary frequency provisions of PLL-synchronized VSC HVDC when attached to islanded AC system. IEEE Trans. Power Syst. **33**(4), 4179–4188 (2018)
11. Y. Sun et al., New perspectives on droop control in AC microgrid. IEEE Trans. Ind Electron. **64**(7), 5741–5745 (2017)

Part IV
Self-organized Control of Microgrid Clusters

Chapter 15
Distributed Coordination Control for Hybrid AC/DC Microgrid

15.1 Hybrid AC/DC Microgrid Configuration

HYHRID AC/DC microgrid is considered as the ideal structure of microgrid (MG) since it not only integrates advantages of both AC and DC MGs, but also can improve the power efficiency without extra energy conversion [1–4]. In hybrid MG, AC DGs and loads are connected to AC bus, while DC sub-grid contains batteries, DC renewable energy sources (RESs) and DC loads. To achieve power sharing, DGs in AC sub-grids are controlled by f/P droop while their counterparts in DC sub-grids are controlled by V/I droop [5–8]. Different sub-grids are connected by interlinked converters (ILCs) [9]. To avoid the heavy load condition of one sub-grid and light load condition of another, the main task of ILCs is to achieve the global power sharing (GPS) between DC and AC sub-grids.

To achieve better performance of power sharing accuracy with frequency/voltage restoration, distributed control is adopted in MGs with only sparse communication. For AC and DC MGs, many secondary control schemes like distributed averaged proportional-integral (DAPI) control, model-based predictive control (MPC) are introduced [10, 11]. However, for the hybrid AC/DC MG, the main challenge of distributed control is how to obtain the load conditions in the situation of frequency recovery and voltage restoration. Lack of inertia is also an urgent problem for hybrid MG [12]. Meanwhile, as for power control of ILC, there are two types of controls: PLL-based current control and VSG (or droop)-based voltage control. It's crucial to choose suitable power control of ILC for the stable operation of hybrid MG.

X. Hou et al., *Self-organized Operation and Stability Control of Electronized Power Networks*, https://doi.org/10.1007/978-981-96-9067-1_15

15.2 Analysis of Multiobjective Integration

The studied islanded hybrid MG structure is shown in Fig. 15.1. The feeders are considered dominantly inductive in AC sub-grid and resistive in DC sub-grid [4–7]. AC/DC generators and energy storage systems (ESSs) are considered as dispatchable DGs. Therefore, these DGs can support the hybrid MG and achieve power sharing in decentralized manners.

The control objectives of hybrid MG can be summarized in the following three points:

Obj. 1. Proper power coordination (i.e. the power sharing among all ac and dc DGs) and frequency/ voltage recovery.

Obj. 2. Inertia support for hybrid MG to avoid frequency and power fluctuation.

So far, several researches have been done to achieve these objectives whose methods are mainly divided into droop-based control methods and distributed control methods. In this section, the features and drawbacks of these control methods will be discussed in detail.

15.2.1 Droop-Based Control for Hybrid Microgrid

Droop control is used to achieve power sharing in sub-grids, and the mathematical expression can be described by

$$\text{AC sub - grid:} \begin{cases} f_i = f_0 - m_i P_{aci}; \\ E_i = E_0 - n_i Q_i; \end{cases} \tag{15.1}$$

$$\text{DC sub - grid: } V_{dcm} = V_{dc0} - k_m P_{dcm} \tag{15.2}$$

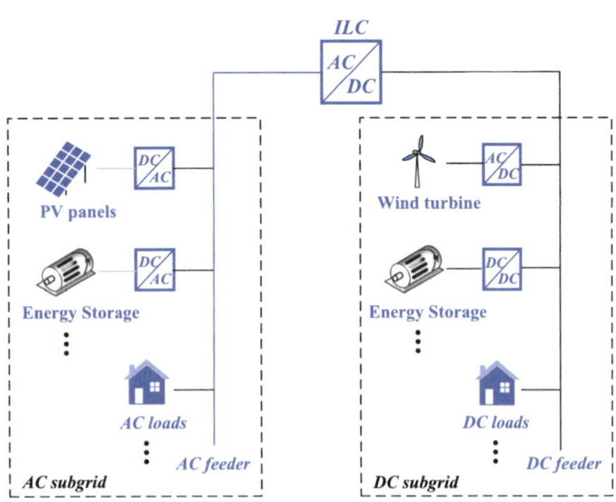

Fig. 15.1 The structure of islanded hybrid microgrid

where f_i, f_0 and P_aci are the output value, nominal value of the frequency and output active power in i-th AC-DG respectively. E_i, E_0 and Q_i are the output value, nominal value of the voltage and output reactive power in i-th AC-DG respectively. mi and n_i are the droop coefficients in i-th AC-DG. V_{dcm}, V_{dc0} and P_{dcm} are the output value, nominal value of the voltage and output active power in m-th DC-DG respectively. k_m is the droop coefficients in m-th DC-DG. When frequency synchronization is realized in the steady state, the proportional power sharing is achieved and can be shown as

$$\frac{P_{ac1}}{P_{ac1.\max}} = \frac{P_{ac2}}{P_{ac2.\max}} \ldots = \frac{P_{aci}}{P_{aci.\max}} \tag{15.3}$$

where $P_{aci.max}$ is the maximum output power of i-th AC-DG. On the other hand, the power sharing accuracy in DC sub-grids is influenced by mismatches of output feeders.

As for ILC control, the information of load conditions of AC and DC sub-grids is important. In droop control, the deviation of frequency and voltage can indicate the load conditions. Therefore, ILC can measure the frequency at AC bus (f) and the voltage at DC bus (V_{dc}) to estimate the load conditions. Firstly, the frequency and voltage deviation are normalized and unified in the same coordinate which is given by

$$f_{pu} = \frac{f - 0.5(f_{\max} + f_{\min})}{0.5(f_{\max} - f_{\min})} \tag{15.4}$$

$$V_{dc.pu} = \frac{V_{dc} - 0.5(V_{dc.\max} + V_{dc.\min})}{0.5(V_{dc.\max} - V_{dc.\min})} \tag{15.5}$$

where subscript ‘pu’ denotes the per unit value of each variable. Secondly, the PI controller is implemented for$V_{dc.pu}$ and f_{pu} as

$$i_{d.ic} = G_{PI(s)}\left(V_{dc.pu} - f_{pu}\right), i_{q.ic} = 0 \tag{15.6}$$

where $i_{d.ic}$ and $i_{q.ic}$ are the current references of ILC in the DQ frame. The current reference of d-axis is controlled by the PI controller of $G_{PI(s)}$ and the current in q-axis is zero. In the steady state, $V_{dc.pu}$ and f_{pu} are controlled to reach the same which means the identical load conditions between AC and DC sub-grids.

In a word, the droop-based method can achieve only GPS in hybrid MG. On the other hand, droop control will lead to frequency deviation, voltage drop and low inertia.

15.2.2 Distributed-Based Control for Hybrid Microgrid

As seen from (15.1) and (15.2), droop control inevitably causes the frequency deviation and voltage drop. Then, the secondary distributed control method is used to achieve frequency recovery and voltage restoration and is illustrated by

$$\begin{cases} f_i = f_0 - m_i P_{aci} + \Omega_i;\ \dot{\Omega}_i = \sum (\Omega_j - \Omega_i) \\ E_i = E_0 - n_i Q_i + \gamma_i;\ \dot{\gamma}_i = \sum (Q_j - Q_i) \\ V_{dcm} = V_{dc0} - k_m P_{dcm} + \sigma_m;\ \dot{\sigma}_m = \sum (P_n - P_m) \end{cases} \tag{15.7}$$

where Ω_i and γ_i are the frequency and voltage regulation terms of i-th AC-DG. σ_m is the voltage regulation terms of m-th DC-DG [13].

By implementing (15.7), the frequency recovery and voltage restoration are achieved. However, when f and V_{dc} are tightly controlled at the nominal value, the load conditions in sub-grids become imperceptible. In other words, ILC can't transfer the proper power to achieve GPS based on (15.4)–(15.6). Therefore, we have

$$f = f_0;\ V_{dc} = V_{dc0} \tag{15.8}$$

$$\frac{P_{ac1}}{P_{ac1.\max}} = \ldots = \frac{P_{aci}}{P_{aci.\max}} \neq \frac{P_{dc1}}{P_{dc1.\max}} = \ldots = \frac{P_{dcm}}{P_{dcm.\max}} \tag{15.9}$$

In a word, the standard distributed methods can achieve power sharing, frequency recovery and voltage restoration in sub-grids. However, they fail to realize the power sharing among DC and AC sub-grids and inertia support for hybrid MG.

15.3 Distributed Control Based on NPC and VSG

The overall structure of the introduced method is designed and shown in Fig. 15.2. The blue dashed lines in Fig. 15.2 denote the distributed cyber links in DC/AC sub-grids and ILC. The distributed control is adopted to achieve frequency recovery and voltage restoration. The NPC control is used to achieve GPS of hybrid MG. In addition, we adopt modified VSG control to enhance the frequency performance and avoid current distortion under high grid-impedance. The control mechanism will be introduced in the following section in detail.

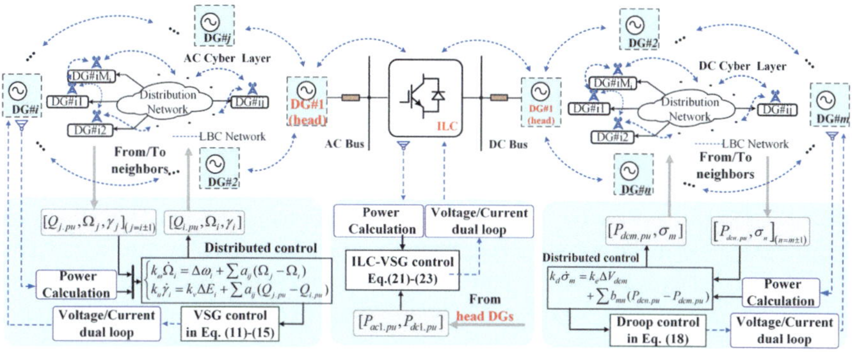

Fig. 15.2 Overall control diagram of hybrid MG

15.3.1 Two-Layer Architecture in AC Subgrid

(1) VSG based primary control

In order to improve the holistic inertia of MG, VSG control is an effective way to emulate the conventional synchronous generator compared to droop control. The swing equation of VSG can be described as follows

$$J\omega_0 \frac{\mathrm{d}\omega_i}{\mathrm{d}t} = P_{ini} - P_{aci} - D(\omega_i - \omega_0) \tag{15.10}$$

where ω_i is the generated angular frequency reference of i-th inverter output voltage, ω_0 is the nominal angular frequency. D is the damping factor and J is the virtual inertia. P_{aci} denotes the output active power of i-th inverter and P_{ini} represents the reference value of active power produced by the controller.

In order to coordinate the DGs of different capacities in hybrid microgrid, the process of transforming signals into per-unit values is introduced and (15.10) is modified as

$$J_i \frac{\mathrm{d}\omega_{i.pu}}{\mathrm{d}t} = P_{ini.pu} - P_{aci.pu} + D_i\left(\omega_{0.pu} - \omega_{i.pu}\right) \tag{15.11}$$

$$\omega_{i.pu} = \frac{\omega_i}{\omega_0};\ P_{aci.pu} = \frac{P_{aci}}{P_{0i}} \tag{15.12}$$

where P_{0i} is the rated active power of i-th DG. The normalization method is used in swing equation of (15.11). The normalized variables are calculated by (15.12) and denoted by subscript 'pu'. In the normalized framework, the damping factor D_i is set the same value for the proportional power sharing among DGs [14]. Meanwhile, the design of the inertia coefficient J_i usually needs to meet the requirements of the system damping ratio. Thus, J_i can be set the same in the normalized system to avoid large oscillation [15].

The control input term '$P_{ini.pu}$' takes the following form.

$$P_{ini.pu} = m_i \left(\omega_{0.pu} - \omega_{i.pu}\right) + \Omega_i \tag{15.13}$$

where m_i is the proportional coefficient of the governor and is designed with the same value among DGs in the normalized framework. Ω_i is the frequency restoration term from the secondary layer (which will be explained in the following subsection). In the meanwhile, the reactive power loop adopts the V/Q droop and is designed as

$$E_{i.pu} = E_{0.pu} - nQ_{i.pu} + \gamma_i \tag{15.14}$$

$$E_{i.pu} = \frac{E_i}{E_0};\ Q_{i.pu} = \frac{Q_i}{Q_0} \tag{15.15}$$

where Q_0 is the rated reactive power of DGs and γ_i is the voltage restoration term from the secondary layer (which will be explained in the following subsection).
(2) Distributed secondary control in AC sub-grid
The secondary control structure is based on the distributed network and all VSGs are designed to merely communicate with their neighbors. The frequency and voltage restoration terms in (15.13)–(15.14) are obtained by

$$k_{\omega i}\dot{\Omega}_i = \left(\omega_{0.pu} - \omega_{i.pu}\right) + \sum_{j \in N_i} a_{ij}(\Omega_j - \Omega_i) \tag{15.16}$$

$$k_{ui}\dot{\gamma_i} = k_{vi}\left(E_{0.pu} - E_{i.pu}\right) + \sum_{j \in N_i} a_{ij}(Q_{j.pu} - Q_{i.pu}) \tag{15.17}$$

where $k_{\omega i}$ is the positive gain of secondary frequency loop. k_{ui} and k_{vi} are the positive gain of secondary voltage loop. The communication network among DGs is designed in the distributed manner. The topology is described by a weighted graph $G = (\upsilon, \varepsilon, A)$ where $\upsilon = \{1, 2, \ldots, k\}$ is the DG node set, $\varepsilon \subseteq \upsilon \times \upsilon$ is the set of communication links and A is the adjacency matrix of the graph with its elements a_{ij}=0,1. The circular communication graph is considered as a reliable structure with enough redundance to endure the case of any single point failure [13].

The frequency restoration of ω_{pu} is achieved in the steady state according to (15.11) and (15.16). On the other hand, there is a trade-off between voltage restoration and reactive power sharing. In order to achieve bus-voltage regulation and reactive power sharing simultaneously, we label the nearest AC-DG to AC-Bus as the head DG1 to control the bus voltage of AC sub-grids. $k_{\upsilon 1}$ is designed positive for voltage restoration and a_{1j} is zero. Afterwards, the other DGs are set as follower DGs. They are controlled to realize accurate reactive power sharing. The parameters of followers $a_{ij(i \neq 0)}$ is positive and $k_{\upsilon i(i \neq 0)}$ is zero.

15.3.2 Two-Layer Architecture in DC Subgrid

(1) Droop based primary control
The V/P droop is adopted in primary control. Similar to the normalized control in AC sub-grid, DGs of different capacities in DC sub-grid are also transformed into per-unit and the modified droop is described as follows

$$V_{dcm.pu} = V_{dc0.pu} - kP_{dcm.pu} + \sigma_m \tag{15.18}$$

$$V_{dcm.pu} = \frac{V_{dcm}}{V_{dc0}} \tag{15.19}$$

where k is the droop coefficient of DC sub-grid. σ_m is voltage restoration term from the secondary layer (which will be illustrated in the following subsection).
(2) Distributed secondary control in DC sub-grid
The secondary distributed control structure of DC sub-grid is similar to AC sub-grid and the circular network is adopted as well. The voltage restoration term σ is calculated by

$$k_{dm}\dot{\sigma}_m = k_{em}\left(V_{dc0.pu} - V_{dcm.pu}\right) + \sum_{n\in N_m} b_{mn}(P_{dcn.pu} - P_{dcm.pu}) \tag{15.20}$$

where k_{dm} and k_{em} are the positive gain of secondary voltage loop. Similar to the design in AC sub-grid, we label the nearest DC-DG to DC-Bus as the head DG_1 to control the bus voltage of DC sub-grids. k_{e1} is set positive for voltage restoration and b_{1n} is zero. Afterwards, the other DGs are seen as follower DGs. They are controlled to achieve accurate power sharing. The parameters of followers $b_{mn(m\neq 0)}$ =1 and $k_{em(m\neq 0)}$ =0.

15.3.3 VSG-Based NPC Control Under a Single ILC

In order to achieve the GPS of hybrid MG, ILC needs to estimate the load conditions and control the power flow between two sub-grids. Therefore, the head DGs in sub-grids transmit the output-power information to ILC. The normalized active power of AC and DC head sources are represented by $P_{ac1.pu}$ and $P_{dc1.pu}$, respectively. Define that the positive direction of P_{ic} is from DC side to AC side. Then $P_{ac1.pu}$ and $P_{dc1.pu}$ are obtained by ILC through communication links and are sent to NPC controller, which can be realized by a PI controller.

$$\delta P_{ic} = \left(k_{pp} + \frac{k_{pi}}{s}\right)\left(P_{ac1.pu} - P_{dc1.pu}\right) \tag{15.21}$$

where k_{pp} and k_{pi} are the PI coefficients of NPC controller. In addition, the adaptive VSG control is embedded with NPC controller and the overall control of ILC is illustrated by

$$J_{ic}\frac{\mathrm{d}\omega_{ic.pu}}{\mathrm{d}t} = P_{in\_ic.pu} - P_{ic.pu} + D\left(\omega_{0.pu} - \omega_{ic.pu}\right) \tag{15.22}$$

$$P_{in\_ic.pu} = \delta P_{ic} + m\left(\omega_{0.pu} - \omega_{pu}\right) \tag{15.23}$$

where $\omega_{ic.pu}$ is the normalized value of ILC's output frequency. $P_{ic.pu}$ is the normalized value of ILC's output power and J_{ic} is the virtual inertia of ILC. According to (15.21)–(15.23), the frequency synchronization between ILC and AC sub-grid can be achieved without PLL. Besides, $P_{ac1.pu}$ and $P_{dc1.pu}$ are controlled at the same values in the steady state by NPC controller and are shown as follows:

$$\frac{P_{ac1}}{P_{ac1.0}} = P_{ac1.pu} = P_{dc1.pu} = \frac{P_{dc1}}{P_{dc1.0}} \tag{15.24}$$

In the meanwhile, by adopting (15.11) and (15.18) in AC and DC sub-grids, the accurate power sharing in AC and DC sub-grids can be realized in the steady state and are given by

$$\frac{P_{ac1}}{P_{ac1.0}} = \frac{P_{ac2}}{P_{ac2.0}} = \ldots = \frac{P_{aci}}{P_{aci.0}} \tag{15.25}$$

$$\frac{P_{dc1}}{P_{dc1.0}} = \frac{P_{dc2}}{P_{dc2.0}} = \ldots = \frac{P_{dcj}}{P_{dcj.0}} \tag{15.26}$$

Combing (15.24)–(15.26), the GPS of hybrid MG is achieved in the steady state and can be shown as

$$\frac{P_{ac1}}{P_{ac1.0}} = \ldots = \frac{P_{aci}}{P_{aci.0}} = \frac{P_{dc1}}{P_{dc1.0}} = \ldots = \frac{P_{dcj}}{P_{dcj.0}} \tag{15.27}$$

Besides, the equivalent inertia of hybrid MG in the introduced method can be evaluated by [16]

$$J_{eq} = \frac{\sum_{n=1}^{N} J_n + \sum_{l=1}^{L} J_{ic,l}}{N + L} \tag{15.28}$$

where N is the number of ac-VSG sources and L is the number of ILC-VSG sources. From (15.28), the introduced VSG-based NPC control provide inertia support from J, J_{ic}. In contrast, the droop and PLL-based methods have little inertia and their inertia can be calculated by $1/(m_i\omega_c)$ [17]. ω_c is the cut-off frequency of power filter. Thus, the equivalent inertia of the system increases and the frequency performance can be improved by the introduced method.

Remark 1: The introduced NPC controller of ILC obtains the power information from the head DGs in sub-grids. It helps ILC to estimate the load conditions and coordinate power flow properly among AC and DC sub-grids when frequency and DC voltage are controlled at the nominal value.

Remark 2: The VSG controller is embedded with NPC controller and the voltage control is adopted for inner loop of ILC. Compared to the conventional PLL-based current control for ILCs in [13, 18–24], the VSG-based voltage control for ILC has positive merits like provision of inertia, voltage support, islanded operation and good control performance in weak grids [3, 14, 25]. In particular, the introduced method is more suitable for islanded microgrid operation since it can provide voltage regulation and improve the holistic inertia level of the system compared with PLL-based method.

15.3.4 VSG-Based NPC Under Multiple ILCs

In order to improve the reliability of hybrid MG, the structure with multiple ILCs is recommended for enough redundance [9]. Multiple ILCs are aimed to proportionally share the transmission power between ac and dc subgrids, namely ILC power sharing (IPS). Based on the NPC control of (15.21), we extend it to multi-ILC cases.

$$\begin{cases} \delta P_{ic,j} = \left(k_{pp} + \frac{k_{pi}}{s}\right) e_{ic,j} \\ e_{ic,j} = \alpha\left(P_{ac1.pu} - P_{dc1.pu}\right) + \beta \sum\limits_{k \in N_{ic,j}} \left(P_{ic,k.pu} - P_{ic,j.pu}\right) \end{cases} \tag{15.29}$$

where $N_{ic,j}$ denotes the set of neighbors of j-th ILC. $P_{ic,k.pu}$ is the normalized output power of k-th ILC. α, β are positive coefficients related to the transient response of GPS and IPS, respectively. (15.29) can be considered as a kind of DAPI controller. The second term of $e_{ic,j}$ is the power sharing error among ILCs. It will converge to zero when the system reaches an equilibrium point due to the distributed consensus. The first term of (15.29) is the power sharing error between ac and dc subgrids. It will also converge to zero by PI regulator when second term of $e_{ic,j}$ converges.

15.4 Stability Analysis

15.4.1 Small-Signal Modeling of Hybrid MG with NPC Control

The model of the closed-loop system with the introduced control method is developed for small-signal stability analysis and parameter design.

Define that the positive direction of P_{ic} is from DC side to AC side. Then, the active power at node i in ac subgrid can be expressed by

$$P_{ac,i} = -s_i P_{ic} + P_{l,i} + E_i \sum_{j=1}^{N} E_j \left(G_{ij} \cos \delta_{ij} + B_{ij} \sin \delta_{ij}\right) \tag{15.30}$$

where G_{ij} and B_{ij} are the conductance and susceptance between node i and j, respectively. δ_{ij} is the power angle between node i and j. s_i is the connection status between node i and ILC and $P_{l,i}$ is the constant power load at node i. In dc subgrid, the nodal power at node i is given by

$$P_{dc,i} = s_i P_{ic} + P_{l,i} + V_{dc,i} \sum_{j=1}^{M} \left(V_{dc,i} - V_{dc,j}\right) Y_{ij} \tag{15.31}$$

where Y_{ij} is the conductance between node i and j in dc subgrid. Afterwards, assuming the *V/Q* loop is stable at the rated value and δ_{ij} is small [18], the small-signal model of the system can be linearized as

$$\Delta P_{ac,i} = -s_i \Delta P_{ic} + E_i^s \sum_{j=1}^{N} E_j^s B_{ij} \cos \delta_{ij}^s \left(\Delta \delta_i - \Delta \delta_j\right) \tag{15.32}$$

$$\Delta P_{dc,i} = s_i \Delta P_{ic} + V_{dc,i}^s \sum_{j=1}^{M} \left(\Delta V_{dc,i} - \Delta V_{dc,j}\right) Y_{ij} \tag{15.33}$$

$$\begin{cases} J_i \Delta \dot{\omega}_i = -(m + D)\, \Delta \omega_i - \Delta P_{ac,i} + \Delta \Omega_i \\ k_{\omega i} \Delta \dot{\Omega}_i = -\Delta \omega_{i.pu} + \sum\limits_{j \in N_i} a_{ij} (\Delta \Omega_j - \Delta \Omega_i) \end{cases} \tag{15.34}$$

$$\begin{cases} \Delta V_{dc,m} = -k \Delta P_{dc,m} + \Delta \sigma_m \\ k_{dm} \Delta \dot{\sigma}_m = -k_{em} \Delta V_{dc,m} + \sum\limits_{n \in N_m} b_{mn} (\Delta P_{dc,n} - \Delta P_{dc,m}) \end{cases} \tag{15.35}$$

$$\begin{cases} J_{ic} \Delta \dot{\omega}_{ic} = -(m + D)\, \Delta \omega_{ic} - \Delta P_{ic} + k_{pp} (\Delta P_{ac1} - \Delta P_{dc1}) + k_{pi}\, p \\ \dot{p} = (\Delta P_{ac1} - \Delta P_{dc1}) \end{cases} \tag{15.36}$$

The subscript 'pu' is omitted for simplicity and superscript 's' means the steady-state value. Further, to explore the impact of communication time delay on system's stability, the delay term $G_d = 1/(\tau_d s + 1)$ is considered in the distributed control terms. Finally, by integrating (15.32)–(15.36), the whole state-space model of the system is illustrated as

$$\Delta\dot{x} = \mathbf{A}_{(4N+3+2M)\times(4N+3+2M)}\Delta\mathbf{x} + \mathbf{Bu} \tag{15.37}$$

where $\mathbf{x}=[\omega_{\mathbf{ac}}\delta_{\mathbf{ac}}\mathbf{P}_{\mathbf{ac}}\Omega\omega_{ic}\delta_{ic}p\mathbf{P}_{\mathbf{dc}}\sigma]$ and the system matrix $\boldsymbol{A}$ is given in Appendix.

15.4.2 Participation Factor Analysis

In general, Participation Factor (PF) analysis has been introduced to analyze the connection between a system variable and some modes and to quantify the corresponding participation degree of a system variable in a mode and vice versa [26]. The calculation of PF is defined as

$$p_{ki} \triangleq u_i^k v_i^k \tag{15.38}$$

where u_i^k and v_i^k are the right and the left eigenvector with the k-th component of the state $\mathbf{x}$.

Further, a hybrid MG with 4 ac DGs, 4 dc DGs and 1 ILC is taken as an example. The nominal parameters and control coefficients of hybrid MG are given in Table 15.2 in Appendix. Table 15.1 give the result of participation factors analysis.

In Table 15.1, the dominant oscillatory modes eigenvalues of hybrid MG are selected for parameter guidance and stability analysis. λ_{5-10} are mainly related with ω_{ac}, δ_{ac}, P_{ac} , $\lambda_{17,18}$ are associated with ω_{ic}, δ_{ic}, P_{ac1}, P_{dc1} and λ_{20-23} are relevant to P_{dc}, σ . It means that the related states can be chosen as feedback information to raise the damping ratio (ζ) of the dominant modes and improve the stability of the system. From (15.36), k_{pp}, k_{pi} and τ_d are the key coefficients of power feedback. Besides, J, D are related coefficients of frequency feedback. Thus, these coefficients should be tuned carefully for stable operation of hybrid MG.

15.4.3 Sensitivity Analysis

To properly design the key control parameters of hybrid MG, sensitivity analysis is implemented. The root locus of inertia of ILC is shown in Fig. 15.3a. When J_{ic} is

Table 15.1 Parameters of hybrid microgrid

Eigenvalue	Oscillation frequency/Hz	Damping ratio	Mainly related state variable
λ_{5-10}	1.6 ~ 2.1	0.24 ~ 0.35	ω_{ac}, δ_{ac}, P_{ac}
$\lambda_{17,18}$	5.9	0.26	ω_{ic}, P_{ac1}, P_{dc1}
λ_{20-23}	119	0.31	P_{dc}, σ

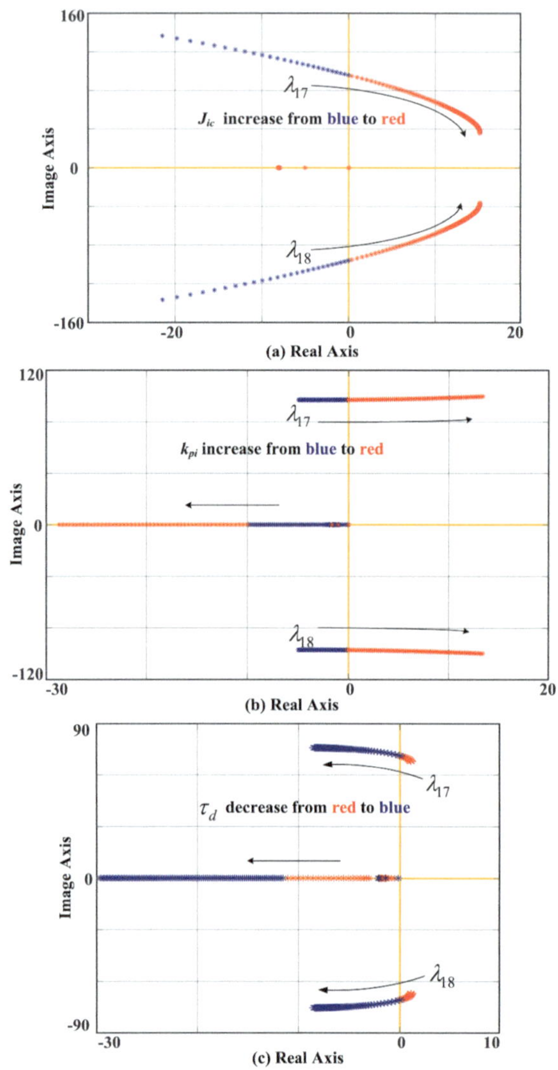

Fig. 15.3 Dominant poles $\lambda_{17,18}$ as **a** inertia J_{ic} increases from 10 to 400, **b** k_{pi} increases from 0.5 to 50, **c** τ_d decreases from 10 to 1e-2

increasing to 130, $\lambda_{17,18}$ lie on the right half-plane which indicates the instability of the system. Therefore, J_{ic} should not be too large. Figure 15.3b illustrate the influence of coefficients k_{pi}. When coefficient k_{pi} increases to 20, $\lambda_{17,18}$ become unsTable Fig. 15.3(c) shows the track of the $\lambda_{17,18}$ as time delay τ_d decreases from 10 to 1e-2 and when τ_d is smaller than 2s, the system become sTable The impact of J_{ic}, k_{pi} and τ_d on the system is similar since the increase of them reduces the damping ratio and even causes the instability of the system.

Based on the above analysis, the smart tuning can be summarized as follows. The trade-off between GPS response and time delay τ_d should be made since both of them reduce the damping ratio of the system. As for inertia placement, it will be discussed in detail in the following subsection.

15.4.4 Inertia Placement for Oscillation Suppression

The inertia placement is important since it is strongly related to oscillatory modes eigenvalues from PF analysis. Assume that the equivalent inertia in (15.28) of hybrid MG is fixed. Then we define the index of inertia deviation σ_{ac-ac} for ac subgrid and σ_{ic-ac} for hybrid MG as follows.

$$\begin{cases} \sigma_{ac-ac} = \sqrt{\frac{\sum_{n=1}^{N}\left(J_n - \bar{J}_{ac}\right)^2}{N}} \\ \sigma_{ic-ac} = \sqrt{\frac{\sum_{n=1}^{N}\left(J_n - J_{eq}\right)^2 + \sum_{l=1}^{L}\left(J_{ic,l} - J_{eq}\right)^2}{N+L}} \end{cases} \tag{15.39}$$

where $\bar{J}_{ac} = \sum_{n=1}^{N} J_n/N$ is the equivalent inertia in ac subgrid. (15.39) indicates the standard deviation of inertia among ac DGs and ILCs at a fixed inertia level. Furthermore, based on small-signal model of (15.37) and PF analysis, the correlations between minimum damping ratio of oscillatory poles and σ_{ac-ac} , σ_{ic-ac} are shown in Fig. 15.4.

We can obtain that the minimum damping ratio of dominant oscillatory poles will decrease as σ_{ac-ac} increases. Thus, the inertia in ac subgrids can be set the same to suppress the oscillation of oscillatory poles. Besides, the minimum damping ratio of dominant oscillatory poles will increase to the maximum first and then it will reduce with σ_{ic-ac} increasing gradually. Meanwhile, the increase of J_{ic} might cause

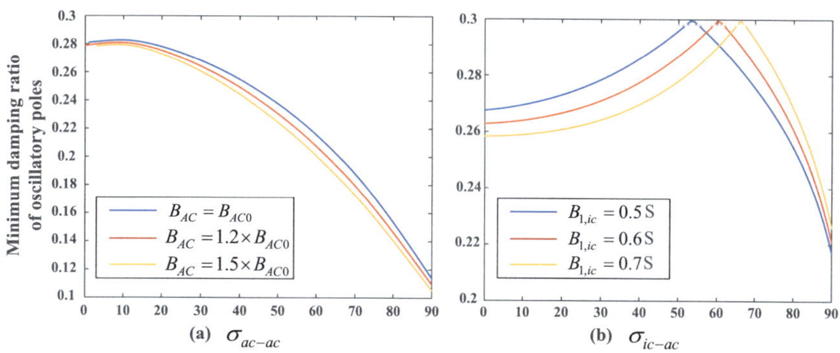

Fig. 15.4 The correlations between minimum damping ratio of oscillatory poles and **a** σ_{ac-ac}, **b** σ_{ic-ac}

instability of system from Fig. 15.3a. Thereby, it is suggested that J_{ic} is smaller than J to suppress oscillation. In this chapte, the equivalent inertia level J_{eq} in the following tests is 90. J is 100 for all ac DGs and J_{ic} is equal to 50.

15.5 Simulation and Hardware-in-Loop Results

To examine the introduced coordination control, a tested hybrid MG with 4 ac DGs, an ILC and 4 dc DGs is built in MATLAB/Simulink, as shown in Fig. 15.5. The three-phase inverters in AC sub-grid use LC filters and are controlled by PIP controller in dq-frame. DC-DC boost converters in DC sub-grid use LC filters and are controller by PIP controller. ILC converter uses LC filter in the introduced VSG-based method and adopts LCL filter in conventional PQ-based method. The parameters of hybrid MG are listed in Table 15.2 in Appendix. The line impedance of L_1 to L_4 and L_{ic} is 2.5, 3, 3.5, 4 and 5 mH in AC sub-grid and R_1 to R_4 and R_{ic} is 0.5, 0.8, 1, 1.2 and 1.6 Ω in DC sub-grid.

15.5.1 Comparison of the Distributed Coordination Control

Firstly, we conduct the comparison tests with method in [21], which used conventional DAPI droop control and PLL-based ILC control. The parameters of the tested system are given in Table 15.2 in Appendix. Figure 15.6 shows the waveforms of hybrid MG under the introduced control and method in [21]. The AC impact load is connected at 3s and removed at 9s. The DC impact load is connected at 6s and

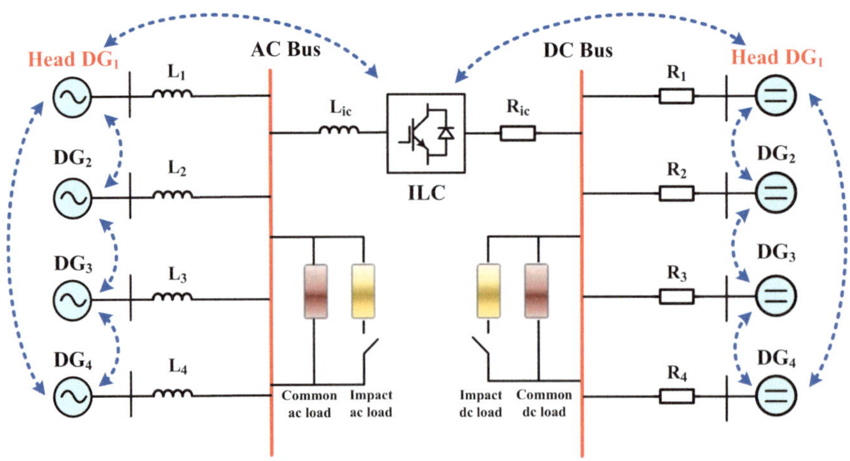

Fig. 15.5 The structure of the hybrid microgrid

Table 15.2 Key parameters of hybrid microgrid

Parameter	Symbol	Value
	System parameters	
Frequency	ω_0	314 rad/s
AC-Voltage	V_{d0}	311 V
DC-Voltage	V_{dc0}	400 V
LC filter of inverters	L_f, C_f	1 mH, 25 μF
LC filter of DC converter	L_b, C_b	0.5 mH, 1.5 mF
VSG-based LC filter of ILC	L_{ic}, C_{ic}	1 mH, 25 μF
PQ-based LCL filter of ILC	L_{ic}, C_{ic}	1.5 mH, 25 μF
Rated power of DGs in AC MG	$P_{0,1-2}/P_{0,3-4}$	4/6 kW
Rated power of DGs in DC MG	$P_{0,1-2}/P_{0,3-4}$	4/6 kW
Rated power of ILC-DG	P_{ic}/P_0	8 kW
Initial AC/DC load	P_{ac1}/P_{dc1}	12/6 kW
Impact AC/DC load		6/3 kW
	Inner loop control	
AC-Voltage loop PI	k_{vp1}, k_{vi1}	0.5, 250
AC-Current loop P	k_{cp1}	20
DC-Voltage loop PI	k_{vp2}, k_{vi2}	1, 10
DC- Current loop P	k_{cp2}	200
	Power loop control	
Control coefficient of AC-DG	m/n	$1 \times 10^3/1 \times 10^{-3}$
	J_{ac}/D	100/20
	$k_\omega/k_p/k_v$	0.2/1/10
	ω_c	50
Control coefficient of ILC	J_{ic}	50
	k_{pp}/k_{pi}	0.5/5
Control coefficient of DC-DG	k	1×10^{-3}
	k_d/k_e	0.1/5

removed at 12s. The GPS of hybrid MG and frequency recovery are achieved under both two methods. Further, we can find that the response time of GPS under the introduced method is smaller than 1 s which faster than method in [21]. Meanwhile, the frequency performance of the introduced method is much better than that of method in [21].

To further compare the frequency performance under these two methods in detail, we introduce the concept of the frequency of the center of inertia (COI), which is calculated by

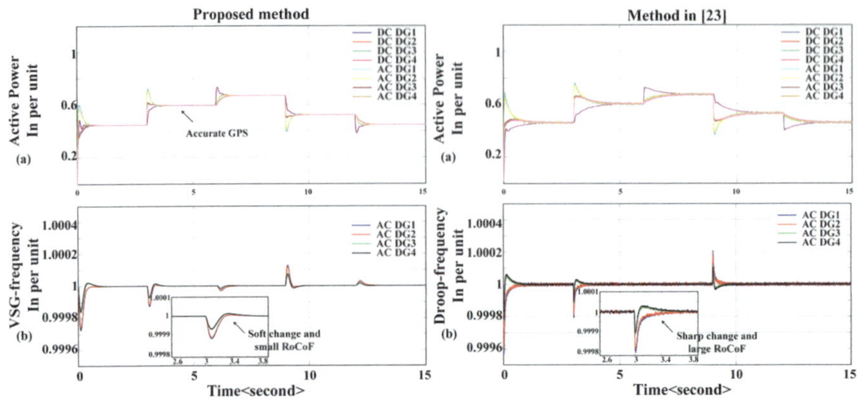

Fig. 15.6 The comparison results of the introduced method and method in [21]. **a** Active power of DC and AC DGs, **b** frequency of DGs

$$f_{COI} = \frac{\sum_{i \in N_{SG}} J_i f_i}{\sum_{i \in N_{SG}} J_i} \tag{15.40}$$

where N_{SG} is the set of synchronous DGs, including VSG-based DGs and ILCs, droop-based DGs. It is considered that f_{COI} can stand for the uniform/coherent motion shared by all synchronous DGs of the energy systems [27]. The comparison results of f_{COI} and its RoCoF are shown in Fig. 15.7. Observing Fig. 15.7, the VSG-based DGs and ILCs can improve the holistic inertia of MG. Therefore, the frequency performance is enhanced with small RoCoF (within 0.2 Hz/s). In contrast, the MG controlled by method [21] has low inertia and encounters large RoCoF which is harmful to the system.

15.5.2 Comparison of Different Inertia Placement

In this subsection, the comparison of different inertia placement is carried out to verify the analysis of Sect. 15.4.4. The results are given in Fig. 15.8.

We can obtain from Fig. 15.8 that when the inertia of all VSGs is set the same, the oscillation of ILC is large since $\lambda_{17,18}$ are the oscillatory poles with minimum damping ratio. With the decrease of J_{ic} , σ_{ic-ac} increases and the damping ratio of $\lambda_{17,18}$ become larger. When J_{ic} decreases to 10, $\lambda_{17,18}$ are not oscillatory poles with minimum damping ratio. The oscillation of ILC is smaller but ILC has little inertia support. Generally, $J_{ic} = 50$ is the optimal choice for ILC considering the frequency oscillation and inertia support, which verifies the analysis of Sect. 15.4.4.

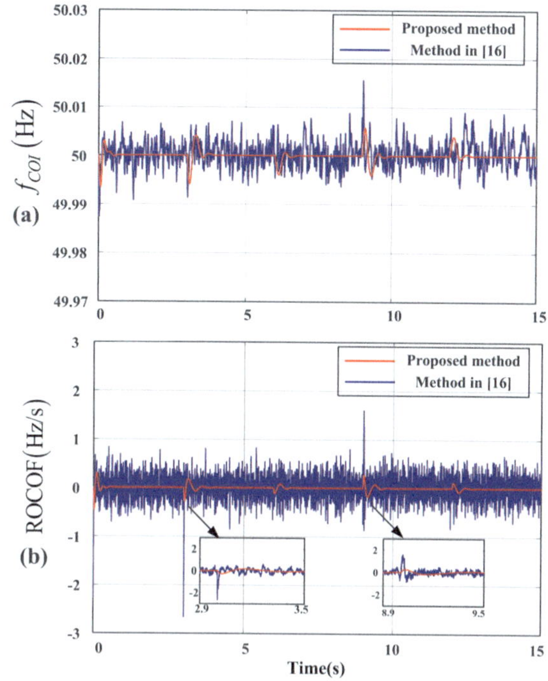

Fig. 15.7 The comparison results of the introduced method and method in [21]. **a** Frequency of COI, **b** RoCoF

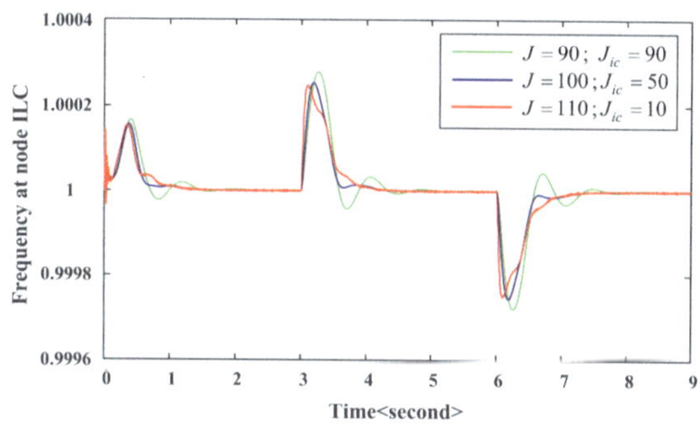

Fig. 15.8 The comparison of frequency response at node ILC with different inertia placement

15.5.3 Method Under Multi-ILC Cases

In this subsection, the introduced NPC control with 3 ILCs is tested. The rated power of 3 ILCs is 8 KW, 4 KW and 4 KW, respectively. The control algorithm is based on (15.29) and the distributed circular communication is used among ILCs. α is equal to 0.8 and β is 0.2.

The results are shown in Fig. 15.9 and GPS of DGs and power sharing among ILCs are both achieved. Comparing multi-ILC in Fig. 15.9 with only one ILC in Fig. 15.7, we can find that the transient performance of GPS is slower under multi-

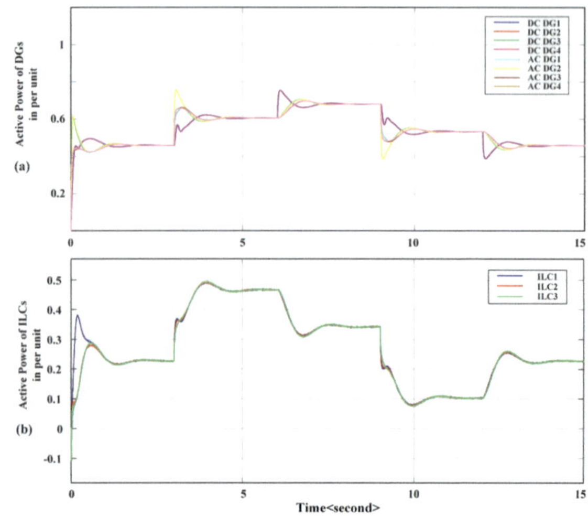

Fig. 15.9 Results of the introduced method under multi-ILC cases **a** global power sharing among DGs, **b** power sharing among ILCs

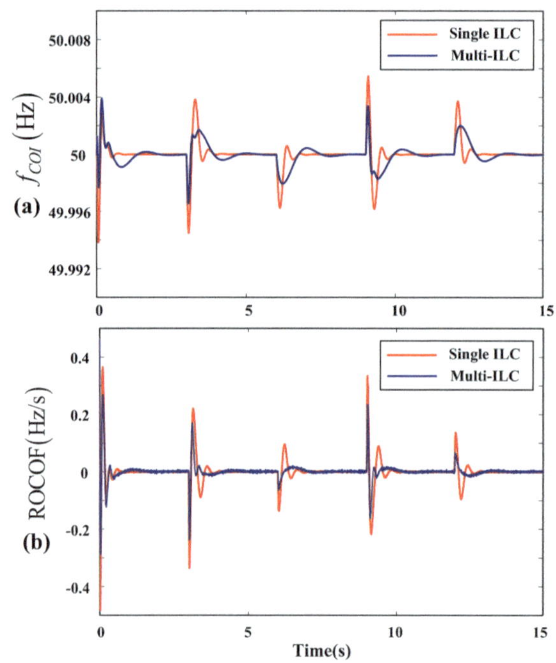

Fig. 15.10 The comparison results of a single ILC and multiple ILCs under introduced method. **a** Frequency of COI, **b** RoCoF

ILC. This is because the synchronization process among multiple ILCs and ac DGs is longer. Furthermore, the frequency performance under these two cases is compared and shown in Fig. 15.10. The RoCoF with multiple ILCs is reduced. Therefore, they can provide more inertia support for hybrid MG and enhance the reliability of the system.

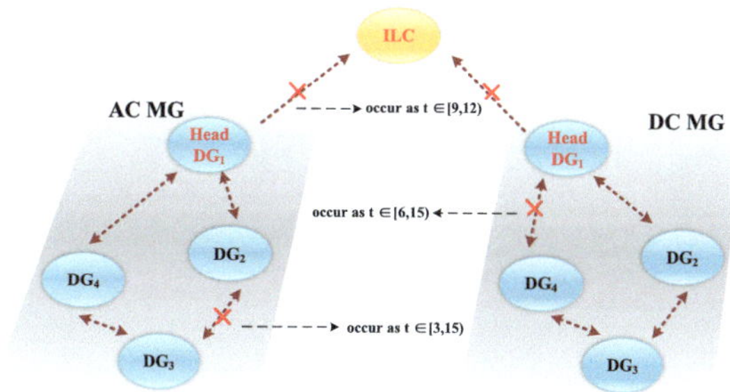

Fig. 15.11 Cyber networks of hybrid microgrid with possible link failure

15.5.4 Resilience on Failure and Time Delay

We proceed the test system in the situations of load change as the aforementioned case. The failure settings of links in hybrid MG are shown in Fig. 15.11 and the time delay is set to 0.1s. The result is shown in Fig. 15.12.

Firstly, the frequency recovery and voltage restoration are achieved. As for the GPS of hybrid MG, the introduced method is almost free from the link failures occurring in AC and DC sub-grids. The GPS of hybrid MG can be maintained by the introduced method. At 9s, the failure of ILC links happens with a DC-load decrease. In this situation, the GPS of MG is failed because the NPC controller stops to update. However, the self-synchronization and the stability of MG are maintained. The links are recovered at 12 s and the GPS is obtained which shows the robustness of the introduced method.

15.5.5 Method Under Grid-On/Off Operation

Generally, MG is required to operate both in islanded mode and grid-connected mode. To test the flexibility of the introduced method, the grid-on/off operation is implemented. In grid-connected mode, ac side sources (ac DGs and ILC) are controlled to output rated power to main grid and dc side sources to maintain the DC-link voltage [28]. The results are shown in Fig. 15.13.

Observing Fig. 15.13, the transient performance of hybrid MG under introduced method is good and fast. The GPS in the steady state is achieved and all DGs output rated power to the grid. It should be noted that there's oscillation during grid-on and off switching due to the conjugate complex poles of VSGs [29].

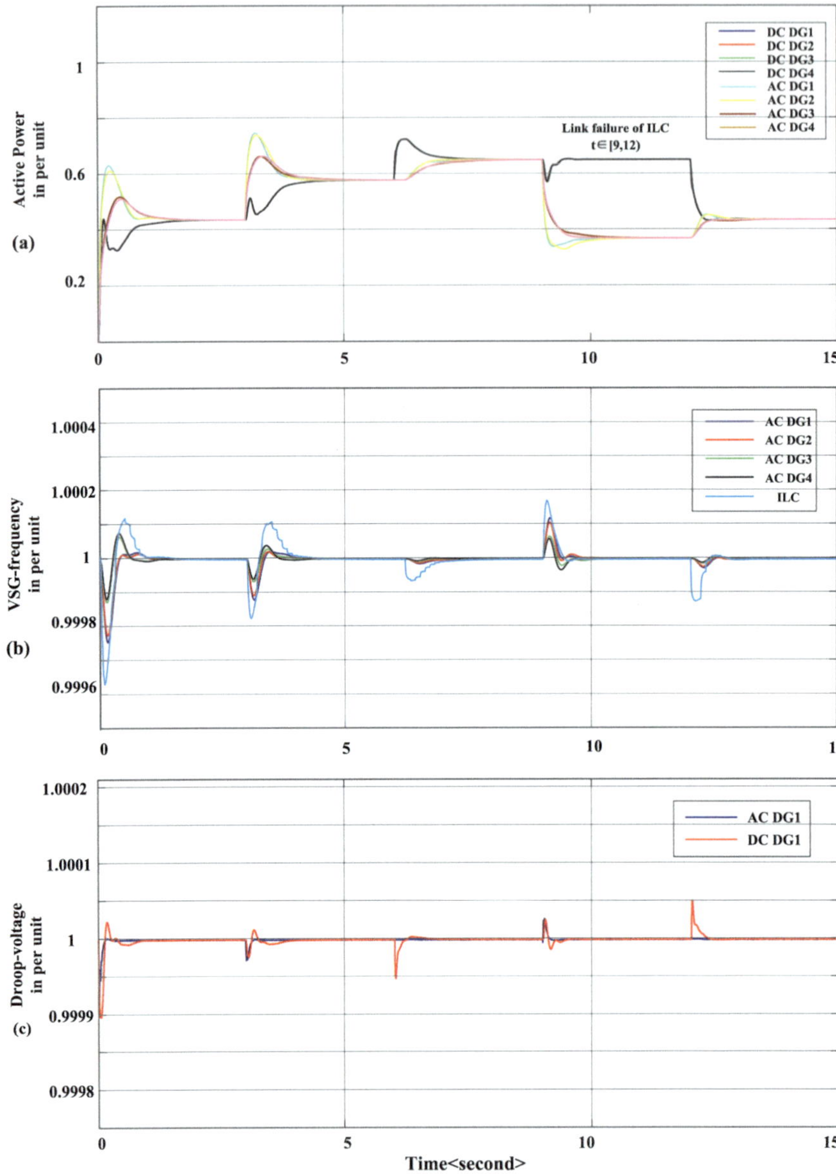

Fig. 15.12 Performance of the introduced control method. **a** Active power of DC and AC DGs, **b** Frequency of DGs, **c** Bus voltage of AC and DC sub-grids

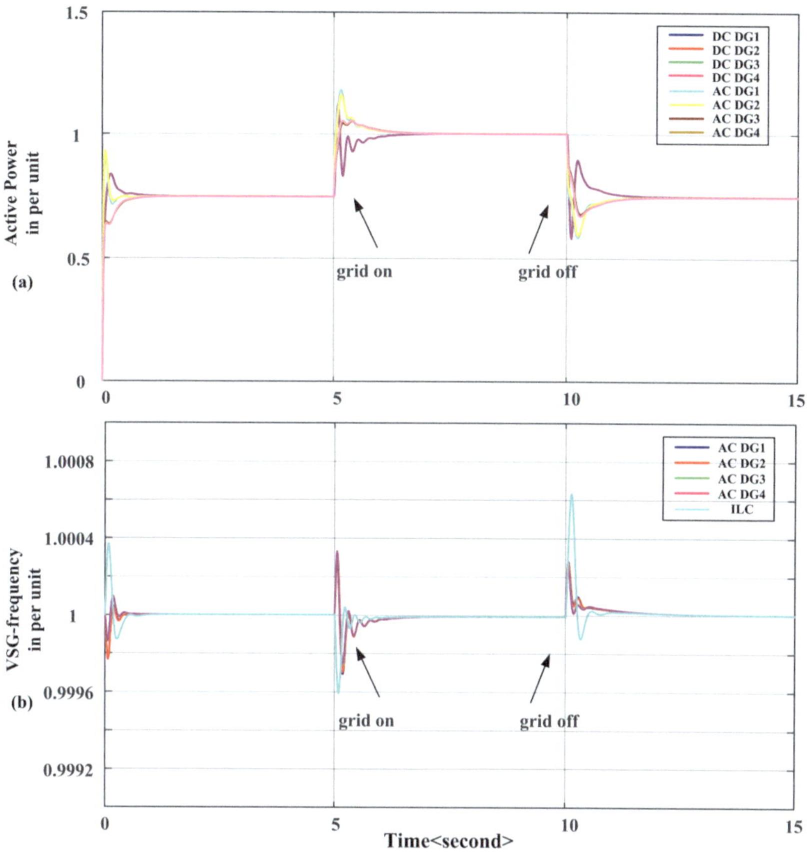

Fig. 15.13 The grid-on/off operation under introduced method

15.6 Hardware-in-Loop Validation of Multi-Subgrid

In this section, the introduced NPC control is verified by real-time HIL test-based OPAL-RT (OP4510). The topology of HIL test is shown in Fig. 15.14 which contains multi-subgrid. The parameters of DC Sub-grid and AC Subgrid-1 are equal to their counterparts in Fig. 15.5. The impedance of AC Subgrid-2 L_5 to L_7 is 2, 2.5 and 3 mH.

Before 5 s, the distributed control in sub-grids is not activated. The power sharing is accurate in two AC sub-grids and is poor in DC sub-grid in Fig. 15.15. In addition, the frequency of DGs and ILCs deviates from the nominal value.

During 5–25 s, the introduced control is activated and the communication delay is set to 0.1 s. The impact load occurs in DC Subgrid-1, AC Subgrid-1 and AC Subgrid-2 at 10, 15 and 20 s respectively. The GPS of hybrid MG is achieved quickly shown

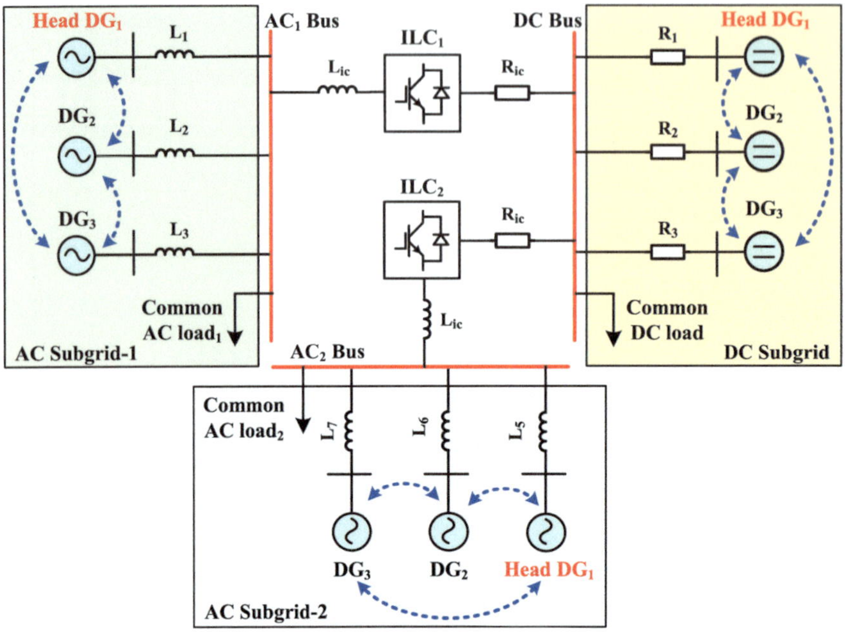

Fig. 15.14 The structure of the hybrid microgrid

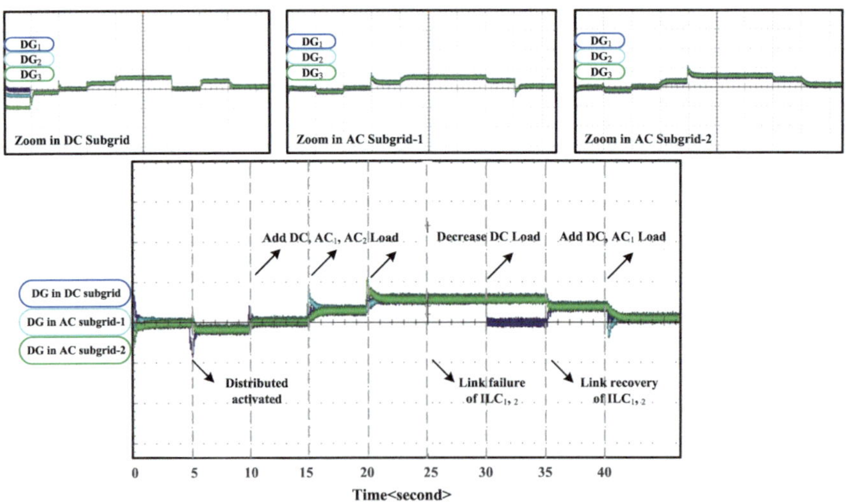

Fig. 15.15 Active power sharing of hybrid MG in per unit

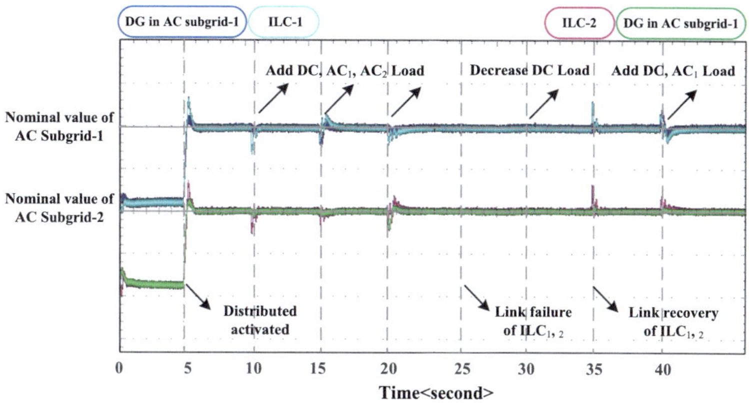

Fig. 15.16 Frequency performance of DGs and ILCs in two AC sub-grids

in Fig. 15.15. The frequency is regulated at the nominal value with little oscillation when load fluctuation happens.

During 25–35 s, the link failure happens at 25 s between ILCs and sub-grids. As shown in Figs. 15.15 and 15.16, the link failure does not affect the hybrid MG which shows the resilience on link failure.

The DC load decrease occurs at 30 s. The output power of DC sub-grid decreases but the output power of two AC sub-grids remains unchanged. After 35 s, the link is recovered and the GPS is obtained again. The AC load shedding occurs at 40 s and the output power of DGs decreases to meet the load demand.

15.7 Conclusion

In this section, VSG-based normalized power coordination (NPC) is introduced for hybrid MG, which achieves accurate GPS among all DGs and frequency/voltage recovery simultaneously. Besides, the cross inertia support is achieved for hybrid MG without frequency derivative calculation. Therefore, the transient performance of frequency and GPS is improved with small RoCoF. To suppress frequency and power oscillation, the index of inertia deviation is introduced to guide the inertia placement of hybrid MG based on PF analysis. Finally, simulation and HIL experiments are implemented to verify the effectiveness and expandability of the introduced method.

Appendix

The system matrix A is given by

$$A = \begin{bmatrix} A_{AC} & A_{AC,IC} & 0 \\ A_{IC,AC} & A_{IC} & A_{IC,DC} \\ 0 & A_{DC,IC} & A_{DC} \end{bmatrix} \tag{15.41}$$

$$A_{AC} = \begin{bmatrix} -\mathbf{JD} & 0 & -\mathbf{J}E & \mathbf{J}E \\ E & 0 & 0 & 0 \\ 0 & \frac{1}{\tau_d}B_{AC} & \frac{-1}{\tau_d}E & 0 \\ \frac{-1}{k_\omega}E & 0 & 0 & -L \end{bmatrix} \tag{15.42}$$

where $\mathbf{J} = \text{diag}\,(1/J_1, \ldots, 1/J_n)$,$\mathbf{D} = \text{diag}\,(m_1 + D_1, \ldots, m_n + D_n)$ .L is Laplace matrix, E is identity matrix and

$$B_{AC} = \begin{bmatrix} \sum_{j=0,j\neq 1}^{n} E_1 E_j B_{1j}\cos\delta_{1j} & -E_1E_2B_{12}\cos\delta_{12} & \cdots & -E_1E_nB_{1n}\cos\delta_{1j} \\ -E_2E_1B_{21}\cos\delta_{21} & \sum_{j=0,j\neq 2}^{n} E_2E_jB_{2j}\cos\delta_{2j} & \cdots & -E_2E_nB_{2n}\cos\delta_{2n} \\ \vdots & \vdots & \ddots & \vdots \\ -E_nE_1B_{n1}\cos\delta_{n1} & -E_nE_2B_{n2}\cos\delta_{n2} & \cdots & \sum_{j=0}^{n-1} E_nE_jB_{nj}\cos\delta_{nj} \end{bmatrix}$$

$$A_{DC} = \begin{bmatrix} \frac{-1}{\tau_d}(E + kY_{DC}) & \frac{1}{\tau_d}Y_{dc} \\ \frac{-1}{k_d}L & 0 \end{bmatrix} \tag{15.43}$$

where the dc conductance matrix is

$$Y_{dc} = \begin{bmatrix} \sum_{j=0,j\neq 1}^{n} E_1Y_{1j} & -E_1Y_{12} & \cdots & -E_1Y_{1n} \\ -E_2Y_{21} & \sum_{j=0,j\neq 2}^{n} E_2Y_{2j} & \cdots & -E_2Y_{2n} \\ \vdots & \vdots & \ddots & \vdots \\ -E_nY_{n1} & -E_nY_{n2} & \cdots & \sum_{j=0}^{n-1} E_nY_{nj} \end{bmatrix} \tag{15.44}$$

$$A_{IC} = \begin{bmatrix} \frac{(m+D)}{J_{ic}} & \frac{-E_1E_{ic}B_{1,ic}}{J_{ic}} & k_{pi} \\ 1 & 0 & 0 \\ 0 & 0 & 0 \end{bmatrix}$$

References

1. Y. Xu, H. Sun, W. Gu, Y. Xu, Z. Li, Optimal distributed control for secondary frequency and voltage regulation in an islanded microgrid. IEEE Trans. Ind. Informat. **15**(1), 225–235 (2019). (Jan.)
2. P. Lin, C. Jin, J. Xiao, X. Li, D. Shi, Y. Tang, P. Wang, A distributed control architecture for global system economic operation in autonomous hybrid AC/DC microgrids. IEEE Trans. Smart Grid **10**(3), 2603–2617 (2019)

3. X. Hou, K. Sun, N. Zhang, F. Teng, X. Zhang, T.C. Green, Priority-driven self-optimizing power control scheme for interlinking converters of hybrid AC/DC microgrid clusters in decentralized manner. IEEE Trans. Power Electron. **37**(5), 5970–5983 (2022). (May)
4. M. Su, K. Zhang, Y. Sun, H. Han, G. Shi, S. Fu, Coordinated control for PV-ESS islanded microgrid without communication. Int. J. Electrical Power Energy Syst. **136** (2021)
5. S. Fu, Y. Sun, L. Li, Z. Liu, H. Han, M. Su, Power oscillation suppression of multi-VSG grid via decentralized mutual damping control. IEEE Trans. Ind. Electron. **69**(10), 10202–10214 (2022). (Oct.)
6. Y. Sun, X. Hou, J. Yang, H. Han, M. Su, J.M. Guerrero, New perspectives on droop control in AC microgrid. IEEE Trans. Ind. Electron. **64**, 5741–5745 (2017)
7. M. Su, Z. Liu, Y. Sun, H. Han, X. Hou, Stability analysis and stabilization methods of DC microgrid with multiple parallel-connected DC-DC converters loaded by CPLs. IEEE Trans. Smart Grid **9**, 132–142 (2018)
8. K. Zhang et al., *New Perspectives on Transient Stability Between Grid-Following and Grid-Forming VSCs* arXiv preprint arXiv:2207.03273 (2022)
9. P. Lin, P. Wang, C. Jin, J. Xiao, X. Li, F. Guo, C. Zhang, A distributed power management strategy for multi-paralleled bidirectional interlinking converters in hybrid AC/DC microgrids. IEEE Trans. Smart Grid **10**(5), 5696–5711 (2019)
10. J.W. Simpson-Porco, Q. Shafiee, F. Dörfler, J.M. Guerrero, F. Bullo, Secondary frequency and voltage control of islanded microgrids via distributed averaging. IEEE Trans. Ind. Electron. **62**(11), 7025–7038 (2015). (Nov.)
11. J.S. Gomez, D. Saez, J.W. Simpson-Porco, R. Cardenas, Distributed predictive control for frequency and voltage regulation in microgrids. IEEE Trans. Smart Grid **11**(2), 1319–1329 (2020). (Mar.)
12. L. He, Yong Li, J. M. Guerrero,, Y. Cao, A comprehensive inertial control strategy for hybrid AC/DC microgrid with distributed generations. IEEE Trans. Smart Grid **11**(2), 1737–1747 (2020)
13. E. Espina, R. Cárdenas-Dobson, J.W. Simpson-Porco, D. Sáez, M. Kazerani, A consensus-based secondary control strategy for hybrid AC/DC microgrids with experimental validation. IEEE Trans. Power Electron. **36**(5), 5971–5984 (2021). (May)
14. M. Schweizer, S. Almér, S. Pettersson, A. Merkert, V. Bergemann, L. Harnefors, Grid-forming vector current control. IEEE Trans. Power Electron. **37**(11), 13091–13106 (2022). (Nov.)
15. S. Fu, Y. Sun, L. Li, Z. Liu, H. Han, M. Su, Power oscillation suppression in multi-VSG grid by adaptive virtual impedance control. IEEE Syst. J. https://doi.org/10.1109/JSYST.2022.3173666
16. J. Liu, Y. Miura, T. Ise, Comparison of dynamic characteristics between virtual synchronous generator and droop control in inverter-based distributed generators. IEEE Trans. Power Electron. **31**(5), 3600–3611 (2016). (May)
17. J. Fang, R. Zhang, H. Li, Y. Tang, Frequency derivative-based inertia enhancement by grid-connected power converters with a frequency-locked-loop. IEEE Trans. Smart Grid **10**(5), 4918–4927 (2019). (Sept.)
18. K. Zhang, M. Su, Y. Sun, P. Wu, Z. Luo, H. Han, A novel distributed control for hybrid AC/DC microgrid with consideration of power limit, in *2021 IEEE 12th Energy Conversion Congress & Exposition-Asia (ECCE-Asia)* (2021), pp. 1856–1859
19. J.-W. Chang, G.-S. Lee, S.-I. Moon, P.-I. Hwang, A novel distributed control method for interlinking converters in an islanded hybrid AC/DC microgrid. IEEE Trans. Smart Grid **12**(5), 3765–3779 (2021). (Sept.)
20. H. Yoo, T. Nguyen, H. Kim, Consensus-based distributed coordination control of hybrid AC/DC microgrids. IEEE Trans. Sustain. Energy **11**(2), 629–939 (2020). (April)
21. E. Rute-Luengo et al., Distributed model-based predictive secondary control for hybrid AC/DC microgrids. IEEE Trans. Emerg. Sel. Top. Power Electron. https://doi.org/10.1109/JESTPE.2022.3157979
22. A. A. Eajal, A. H. Yazdavar, E. F. El-Saadany, M. M. A. Salama, Optimizing the droop characteristics of AC/DC hybrid microgrids for precise power sharing. IEEE Syst. J. **15**(1), 560–569 (2021)

23. Y. Xia, Y. Peng, P. Yang, M. Yu, W. Wei, Distributed coordination control for multiple bidirectional power converters in a hybrid AC/DC microgrid. IEEE Trans. Smart Grid **32**(6), 4949–4959 (2017). (June)
24. J.-W. Chang, S. Chae, G.-S. Lee, Distributed optimal power sharing strategy in an islanded hybrid AC/DC microgrid to improve efficiency. IEEE Trans. Power Deliv. (2022). https://doi.org/10.1109/TPWRD.2022.3197434
25. Y. Li, Y. Gu, T.C. Green, Revisiting grid-forming and grid-following inverters: a duality theory. IEEE Trans. Power Syst. **37**(6), 4541–4554 (2022). (Nov.)
26. U. Markovic, Z. Chu, P. Aristidou, LQR-based adaptive virtual synchronous machine for power systems with high inverter penetration. IEEE Trans. Sustain. Energy **10**(3), 1501–1512 (2019). (Jul.)
27. F. Garofalo, L. Iannelli, F. Vasca, Participation factors and their connections to residues and relative gain array, in *Proceedings of 15th Triennal World Congr. Int. Fed. Automat. Control, Barcelona, Spain*, 21–26 Jul 2002, p. 180
28. D. Ke, C.Y. Chung, J. Xu, Y. Shen, Y. Sun, Inertia emulation uncorrelated with electromechanical dynamics to improve frequency transients using center of inertia (COI) frequency signal. IEEE Trans. Power Syst. **36**(3), 1736–1749 (2021). (May)
29. F. Nejabatkhah, Y.W. Li, Overview of power management strategies of hybrid AC/DC microgrid. IEEE Trans. Power Electron. **30**(12), 7072–7089 (2015). (Dec.)

Chapter 16
Self-optimizing Power Control for Microgrid Clusters

16.1 Microgrid Clusters

Microgrids have become a cost-efficient solution to locally integrate the high penetration of distributed generations (DGs) and provide reliable power to customers in future urban power systems [1]. It can maintain the emergent operations and protect critical infrastructure from power outages in the event of physical/cyber disruptions in the utility distribution grid [2].

According to the voltage type of bus supply, microgrids can be classified into three categories: AC microgrid, DC microgrid, and hybrid AC/DC microgrid [3, 4].In addition, interlinking converters (ILCs) are implemented to interlink the DC microgrid and AC microgrid together to form a hybrid microgrid [5], which has been regarded as the most likely future microgrid structures by integrating the advantages of both AC and DC microgrid [6]. Networked hybrid microgrid clusters are formed by interlinking the microgrids with their individual neighbors through ILCs, as shown in Fig. 16.1. Rather than focusing on a single microgrid, multiple microgrid clusters can further enhance the systematic and coordinated operation [7].

In hybrid microgrid clusters, the role and control methods of ILCs significantly affect the optimal power exchange [8], power quality [9–11], system stability [12], resilience capability [13], and ancillary services [14–16]. With regard to the control framework, the power control strategies of ILCs are classified into three categories: centralized [17–19], distributed [20–22] and decentralized [23–29], according to the communication dependency.

X. Hou et al., *Self-organized Operation and Stability Control of Electronized Power Networks*, https://doi.org/10.1007/978-981-96-9067-1_16

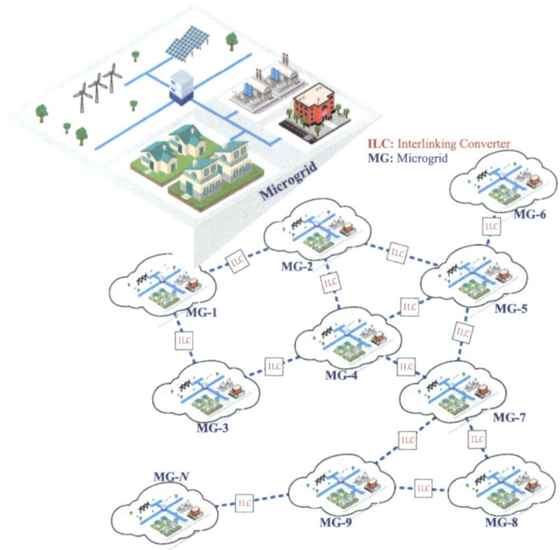

Fig. 16.1 Schematic representation of networked microgrid clusters

16.2 Hybrid Microgrid Clusters and Basic Control

16.2.1 Configuration of Hybrid Microgrid Clusters

Figure 16.2 shows the typical configuration of hybrid microgrid clusters with four AC and DC microgrids. These subgrids are composed of various loads, dispatchable DGs, including storages, fuel cell, micro-turbine, etc., and non-dispatchable renewable generations. Four subgrids are interlinked by bidirectional ILCs, which includes DC/DC, DC/AC, and AC/AC ILCs for linking different microgrid types. Without loss of generality, this study focuses on the control technologies of **DC/AC ILCs** for hybrid AC/DC microgrid clusters.

Practically, in networked microgrid clusters, the subgrid load-priorities and power quality requirements for different areas can be different as shown in Table 16.1. In our work, the microgrids are divided into three categories according to their individual fed-load features and priorities, namely **Class A**-microgrid with critical load, **Class B**-microgrid with fixed non-critical load, and **Class C**-microgrid with flexible non-critical load [11, 30, 31]. On this basis, the concept of subgrid priority index (important weight) is introduced to evaluate the importance of each subgrid, which lays a solid foundation of the optimal power exchange model in hybrid microgrid clusters. It is noted that the flexible load part of the low-priority subgrid could be curtailed or shifted according to the hierarchical segment of AC-frequency and DC-voltage deviations under the power shortage.

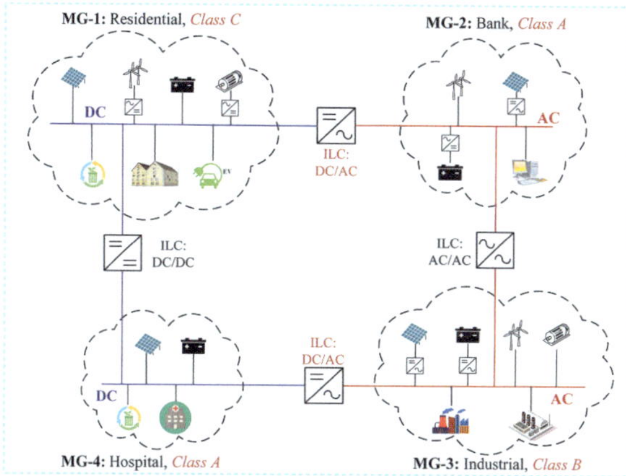

Fig. 16.2 A typical configuration of hybrid AC/DC microgrid clusters with different capacities and priorities

Table 16.1 Operation practice and requisite of different real-world microgrids [11, 30, 31]

Load feature within each microgrid	Microgrid scenario examples	Voltage deviation of Power quality requirement	Sub-grid priority index
Class A: Critical load	Hospital/bank/police/ data center microgrid	$\|V/f_{error}\| < 0.5\%$	High ★★★
Class B: Fixed non-critical load	Industrial/commercial microgrid	$\|V/f_{error}\| < 3\%$	Medium ★★
Class C: Flexible non-critical load	Residential microgrid	$\|V/f_{error}\| < 5\%$	Low ★

16.2.2 *Self-Organizing Operation Mechanism*

To control and manage the complex hybrid microgrid clusters, a self-organizing operation mechanism is necessary for a transition from a centralized system to a decentralized system [32]. In this study, a simple self-organizing operation rule is formed based on the introduced self-optimizing power flow principle. Some main features of self-organizing resilience operation are as follows:

(1) ***Decentralized self-optimizing manner.*** The decentralized self-optimizing power exchange control scheme of ILCs should be achieved by only local information, which is off-line designed without requiring a central controller and has high system resilience to communication failure.

(2) ***Reserved dispatchable source sharing***. All dispatchable sources of microgrid clusters should be shared with each other according to the predesigned subgrid priorities.
(3) ***Mutual uncertainty complementarity***. Due to the intermittency of wind/solar-based renewable generations and the load uncertainty in each subgrid, it is likely that the microgrid experiences power sufficiency and deficiency. To achieve mutual uncertainty complementarity, we should control the power flows of ILCs from the power-sufficiency subgrid to the power-deficiency subgrid.
(4) ***Autonomous load shedding***. The less-priority microgrid should shed load automatically by judging real-time frequency/voltage deviation when total power is in deficit.
(5) ***One-ILC-fault redundancy***. The introduced control principle can accommodate one-ILC-failure.

16.2.3 Basic Control of AC and DC Subgrids

Many advanced control strategies for individual AC or DC microgrid have been maturely introduced in [33–35]. The normal operations of AC and DC subgrids mainly rely on parallel converters as well as the corresponding controller designs, which ensure the sufficient stability margin and dynamic responses. However, these damping controller designs just affect the dynamic operation without interfacing with the steady-state performance. For this chapte, the focus is on the introduced priority-driven self-optimizing power control for ILCs in microgrid clusters, thus the compensation controller designs for AC and DC subgrids would not be discussed here in detail.

(1) *Control of non-dispatchable sources*: To maximize the renewable energy harvest, power conversion control mode of renewable generations always works in maximum-power-point-tracking (MPPT), except for the extreme light-load conditions. Various MPPT control methods for different renewable power generations have been maturely developed. For the simplicity of demonstration, MPPT-controlled renewable power generations are considered as negative loads in this study.

(2) *Control for dispatchable sources*: To establish AC and DC subgrid voltage and support MPPT-controlled renewable power generations, some dispatchable generators and energy storages are mostly needed to work in voltage control mode. Especially in islanded microgrid, one key challenge is the demand power sharing among dispatchable sources. To this issue, a classic droop control method has been introduced to realize decentralized power sharing objective based on the droop gains design [3, 4]. Similarly, the decentralized droop control structure for proportional power sharing among all dispatchable resources is also applied in this study. To focus on the power exchange control scheme of ILCs for hybrid microgrid clusters, the droop-controlled dispatchable power sources in AC and DC subgrids are combined as one AC and one DC main voltage source for each subgrid.

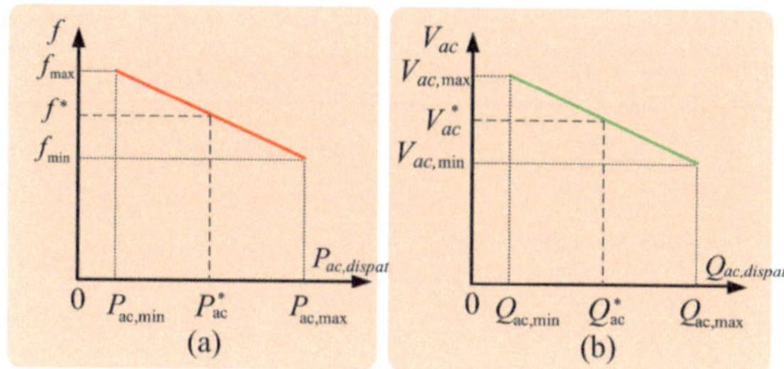

Fig. 16.3 Classic droop control for AC subgrid. **a** *P-f* droop control. **b** *Q-V* droop control

In this section, classic droop control strategies for AC and DC subgrids are firstly introduced to regulate the power flow within each-subgrid separately. For AC subgrid, the droop control characteristic is shown in Fig. 16.3. The frequency reference f_{ac} and the voltage amplitude reference V_{ac} are presented as follows

$$f_{ac} = f_{\max} - mP_{ac,dispat}, m = \frac{f_{\max} - f_{\min}}{P_{ac,\max} - P_{ac,\min}} \tag{16.1}$$

$$V_{ac} = V_{ac,\max} - nQ_{ac,dispat}, n = \frac{V_{ac,\max} - V_{ac,\min}}{Q_{ac,\max} - Q_{ac,\min}} \tag{16.2}$$

where m and n are the $P - f$ and $Q - V$ droop coefficients, respectively. $P_{ac,dispat}$ and $Q_{ac,dispat}$ are the output active and reactive power of dispatchable source in AC subgrid. From Fig. 16.3, f_{max} and f_{min} are the maximum and minimum values of permissible frequency. $V_{ac,max}$ and $V_{ac,min}$ are the maximum and minimum values of permissible voltage amplitude.

As shown in Fig. 16.4, the P-V droop control characteristic for DC subgrid is given by

$$V_{dc} = V_{dc,\max} - kP_{dc,dispat}, k = \frac{V_{dc,\max} - V_{dc,\min}}{P_{dc,\max} - P_{dc,\min}} \tag{16.3}$$

Fig. 16.4 Classic $P - V$ droop control for DC subgrid

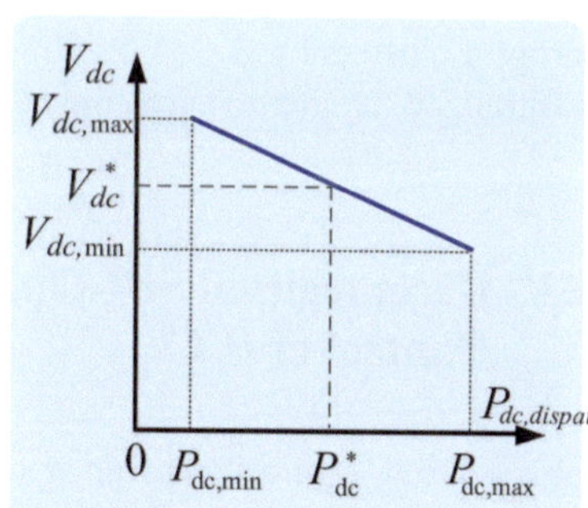

where $P_{dc,dispat}$ is the output active power of dispatchable source in DC subgrid. k is the $P - V$ droop coefficient. From Fig. 16.4, $V_{dc,max}$ and $V_{dc,min}$ are the maximum and minimum values of permissible bus voltage amplitude.

16.2.4 *Control Problem Description of ILCs*

According to the droop equations (16.1–16.3), the increase of load-demand within each-subgrid will lead to a decrease of the AC-subgrid frequency or DC-subgrid voltage. So, the relative deviations of AC-frequency and DC-voltage with respect to the rated-values are favorable indications of the real-time operation state-deviation in each-subgrid [23–25]. In this way, the normalized AC-frequency and DC-voltage are given by

$$\begin{cases} f_{pu} = \frac{f_{ac}-0.5(f_{\max}+f_{\min})}{0.5(f_{\max}-f_{\min})} \in [-1, 1] \\ V_{dc,pu} = \frac{V_{dc}-0.5(V_{dc,\max}+V_{dc,\min})}{0.5(V_{dc,\max}-V_{dc,\min})} \in [-1, 1] \end{cases} \tag{16.4}$$

Then, a uniform $\eta_{pu,i}$ of operation state deviation index for i-th AC or DC subgrid is given by

$$\eta_{pu,i} = \begin{cases} f_{pu,i}; \text{ForACsubgrid}\#i \\ V_{dc,pu,i}; \text{ForDCsubgrid}\#i \end{cases} \tag{16.5}$$

Equation (16.5) is intended to unify expression forms of power sufficiency-deficiency states of AC-type and DC-type subgrid. For AC-type subgrid#i, $\eta_{pu,i}$ $=f_{pu,i}$. For DC-type subgrid#i, $\eta_{pu,i}$ $=V_{pu,i}$. This unified form can be used for a more convenient expression of the following optimal problem regardless of whether it's AC or DC subgrid. From (16.4)–(16.5), $\eta_{pu,i}$=1 represents power-sufficiency operation state; $\eta_{pu,i} = -1$ represents fully power-deficiency operation state; $\eta_{pu,i}$ =0 represents nominal-load operation state of i-th subgrid.

The power exchange between microgrid clusters is coordinated by the ILC, as shown in Fig. 16.5. Different from the "global-equal sharing" concept in [23–25], this study focuses on two issues from the "priority-driven optimizing" perspective: (1) how to build the optimal power exchange model and manage the power flow among microgrid clusters by ILCs? (2) How to realize the control algorithm for ILCs in a decentralized manner to avoid the need of central controller and communication?

16.3 Decentralized Self-Optimizing Power Exchange Control of ILCs

In this section, a priority-driven optimal power exchange model of ILCs is firstly introduced, and then optimal-oriented primary quasi-droop control of ILCs is implemented. The self-optimizing power control scheme indicates the power exchange

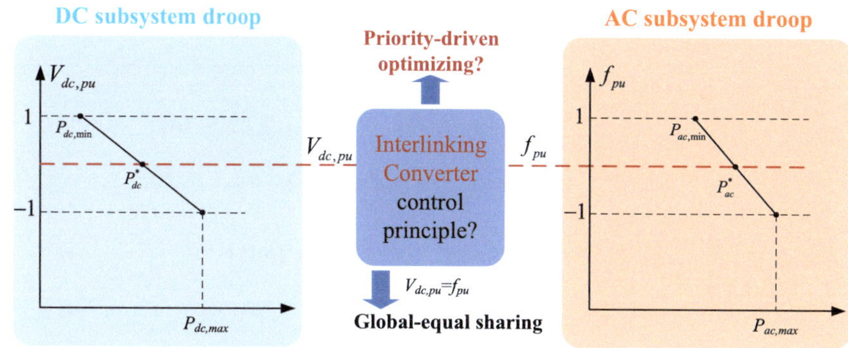

Fig. 16.5 Key control problem description of ILCs

reference value for the primary control of the ILC. And the primary control is utilized to achieve the accurate tracking performances by dual-voltage-current closed loops.

16.3.1 Flexible Priority-Driven Optimal Power Exchange Model

For this first issue, the question is how much power should be exchanged through ILC from the one subgrid to the other subgrid. Herein, a flexible priority-driven principle is introduced to address the optimal power exchange problem through ILCs. This optimal problem is summarized in (16.6).

In (16.6), $P_{ILC,ij}$ is the real power exchange of the ILC from the i-th subgrid to j-th subgrid. w_i is the important weight coefficient of i-th subgrid priority. $\eta_{pu,i}$ is the deviation index of operation state for i-th subgrid, which implies the load-levels of light-load or over-load. N is the total subgrid number in hybrid microgrid clusters. Ω_i represents the physically interlinked subgrids with the i-th subgrid. $P_{load,i}$ is the actual load demand of the i-th subgrid. $P_{dispat,i}$ is output power of droop-controlled dispatchable sources in i-th subgrid. $P^{\max}_{dispat,i}$ is maximum power capacity of dispatchable sources in i-th subgrid. $P_{non-dispat,i}$ is output power of MPPT-controlled non-dispatchable sources in i-th subgrid, which can be treated as negative loads in the optimal model.

$$
\begin{aligned}
&\underset{P_{ILC,ij}}{\text{minimize}}\ J = \sum_{i=1}^{N} w_i \eta_{pu,i}^2 \\
&subjectto \\
&\begin{cases}
P_{dispat,i} - \sum\limits_{j\in\Omega_i} P_{ILC,ij} = P_{load,i} - P_{non-dispat,i};\ (a) \\
P_{dispat,j} + \sum\limits_{i\in\Omega_j} P_{ILC,ij} = P_{load,j} - P_{non-dispat,j};\ (b) \\
\sum\limits_{i=1}^{N} P_{dispat,i} = \sum\limits_{i=1}^{N} (P_{load,i} - P_{non-dispat,i});(c) \\
\eta_{pu,i} = 1 - \frac{2P_{dispat,i}}{P_{dispat,i}^{\max}};(d)
\end{cases}
\end{aligned}
\tag{16.6}
$$

Equation(16.6) means the optimization objective by minimizing the total normalized DC-voltage/AC-frequency state deviations of microgrid clusters with different priorities. (16.6a) and (16.6b) represent the supply-demand power balance of the i-th subgrid and j-th subgrid. (16.6c) represent the supply-demand power balance of the whole N microgrid clusters. (16.6) are deduced according to the normalization feature of droop-controlled subgrid from (16.1)–(16.5).

Then, a simplified mathematical description of optimal problem is obtained from (16.6)–(16.6d)

$$
\begin{aligned}
&\min(J = \sum_{i=1}^{N} w_i \eta_{pu,i}^2) \\
&\text{s.t.} \sum_{i=1}^{N} \frac{P_{dispat,i}^{\max}(1-\eta_{pu,i})}{2} = P_{\text{uncertainty}}
\end{aligned}
\tag{16.7}
$$

where $f_{uncertainty}$ implies the equivalent load demands due to the uncertainty of renewable sources and load-feeding.

From (16.6)–(16.7), the optimization objective J is the quadratic function of variable $\eta_{pu,i}$, and the second derivative Hessian Matrix $\mathbf{H}(J)=diag[2w_1\ 2w_2\ \cdots\ 2w_N]$ is a positive definite matrix. So, the objective function J is convex. Besides, the optimization constraint of (16.7) is a first-order function of variable $\eta_{pu,i}$, and the feasible region is a plane, which is convex. In brief, the optimization problem is convex, and a single minimum exists. Meanwhile, it is noted that this optimization problem focuses on the optimized power exchanges of ILCs from one subgrid to the other subgrid. The power loss of the line impedance and the node power equation within each subgrid are ignored.

Accordingly, the Lagrange function [36] is adopted to find the optimal solution in (16.7) as follow:

$$
L = \sum_{i=1}^{N} w_i \eta_{pu,i}^2 + \lambda \left(P_{\text{uncertainty}} - \sum_{i=1}^{N} \frac{P_{dispat,i}^{\max}(1-\eta_{pu,i})}{2} \right)
\tag{16.8}
$$

where λ is Lagrange multiplier. Therefore, the optimum solution is obtained by solving the following equation:

$$\begin{cases} \frac{\partial L}{\partial \eta_{pu,1}} = 0, \frac{\partial L}{\partial \eta_{pu,2}} = 0, \ldots, \frac{\partial L}{\partial \eta_{pu,N}} = 0 \\ \frac{\partial L}{\partial \lambda} = P_{\text{uncertainty}} - \sum_{i=1}^{N} \frac{P_{dispat,i}^{\max}(1-\eta_{pu,i})}{2} = 0 \end{cases} \tag{16.9}$$

Simplifying (16.9) yields the necessary condition (16.10) for the optimality.

$$\frac{w_1}{P_{dispat,1}^{\max}}\eta_{pu,1} = \frac{w_2}{P_{dispat,2}^{\max}}\eta_{pu,2} = \cdots = \frac{w_N}{P_{dispat,N}^{\max}}\eta_{pu,N} \tag{16.10}$$

From (16.10), although the value of $P_{ILC,ij}$ is difficult to be directly solved due to the load uncertainty, it can be easily obtained that the minimum of objective function (16.6) can be achieved autonomously when the condition (16.10) is met. As the microgrid clusters are physically connected by all ILCs, (16.10) can be held by controlling each ILC in a decentralized manner. Moreover, the steady-state operation deviation index $\eta_{pu,i}$ is a constant value. Thus, the desired power reference for each ILC can be designed by a proportional-integral (PI) controller with steady-state zero-error to guarantee (16.10) hold.

$$P_{ILC,ij}^{*} = (k_p + \frac{k_i}{s})(\frac{w_i}{P_{dispat,i}^{\max}}\eta_{pu,i} - \frac{w_j}{P_{dispat,j}^{\max}}\eta_{pu,j}) \tag{16.11}$$

where $P_{ILC,ij}^{*}$ implies the optimized power exchange value of the ILC_{ij} from the i-th subgrid to j-th subgrid.

From (16.10)–(16.11), it is noted that the priority weight w_i and the maximum dispatchable power capacity $P_{dispat,i}^{\max}$ of the i-th subgrid can be calculated off-line according to their individual subgrid priority and capacity. Hereby, the introduced power flow principle is flexible and can be designed according to the application scenarios of hybrid microgrid clusters.

Practically, the concept of subgrids' priority weights are introduced to evaluate the load importance classes of each subgrid, which can be defined as the proportion of important customers to total customers in a certain subgrid [37, 38]. And these weight values are normally designed according to the voltage deviation criteria of power quality requirement for different subgrids, as shown in Table 16.1. For a better understanding, a simple design procedure is illustrated in Fig. 16.6. For simplicity, we assume three subgrids with same power capacities. From the optimum solution of (16.10), the steady-state equation $w_1\eta_{pu,1} = w_2\eta_{pu,2} = w_3\eta_{pu,3}$ must be firstly guaranteed. Then, according to the voltage deviation requirement of power quality in Table 16.1, we set the high-priority subgrid-1 with $|V/f_{error,1}| < 0.5\%$ the medium-priority subgrid-2 with $|V/f_{error,2}| < 3\%$, and the low-priority subgrid-3 with $|V/f_{error,3}| < 5\%$. As a result, $\eta_{pu,1} : \eta_{pu,2} : \eta_{pu,3} = \left|V/f_{error,1}\right| : \left|V/f_{error,2}\right| : \left|V/f_{error,3}\right| = 0.5 : 3 : 5$. And $w_1 : w_2 : w_3 = 10 : 1.66 : 1$ is derived.

From Fig. 16.6, a higher weight w_i implies a higher priority of power demand and higher voltage quality. Specially, the power supply and voltage quality of the high-priority subgrid-1 would be guaranteed while the low-priority subgrid-3 can

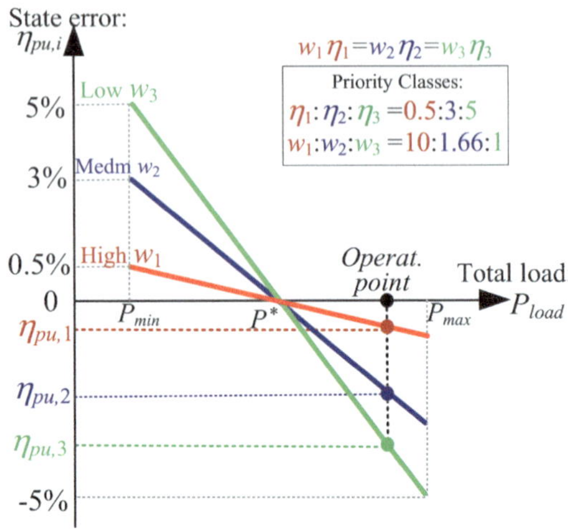

Fig. 16.6 Simplified design diagram of subgrids with different priority weights

be compromised. Under the power over-shortage, the voltage/frequency state deviation of the low-priority subgrid-3 is relatively large, and the flexible load part of the subgrid-3 could be curtailed or shifted by under-frequency/voltage load-shedding. It is noted that the autonomous load shedding is carried out by judging local frequency/voltage deviation within each subgrid, and the deviations are not shared among subgrids. For overall microgrid clusters, the autonomous load shedding can be equivalent to cutting off part of the load to guarantee the priority-driven power supply, which does not affect the effectiveness of our introduced method.

16.3.2 Optimal-Oriented Decentralized Quasi-Droop Control

To achieve the control objective in (16.10)–(16.11), a quasi-droop control is introduced for the ILC_{ij}.

$$f_{ILC,ij} = f_{ILC}^{*} + (k_{ILCp,ij} + \frac{k_{ILCi,ij}}{s})(P_{ILC,ij}^{*} - P_{ILC,ij}) \tag{16.12}$$

$$V_{ILC,ij} = V_{ILC,\max} - n_{ILC,ij} Q_{ILC,ij} \tag{16.13}$$

where $k_{ILCp,ij}$ and $k_{ILCi,ij}$ are the PI coefficients of $P - f$ quasi-droop, respectively. $n_{ILC,ij}$ is the $Q - V$ quasi-droop coefficient. $P_{ILC,ij}$ is the active power exchange of the ILC from the i-th subgrid to j-th subgrid, and $P_{ILC,ij}= -P_{ILC,ji}$. The power flow direction of a positive active power is assumed from i-th DC side to j-th AC side. When the power value is more than zero, the ILC works in inverter mode, and when less than zero, the ILC works in rectifier mode. $Q_{ILC,ij}$ is the output reactive power of the ILC

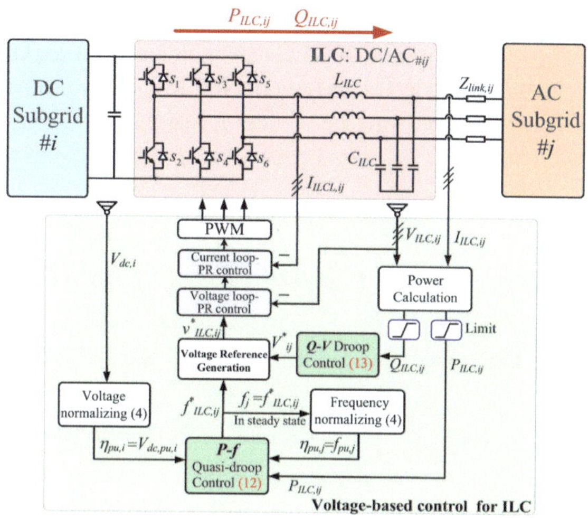

Fig. 16.7 Introduced voltage-based quasi-droop control for ILC_{ij}

from the i-th DC subgrid to j-th AC subgrid. $V_{ILC,ij}$ is the voltage amplitude reference of the ILC_{ij}. In (16.12), the frequency reference is generated according to the real-time operation states of interlinked AC and DC subgrids. Due to the steady state zero-error of PI controller, $P_{ILC,ij}^* = P_{ILC,ij}$ and $\frac{w_i}{P_{dispat,i}^{max}}\eta_{pu,i} = \frac{w_j}{P_{dispat,j}^{max}}\eta_{pu,j}$ in steady-state, and the above optimal principle (16.6)–(16.10) is achieved autonomously. In (16.13), the $Q - V$ droop control is adopted for the voltage control and reactive power compensation to AC subgrid. The detailed voltage-based control scheme for the ILC_{ij} is shown in Fig. 16.7. The introduced control (16.12)–(16.13) is used to indicate the inner voltage reference for dual-voltage-current closed loops. The voltage and current references are tracked by the proportional-resonant (PR) control.

Moreover, it is noted that the interlinking converters can operate as a voltage-controlled-source or current-controlled source. The control mode is always chosen according to the practical requirements of current dynamic response, power quality and the stability of grid strengths [39]. Comparatively, the voltage-controlled mode for ILCs can work in the weak grid condition, and provide an ancillary function of voltage regulation.

16.4 Stability Analysis

Stability analysis and controller designs are essential for the normal operation of the interlinking converters. Thus the small signal stability analysis for the ILCs is carried out in this section. Several basic assumptions are highlighted before modeling the ILCs. (i) The dynamics of dual-voltage-current closed control loops (hundreds of hertz) are overlooked, considering the timescale of interest in this study is focused on the power exchange control dynamics (typically $2 \sim 15$ hertz). (ii) The resistive component and electromagnetic transients in the line circuit are neglected. (iii) The voltage amplitude of the ILC can be treated as a constant due to the salient time-scale separation between the dynamics of frequency control and voltage control. (iv) The DC-link and AC-link bus dynamics of the ILC are assumed to the droop controlled features of (16.1), (16.3) with feeding constant power loads.

16.4.1 Overall Control Model of the ILC

First, the power transmission model of the ILC is built.

$$P_{ILC} = \frac{V_{ILC} V_{ac}}{Z_{link}} \sin\left(\delta_{ILC} - \delta_{ac}\right) \tag{16.14}$$

where V_{ILC} and δ_{ILC} are the voltage magnitude and angle of the ILC, respectively. V_{ac} and δ_{ac} are the voltage magnitude and angle of the AC-link connection bus, respectively. δ_{ac} is the line reactance between the ILC and AC-link bus.

Then, linearizing (16.1), (16.3)–(16.6), (16.11)–(16.12) and (16.14) around the steady state points yields

$$\begin{cases} \dot{\tilde{\delta}}_{ac} = \tilde{f}_{ac} = -m\tilde{P}_{ac,dispat};\ \tilde{V}_{dc} = -k\tilde{P}_{dc,dispat} \\ \tilde{\eta}_{pu,ac} = \tilde{f}_{pu} = a_f \tilde{f}_{ac};\ \tilde{\eta}_{pu,dc} = \tilde{V}_{dc,pu} = a_V \tilde{V}_{dc} \\ \tilde{P}^*_{I_{LC}} = k_p(w_{dcp}\tilde{\eta}_{pu,dc} - w_{acp}\tilde{\eta}_{pu,ac}) + \tilde{\Phi}_{ILC} \\ \dot{\tilde{\Phi}}_{ILC} = k_i(w_{dcp}\tilde{\eta}_{pu,dc} - w_{acp}\tilde{\eta}_{pu,ac}) \\ \dot{\tilde{\delta}}_{ILC} = \tilde{f}_{ILC} = k_{ILCp}(\tilde{P}^*_{I,LC} - \tilde{P}_{ILC}) + \tilde{\Theta}_{ILC} \\ \dot{\tilde{\Theta}}_{ILC} = k_{ILCi}(\tilde{P}^*_{I_{LC}} - \tilde{P}_{ILC}); \\ \tilde{P}_{ILC} = P^{\max}_{ILC}\left(\tilde{\delta}_{ILC} - \tilde{\delta}_{ac}\right); \\ \tilde{P}_{dc,dispat} - \tilde{P}_{ILC} = \tilde{P}_{dc,load} = 0;\ \tilde{P}_{ac,dispat} + \tilde{P}_{ILC} = \tilde{P}_{ac,load} = 0 \end{cases} \tag{16.15}$$

where '~' denotes small perturbation around equilibrium points. And

$$\begin{cases} a_f = \frac{1}{0.5(f_{\max} - f_{\min})};\ a_V = \frac{1}{0.5(V_{dc,\max} - V_{dc,\min})} \\ w_{acp} = \frac{w_i}{P^{\max}_{dispat,ac}};\ w_{dcp} = \frac{w_i}{P^{\max}_{dispat,dc}} \\ P^{\max}_{ILC} = \frac{V_{ILC} V_{ac}}{Z_{link}} \cos\left(\delta_{ILC} - \delta_{ac}\right) \end{cases} \tag{16.16}$$

Rewriting (16.15) in matrix form, the state-space equations of the power control stability of the ILC is presented as

$$\begin{bmatrix} \dot{\tilde{\delta}}_{ac} \\ \dot{\tilde{\Phi}}_{ILC} \\ \dot{\tilde{\delta}}_{ILC} \\ \dot{\tilde{\Theta}}_{ILC} \end{bmatrix} = [A_{ILC}] \begin{bmatrix} \tilde{\delta}_{ac} \\ \tilde{\Phi}_{ILC} \\ \tilde{\delta}_{ILC} \\ \tilde{\Theta}_{ILC} \end{bmatrix} \tag{16.17}$$

where $a_{ILC} = P_{ILC}^{\max}(w_{dcp}a_V k + w_{acp}a_f m)$.

$$[A_{ILC}] = \begin{bmatrix} -mP_{ILC}^{\max} & 0 & mP_{ILC}^{\max} & 0 \\ k_i a_{ILC} & 0 & -k_i a_{ILC} & 0 \\ k_{ILCp}P_{ILC}^{\max} + k_{ILCp}k_p a_{ILC} & k_{ILCp} & -k_{ILCp}P_{ILC}^{\max} - k_{ILCp}k_p a_{ILC} & 1 \\ k_{ILCi}P_{ILC}^{\max} + k_{ILCi}k_p a_{ILC} & k_{ILCi} & -k_{ILCi}P_{ILC}^{\max} - k_{ILCi}k_p a_{ILC} & 0 \end{bmatrix} \tag{16.18}$$

16.4.2 Eigenvalue Analysis

The eigenvalues of matrix A_{ILC} in (16.18) can be used to study the stability of the ILC around the state of equilibrium. For the HIL test system to be described later in Sect. 16.5, the root-locus plots of matrix A_{ILC} for the ILC#12 are shown in Fig. 16.8 while varying the control parameters, k_{ILCp} and k_{ILCi}.

Figure 16.8a shows the root locus with the proportional parameter k_{ILCp} changing from $1e^{-5}$ to $1e^{-3}$. In Fig. 16.8a, there is a simple eigenvalue at zero corresponding to rotational symmetry, which has been proven in [40]. And the rest eigenvalues are in the left half-plane when $k_{ILCp} > 3e^{-5}$. With k_{ILCp} increases in Fig. 16.8a, the dominant eigenvalues (λ_1 and λ_2) gradually move away from the imaginary axis. Meanwhile, the eigenvalue λ_3 moves close to the imaginary axis, which makes the system more easy to become unstable Therefore, it is important to choose a suitable value of the proportional parameter k_{ILCp} to ensure the stability of the ILC. Figure 16.8b shows the root locus as the integral parameter k_{ILCi} increases from $1e^{-5}$ to $3e^{-3}$. In this case, there also exists an eigenvalue at zero [40]. When the value of k_{ILCi} becomes large, the dominant pole λ_1 moves away from the imaginary axis. And λ_2 and λ_3 gradually move away from the real axis, which decreases the damping ratio of the system. Thus, it is important to choose a suitable value of the integral parameter k_{ILCi} to ensure the satisfactory dynamic response of the system.

16.5 Hardware-in-Loop Results

The introduced priority-driven self-optimizing power flow control scheme of ILCs is verified by real-time HIL tests. The HIL system includes two parts: physical main circuits and algorithm controller. The physical circuits are realized by the real-time

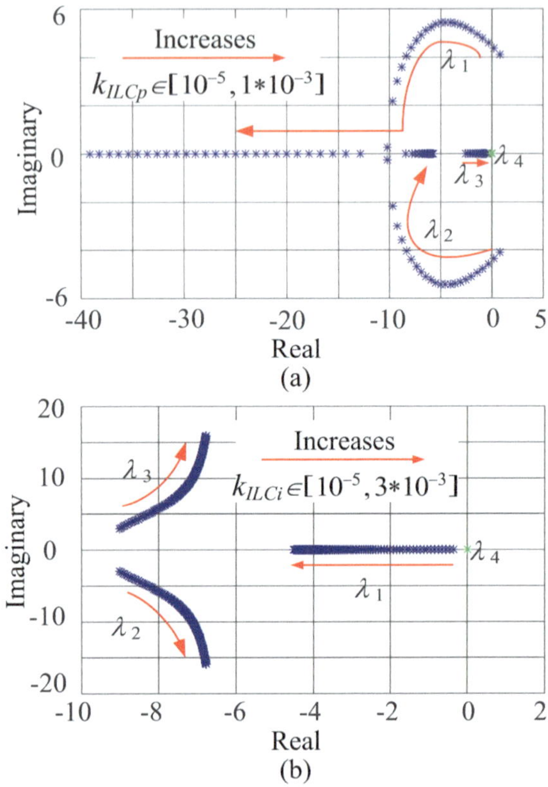

Fig. 16.8 Eigenvalues of state-space matrix A_{ILC} for the ILC#12. **a** $1e^{-5} < k_{ILCp} < 1e^{-3}$, **b** $1e^{-5} < k_{ILCi} < 3e^{-3}$

simulator OP5600 with $20\mu s$ time-step, which can accurately and effectively mimic the dynamics of the power electronic components. The algorithm controller is the real-hardware dSPACE1202 Microlab-box, and the sampling frequency is 20 kHz.

The studied system model of four hybrid AC/DC microgrids is described in Fig. 16.9. Subgrid-1 and subgrid-3 are DC type. Subgrid-2 and subgrid-4 are AC type. There are four H-bridge-based bidirectional ILCs to interlink the four subgrids, ILC#12, ILC#32, ILC#34 and ILC#14. The introduced decentralized quasi-droop control scheme of four ILCs is shown in Fig. 16.8. The power flow direction reference of a positive active power is assumed from DC side to AC side. For example, $P_{ILC,12}$ implies the power exchange value of ILC#12 from subgrid-1 to subgrid-2. When the power value is more than zero, the ILC works in inverter mode, and when less than zero, the ILC works in rectifier mode.

The physical and control parameters in HIL system are listed in Table 16.2. The priority weights w_i of four subgrids are 3, 1, 1, and 1, respectively. The DC subgrid-1 has a high priority of power supply and high requirement of voltage deviation. Besides, the dispacthable power capacity of subgrid-3/subgrid-4 is twice as much as that of subgrid-1/ subgrid-2. The nominal power capacities of four subgrids are 2.5,

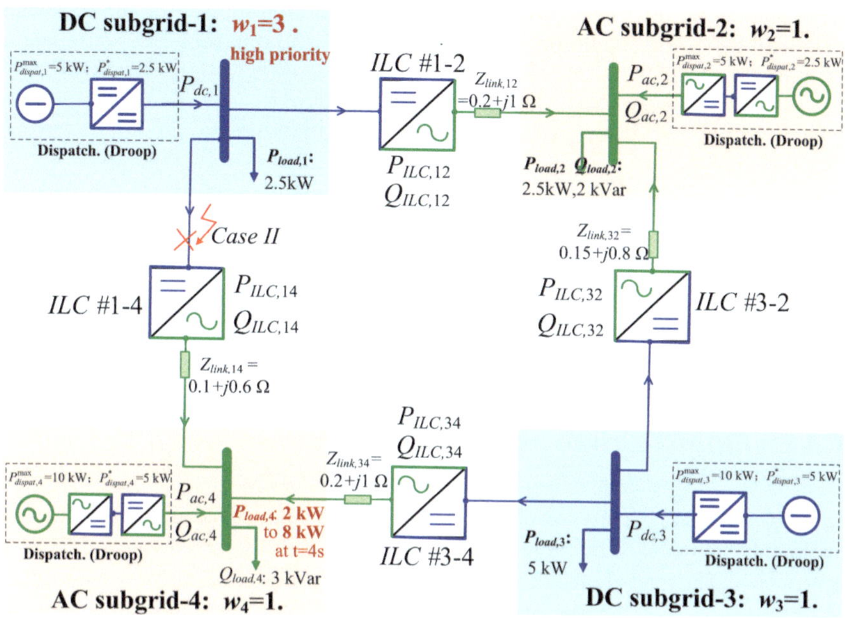

Fig. 16.9 Studied system model of four hybrid AC/DC microgrids

Table 16.2 HIL system parameters

Description	Symbol	Value
Physical parameters of four subgrids		
DC subgrid voltage	$V_{dc,max}/V_{dc}^{*}/V_{dc,min}$	700/ 685/ 670 V
AC subgrid frequency	$f_{max}/f^{*}/f_{min}$	50.5/ 50/ 49.5 HZ
AC subgrid voltage	$V_{ac,max}/V_{ac}^{*}/V_{ac,min}$	321/ 311/ 301 V
Priority weight	$\mathbf{w_1/w_2/w_3/w_4}$	**3/ 1/ 1/ 1**
Max. dispat. power cap.	$\mathbf{P^{max}_{dispat,1} \cdots P^{max}_{dispat,4}}$	**5/ 5/ 10/ 10 kW**
Nom. dispat. power cap.	$P^{*}_{dispat,1} \cdots P^{*}_{dispat,4}$	2.5/ 2.5/ 5/ 5 kW
Nom. load active power	$P^{*}_{load,1} \cdots P^{*}_{load,4}$	2.5/ 2.5/ 5/ 5 kW
Nom. load reactive power	$Q^{*}_{load,1} \cdots Q^{*}_{load,4}$	0/ 2/ 0/ 3 kVar
Control parameters of ILCs		
Ref. voltage/ frequency	$V_{ILC,max}/f^{*}_{ILC}$	321 V/ 50 Hz
Q-V droop coefficient	n_{ILC}	$3e^{-3}$
P-f droop PI coefficient	k_{ILC_p}/k_{ILC_i}	$2e^{-4}/1e^{-4}$
Power control Coefficient	k_p/k_i	$5e^2/1e^4$

2.5, 5, and 5 kW, respectively. From the optimal solution in (16.10), the following theoretical condition should be enabled in test results to minimize the optimization objection (16.6).

$$\begin{cases} \frac{\eta_{pu,1}}{1} = \frac{\eta_{pu,2}}{3} = \frac{\eta_{pu,3}}{6} = \frac{\eta_{pu,4}}{6}; \text{ this study} \\ \frac{\eta_{pu,1}}{1} = \frac{\eta_{pu,2}}{1} = \frac{\eta_{pu,3}}{1} = \frac{\eta_{pu,4}}{1}; \text{ in[23]} - [25] \end{cases} \quad (16.19)$$

To verify the priority-driven optimizing result of this study, a comparison with existing "global equal sharing" method [23–25] is carried out in Case I. Then, to further verify the redundancy of single point of ILC failure, Case II is built under ILC#14 failure. To analysis the impact of a large difference between priority weights on the transient process, Case III with w_1=5 is tested under the load-change condition.

16.5.1 *Priority-Driven Self-Optimizing Power*

Figure 16.10 shows the test waveforms of Case I. The detailed steady-state values are concluded and analyzed in Table 16.3. The load of subgrid-4 increases from 2 kW to 8 kW at t=4s. The load of the other three subgrids are in nominal values, whose values are 2.5 kW, 2.5 kW, and 5 kW, respectively. Before 4s, the total nominal dispatchable power capacity of four subgrids is 15 kW, and actual active power of total load demand is 12 kW. Thus, the total active power is surplus (in power-sufficiency), and the operation state deviation indexes of four subgrids are positive in Fig. 16.10 (f.1)–(f.2). After 4s, the total load power becomes 18 kW, which is more than 15 kW at a power deficit state (in power-dificiency), and then the operation state deviation indexes of four subgrids becomes negative in Fig. 16.10 (f.1)–(f.2). From Table 16.3, the steady-state deviation indexes $\eta_{pu,i}$ of four subgrids are 0.04/ 0.12/ 0.24/ 0.24 at t $\in$[0, 4s] and -0.04/ -0.12/ -0.24/ -0.24 at t $\in$[4s, 10s] under the introduced method. Clearly, two different operation states are opposite and symmetrical based on the nominal operation state. And thus, they have the same optimized object. In the existing global equal-sharing method [23–25], Fig. 16.10 (f.2) and Table IV show that the steady-state deviation indexes $\eta_{pu,i}$ of four subgrids are 0.2/ 0.2/ 0.2/ 0.2 at t $\in$[0, 4s] and −0.2/ −0.2/ −0.2/ −0.2 at t $\in$[4s, 10s]. All indexes of four subgrids are always same because of the same priority weight before and after 4s.

From Table 16.3 and Fig. 16.10 (f.1)–(f.2), some connotations about state indexes are drawn under the introduced method: (1) The ratio$\eta_{pu,1}$:$\eta_{pu,2}$:$\eta_{pu,3}$:$\eta_{pu,4}$ =1:3:6:6, which reveals that Eq. (16.19) always holds, which verify the effectiveness of the above theoretical optimal mode and introduced control scheme. (2) Compared with the global equal-sharing method, we can see that, subgrid-1 with high priority weight has a highest voltage quality index and a high priority of power demand, which verifies the introduced priority-driven optimizing concept. (3) Subgrid-3 and subgrid-4 with same capacity and priority have a same state deviation index. (4) Subgrid-2 with small capacity has a better state deviation index than subgrid-3 and subgrid-4 because the power demand of a small-capacity subgrid can be easily met. (5) When total power is heavily deficit, subgrid-3 and subgrid-4 will shed load automatically by

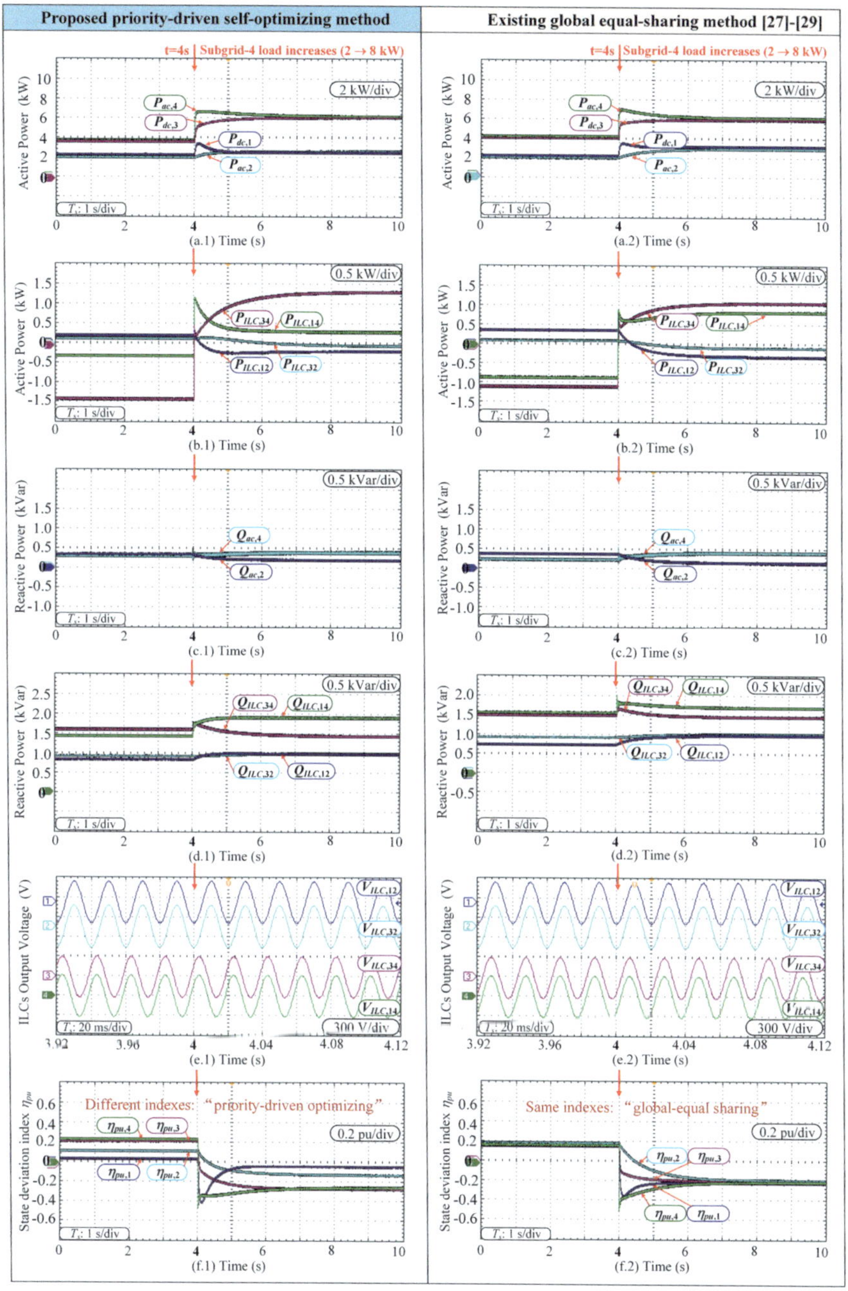

Fig. 16.10 Comparisons of the introduced priority-driven self-optimizing method with the existing global equal-sharing method [23–25]. **a** Output active power of four subgrids, **b** active power of four ILCs, **c** reactive power of two AC subgrids, **d** reactive power of four ILCs, **e** output voltage of four ILCs, and **f** operation state deviation index of four subgrids

Table 16.3 Comparative analysis of introduced method with existing method [23–25]

Steady-state analysis	Power (kW)	introduced method		Existing method	
		t∈[0,4s] Power surplus	t∈[4s, 10s] PPower deficit	t∈[0,4s] Power surplus	t∈[4s, 10s] Power deficit
Subgrid-1	Pdc,1	2.4	2.6	2.0	3.0
	Pload,1	2.5		2.5	
ILC #1-2	$P_{ILC,12}$	0.2	−0.2	0.35	−0.35
Subgrid-2	$P_{ac,2}$	2.2	2.8	2.0	3.0
	$P_{load,2}$	2.5		2.5	
ILC #3-2	$P_{ILC,32}$	0.1	−0.1	0.15	−0.15
Subgrid-3	$P_{dc,3}$	3.8	6.3	4.0	6.0
	$P_{load,3}$	5		5	
ILC #3-4	$P_{ILC,34}$	−1.3	1.4	−1.15	1.15
Subgrid-4	$P_{ac,4}$	3.6	6.3	4.0	6.0
	$P_{load,4}$	2	8	2	8
ILC #3-4	$P_{ILC,14}$	−0.3	0.3	−0.85	0.85
Operation state deviations	$\eta_{pu,1}$-$\eta_{pu,4}$	0.04/0.12/0.24/0.24	−0.04/−0.12/−0.24/−0.24	0.2/0.2/0.2/0.2	−0.2/−0.2/−0.2/−0.2
Optimized Object (6)	J	0.1344	0.1344	0.24	0.24

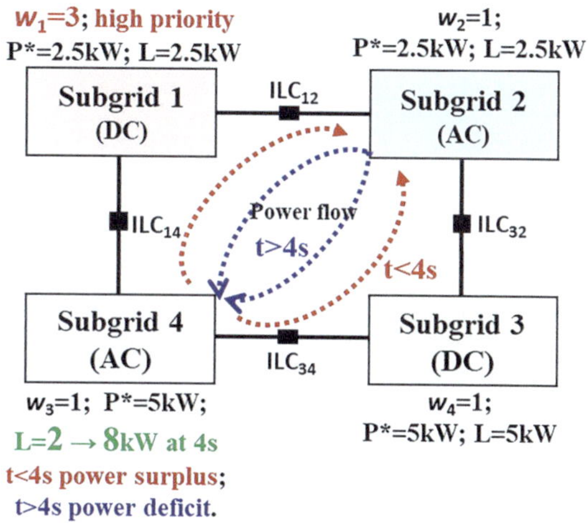

Fig. 16.11 A brief power flow chart of four subgrids by ILCs at different times

judging the large state deviations in order to guarantee the total power balance of high-priority subgrid. (6) The optimized objective value of (16.6) under the introduced method is J=0.1344, which is better than the existing method J=0.24.

$$\begin{cases} P_{dc,1} - P_{ILC,12} - P_{ILC,14} = P_{load,1} \\ P_{ac,2} + P_{ILC,12} + P_{ILC,32} = P_{load,2} \\ P_{dc,3} - P_{ILC,32} - P_{ILC,34} = P_{load,3} \\ P_{ac,4} + P_{ILC,34} + P_{ILC,14} = P_{load,4} \\ P_{dc,1} + P_{ac,2} + P_{dc,3} + P_{ac,4} = P_{load,1} + P_{load,2} + P_{load,3} + P_{load,4} \end{cases} \tag{16.20}$$

Moreover, for the power exchanges among four subgrids in Fig. 16.10 (a.1)–(b.1), Table 16.3 presents the detailed data, and the power flow chart is also briefly given in Fig. 16.11. At t∈[0, 4 s], subgrid-4 is under power sufficiency, then the dispatchable active power in subgrid-4 is transferred by ILCs to the other three subgrids. ILC#34 and ILC#14 work in the rectifier mode, and ILC#12 works in the inverter mode. At t∈[4 s, 10 s], subgrid-4 is under power deficiency, then the dispatchable active power is transferred by ILCs from other three subgrids to the subgrid-4. ILC#34 and ILC#14 work in the inverter mode, and ILC#12 works in the rectifier mode. On the whole, the total dispatchable energy sources are shared among the subgrids according to the priority-driven principle. Besides, the following power balance equation (16.20) is always ensured from Table 16.3.

In addition, Fig. 16.10 (d.1)–(e.1) show the reactive power and voltage waveforms of four ILCs. It illustrates that the quasi-droop controlled ILCs not only achieve a flexible self-optimizing power exchange flow, but ancillary functions of reactive power compensation and voltage support are also provided. Meanwhile, the output voltages of ILCs have a satisfactory dynamic response under load change.

From Case I, the introduced priority-driven self-optimizing power control method can autonomously achieve flexible power exchange management and work well in the load-change condition.

16.5.2 One ILC Unit Fault Redundancy

This case aims to verify the one-ILC fault redundancy. Figure 16.12 shows dynamics of the system with self-optimizing power control when the ILC#14 fails at t=4 s. From Fig. 16.12e, the output current of ILC#14 suddenly changes to 0 at t=4 s. Accordingly, the output active power in Fig. 16.12b and reactive power in Fig. 16.12d of ILC#14 become 0. Because the output power of ILC#14 becomes from 0.3 kW to 0, the power flows of the other three ILCs are re-dispatched.

Although ILC#14 fails, the output active powers of four subgrids are not affected in Fig. 16.12a, and the operation state deviation indexes of four subgrids are unchanged in Fig. 16.12f, which verifies the correctness and uniqueness of priority-driven self-optimizing solutions.

Summarily, the system can still work well after disconnecting the ILC#14 at t=4 s. Meanwhile, ILCs reach the new steady states after about 1 s in Fig. 16.12b. As seen, the results indicate that the introduced method achieves a one-ILC-unit fault redundancy.

16.5.3 Impact of Priority Weights on Transient Process

This case is to analysis the impact of a large difference between priority weights on the transient process of the load-change condition. Compared with the Case I, the priority weight of w_1 increases from 3 to 5 to reduce the operation state deviation of subgrid-1 in the Case III. Figure 16.13 shows dynamics of the system with a higher w_1 when load changes at t=4 s.

From Fig. 16.13a, c, a higher w1 has little impact on the dynamic responses of the output active power and operation state deviation indexes of four subgrids. The transient process time of algorithm convergence is about 2 s, which is always satisfactory. While for Fig. 16.13b, a higher w1 causes a large overshoot of $P_{ILC,14}$. The main reason is that droop-controlled subgrids have less-inertia, and the frequency of subgrid-4 changes instantly when the subgrid-4 load switches, as shown in Fig. 16.13c. As a result, the output active power of the ILC#14 has an instantaneous impact. In the practical microgrid with a certain inertia moment, the power impact of the ILCs would be greatly improved.

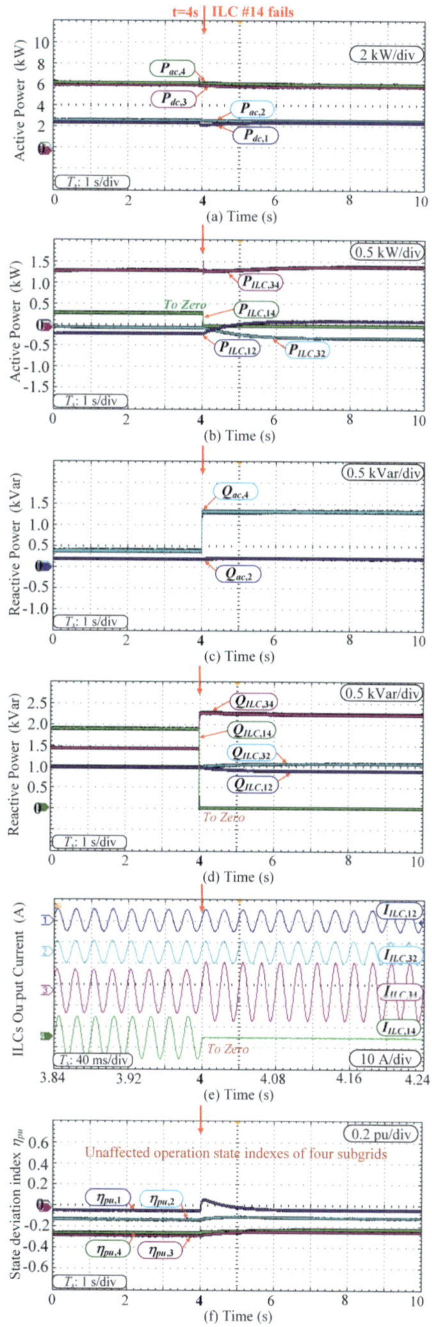

Fig. 16.12 Dynamics of the system with self-optimizing power control when ILC #14 fails at t=4s. **a** Active power of four subgrids, **b** active power of four ILCs, **c** reactive power of two AC subgrids, **d** reactive power of four ILCs, **e** output current of four ILCs, and **f** operation state deviation index of four subgrids

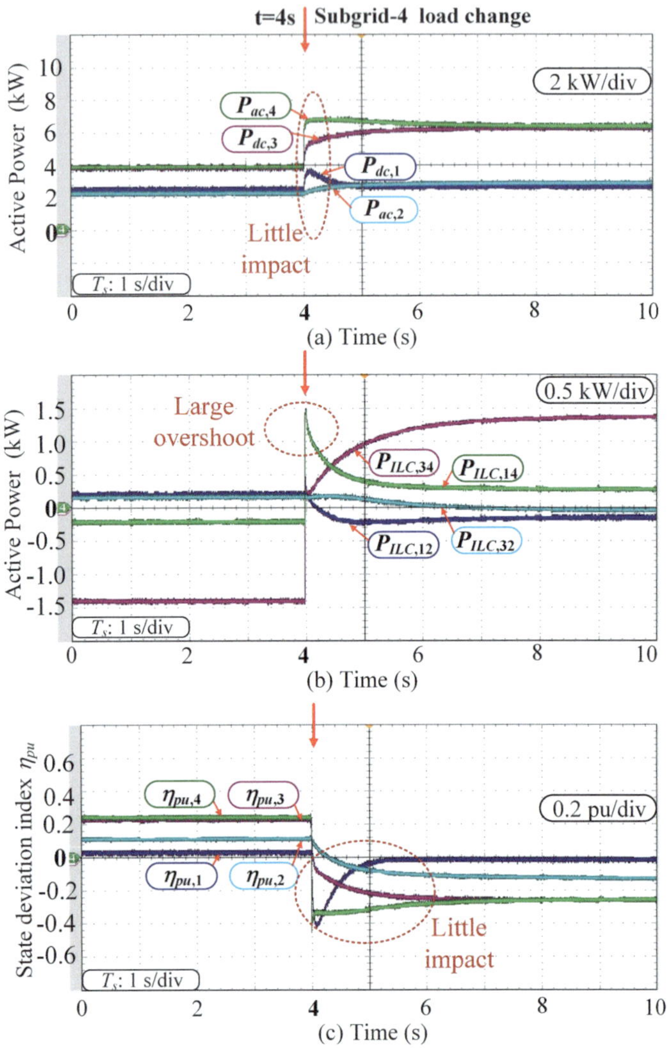

Fig. 16.13 Dynamics of the system with a higher priority weight w_1=5. **a** Active power of four subgrids, **b** active power of four ILCs, and **c** operation state deviation index of four subgrids

16.6 Conclusion

To ensure the optimal power exchanges among microgrid clusters in priority, this study proposes a decentralized self-optimizing power exchange control scheme of ILCs for hybrid microgrid clusters. For a high-priority subgrid with critical loads, its power supply and voltage quality are autonomously prioritized to realize a flexible power flow management. In the introduced strategy, the self-optimizing coordination control of ILCs is executed in a decentralized manner, which controls the power exchanges based on locally monitored AC frequency and DC voltage without a central controller or any communication. Thus, the introduced control is more cost-effective, higher reliable and resilient in terms of communication faults. HIL results show the effectiveness of the introduced method with four interlinked AC/DC microgrids.

Besides, the introduced self-optimizing rule provides a new paradigm to manage complex microgrid clusters towards a decentralized self-organizing grid. It can guarantee the basic operation by avoiding communication dependency, which has some key features to address complicated problems (1) decentralized self-optimizing behavior; (2) reserved dispatchable source sharing among subgrids; (3) mutual PV/wind/load uncertainty complementarity among subgrids; (4) autonomous load shedding of low-priority subgrid; (5) one-ILC-fault redundancy.

Based on this fundamental study, some novel hierarchical control schemes would be explored in future work, where the decentralized self-optimizing power control takes a backup role of primary control and initiates the power exchanges, and the upper-level superior management controller can ensure the economic operation by adjusting the power exchanges. Thus, the introduced self-optimizing method can be compatible with high-level control algorithms. On the other hand, this decentralized self-optimizing backup control can support the resilient operation under extreme events, such as the communication paralysis and central controller failure.

References

1. R.H. Lasseter, Smart distribution: coupled microgrids. Proc. IEEE **99**(6), 1074–1082 (2011). (Jun.)
2. X. Liu, P. Wang, P.C. Loh, A hybrid AC/DC microgrid and its coordination control. IEEE Trans. Smart Grid **2**(2), 278–286 (2011)
3. X. Lu, J.M. Guerrero, K. Sun, J.C. Vasquez, R. Teodorescu, L. Huang, Hierarchical control of parallel AC-DC converter interfaces for hybrid microgrids. IEEE Trans. Smart Grid **5**(2), 683–692 (2014). (Mar.)
4. S.K. Sahoo, A.K. Sinha, N.K. Kishore, Control techniques in AC, DC, and hybrid AC–DC microgrid: a review. IEEE J. Emerg. Sel. Topics Power Electron. **6**(2), 738–759 (2018). (Jun.)
5. X. Shen, D. Tan, Z. Shuai, A. Luo, Control techniques for bidirectional interlinking converters in hybrid microgrids: leveraging the advantages of both AC and DC. IEEE Power Electronics Mag. **6**(3), 39–47 (2019)
6. F. Nejabatkhah, Y.W. Li, Overview of power management strategies of hybrid AC/DC microgrid. IEEE Trans. Power Electron. **30**(12), 7072–7089 (2015). (Dec.)
7. J. Wang, X. Lu, Sustainable and resilient distribution systems with networked microgrids. Proc. IEEE **108**(2), 238–241 (2020). (Feb.)
8. B. Chen, J. Wang, X. Lu, C. Chen, S. Zhao, Networked microgrids for grid resilience, robustness, and efficiency: a review. IEEE Trans. Smart Grid **12**(1), 18–32 (2021). (Jan.)
9. K. Sun, X. Wang, Y.W. Li, F. Nejabatkhah, Y. Mei, X. Lu, Parallel operation of bidirectional interfacing converters in a hybrid AC/DC microgrid under unbalanced grid voltage Conditions. IEEE Trans. Power Electron. **32**(3), 1872–1884 (2017). (Mar.)
10. P. Wang, C. Jin, D. Zhu, Y. Tang, P.C. Loh, F.H. Choo, Distributed control for autonomous operation of a three-port AC/DC/DS hybrid microgrid. IEEE Trans. Ind. Electron. **62**(2), 1279–1290 (2015). (Feb.)

11. L. Meng, F. Tang, M. Savaghebi, J.C. Vasquez, J.M. Guerrero, Tertiary control of voltage unbalance compensation for optimal power quality in islanded microgrids. IEEE Trans. Energy Conversion **29**(4), 802–815 (2014). (Dec.)
12. Y. Zhang, L. Xie, Q. Ding, Interactive control of coupled microgrids for guaranteed system-wide small signal stability. IEEE Trans. Smart Grid **7**(2), 1088–1096 (2016). (Mar.)
13. Q. Zhou, M. Shahidehpour, A. Alabdulwahab, A. Abusorrah, Flexible division and unification control strategies for resilience enhancement in networked microgrids. IEEE Trans. Power Syst. **35**(1), 474–486 (2020). (Jan.)
14. G. Qi, Improved control strategy of interlinking converters with synchronous generator characteristic in islanded hybrid AC/DC microgrid. CPSS Trans. Power Electronics Appl. **2**(2), 149–158 (2017). (Jun.)
15. X. Li, L. Guo, Y. Li, Z. Guo, C. Hong, Y. Zhang, C. Wang, A unified control for the DC-AC interlinking converters in hybrid AC/DC microgrids. IEEE Trans. Smart Grid **9**(6), 6540–6553 (2018)
16. J. Wang, C. Jin, P. Wang, A uniform control strategy for the interlinking converter in hierarchical controlled hybrid AC/DC microgrids. IEEE Trans. Ind. Electron. **65**(8), 6188–6197 (2018). (Aug.)
17. J. Li, Y. Liu, L. Wu, Optimal operation for community-based multi-party microgrid in grid-connected and islanded modes. IEEE Trans. Smart Grid **9**(2), 756–765 (2018). (Mar.)
18. V. Bui, A. Hussain, H. Kim, A strategy for flexible frequency operation of stand-alone multi-microgrids. IEEE Trans. Sustain. Energy **9**(4), 1636–1647 (2018). (Oct.)
19. Z. Wang, B. Chen, J. Wang, M.M. Begovic, C. Chen, Coordinated energy management of networked microgrids in distribution systems. IEEE Trans. Smart Grid **6**(1), 45–53 (2015). (Jan.)
20. P. Lin, P. Wang, C. Jin, J. Xiao, X. Li, F. Guo, C. Zhang, A distributed power management strategy for multi-paralleled bidirectional interlinking converters in hybrid AC/DC microgrids. IEEE Trans. Smart Grid **10**(5), 5696–5711 (2019)
21. P. Lin, C. Jin, J. Xiao, X. Li, D. Shi, Y. Tang, P. Wang, A distributed control architecture for global system economic operation in autonomous hybrid ac/dc microgrids. IEEE Trans. Smart Grid **10**(3), 2603–2617 (2019)
22. Q. Zhou, Z. Li, Q. Wu, M. Shahidehpour, Two-stage load shedding for secondary control in hierarchical operation of islanded microgrids. IEEE Trans. Smart Grid **10**(3), 3103–3111 (2019). (May)
23. C. Jin, P.C. Loh, P. Wang, Y. Mi, F. Blaabjerg, "Autonomous operation of hybrid AC-DC microgrids. IEEE Int. Conf. Sustain. Energy Technol. ICSET **2010** (2010)
24. P. Ch. Loh, D. Li, Y. K. Chai, F. Blaabjerg, Autonomous operation of hybrid microgrid with AC and DC subgrids. IEEE Trans. Power Electron. **28**(5), 2214–2223 (2013)
25. PCh. Loh, D. Li, Y.K. Chai, F. Blaabjerg, Autonomous control of interlinking converter with energy storage in hybrid AC–DC microgrid. IEEE Trans. Ind. App. **49**(3), 1374–1383 (2013). (May)
26. S. Malik, Y. Sun, W. Huang, X. Ai, Z. Shuai, A generalized droop strategy for interlinking converter in a standalone hybrid microgrid. Appl. Energy **226**, 1056–1063 (2018)
27. Y. Xia, W. Wei, M. Yu, Y. Peng, J. Tang, Decentralized multi-time scale power control for a hybrid AC/DC microgrid with multiple subgrids. IEEE Trans. Power Electron. **33**(5), 4061–4072 (2018). (May)
28. Q. Xu, J. Xiao, P. Wang, C. Wen, A decentralized control strategy for economic operation of autonomous AC, DC, and hybrid AC/DC microgrids. IEEE Trans. Energy Conversion **32**(4), 1345–1355 (2017). (Dec.)
29. F. Guo, C. Wen, J. Mao, J. Chen, Y. Song, Hierarchical decentralized optimization architecture for economic dispatch: a new approach for large-scale power system. IEEE Trans. Ind. Informatics **14**(2), 523–534 (2018). (Feb.)
30. M. Tahir, S.K. Mazumder, Event- and priority-driven coordination in next-generation grid. IEEE J. Emerg. Sel. Topics Power Electron. **4**(4), 1186–1194 (2016). (Dec.)

31. J. Li, Y. Liu, L. Wu, Optimal operation for community-based multi-party microgrid in grid-connected and islanded modes. IEEE Trans. Smart Grid **9**(2), 756–765 (2018). (Mar.)
32. B. Kroposki et al., Autonomous energy grids: controlling the future grid with large amounts of distributed energy resources. IEEE Power Energy Mag. **18**(6), 37–46 (2020)
33. S. Bhattacharya, S. Mishra, Stability enhancement of droop controlled PV based microgrid with frequency dependent load, in *2014 North American Power Symposium NAPS* (2014), pp. 0–5
34. A. Kahrobaeian, Y.A.-R.I. Mohamed, Analysis and mitigation of low-frequency instabilities in autonomous medium-voltage converter-based microgrids with dynamic loads. IEEE Trans. Ind. Electron. **61**(4), 1643–1658 (2014). (Apr.)
35. X. Zhang, Q. Gao, Y. Hu, H. Zhang, Z. Guo, Active power reserve photovoltaic virtual synchronization control technology. Chin. J. Elect. Eng. **6**(2), 1–6 (2020). (Jun.)
36. W. Yuan, Y. Wang, D. Liu, F. Deng, Z. Chen, Efficiency-prioritized droop control strategy of AC microgrid. IEEE J. Emerg. Sel. Topics Power Electron. **9**(3), 2936–2950 (2021). (Jun.)
37. T. Ma, J. Wu, X. Niu, Reliability assessment indices and method for urban microgrid, in *Proceedings of 24th International Conference and Exhibition on Electricity Distribution (CIRED), Glasgow, United Kingdom* (2017)
38. K.N. Kumar, B. Sivaneasan, P.L. So, Impact of priority criteria on electric vehicle charge scheduling. IEEE Trans. Transp. Electrification **1**(3), 200–210 (2015). (Oct.)
39. W. Wu et al., Sequence impedance modeling and stability comparative analysis of voltage-controlled VSGs and current-controlled VSGs. IEEE Trans. Ind. Electronics **66**(8), 6460–6472 (2019). (Aug.)
40. J.W. Simpson-Porco, F. Dörfler, F. Bullo, Synchronization and power sharing for droop-controlled inverters in islanded microgrids. Automatica **49**(9), 2603–2611 (2013). (Sep.)

MIX
Papier aus verantwortungsvollen Quellen
Paper from responsible sources
FSC® C105338

If you have any concerns about our products,
you can contact us on
ProductSafety@springernature.com

In case Publisher is established outside the EU,
the EU authorized representative is:
Springer Nature Customer Service Center GmbH
Europaplatz 3, 69115 Heidelberg, Germany

Printed by Libri Plureos GmbH
in Hamburg, Germany